Recent Discoveries and Perspectives in Human Evolution

Papers arising from 'Exploring Human Origins: Exciting Discoveries at the Start of the 21st Century' Manchester 2013

Edited by

Anek R. Sankhyan

BAR International Series 2719
2015

Published in 2016 by
BAR Publishing, Oxford

BAR International Series 2719

Recent Discoveries and Perspectives in Human Evolution

ISBN 978 1 4073 1372 6

BAR Publishing is the trading name of British Archaeological Reports (Oxford) Ltd. British Archaeological Reports was first incorporated in 1974 to publish the BAR Series, International and British. In 1992 Hadrian Books Ltd became part of the BAR group. This volume was originally published by Archaeopress in conjunction with British Archaeological Reports (Oxford) Ltd / Hadrian Books Ltd, the Series principal publisher, in 2015. This present volume is published by BAR Publishing, 2016.

Printed in England

BAR
PUBLISHING

BAR titles are available from:

BAR Publishing
122 Banbury Rd, Oxford, OX2 7BP, UK
EMAIL info@barpublishing.com
PHONE +44 (0)1865 310431
FAX +44 (0)1865 316916
www.barpublishing.com

Table of Contents

List of Figures

List of Tables

RECENT DISCOVERIES AND PERSPECTIVES IN HUMAN EVOLUTION: INTRODUCTION

Anek R. SANKHYAN

The present volume is based on research articles received by me as the convenor of the BH 13 Panel of the international conference on *"Exploring Human Origins: Exciting discoveries at the start of the 21st Century"* held under the auspicious of the 17th Congress of the International Union of the Anthropological and Ethnological Sciences (IUAES) during 5-10 August 2013 at Manchester (U.K.). The prime focus of the conference panel has been to record the more recent fossil, archaeological and genomic discoveries in the field of human origins and evolution, besides a few very significant ones made in 1990s. Some of the scholars who had contributed abstracts to the BH Panel could not send their full papers since they could not attend the conference. One of the main reasons was that the U.K. organisers of the congress could not provide any financial assistance to the contributors of this panel, including the convenor himself, though they provided a good venue for conferencing in Manchester university campus for which the convenor thanks them.

I also have not received any financial support for this publication, such that it is the result of my personal craze and endeavour. I tried hard to make it a fruitful reality by inviting some important contributions from other notable scholars relevant to the theme of the present book to compensate the missing articles figured in the abstracts of the BH 13 panel. In the final present collection there are 22 research articles which fairly cover the intended theme of the conference and represent research areas of anthropology, especially palaeoanthropology, prehistoric archaeology, human molecular genetics and human osteology, besides palaeontology, geology, ancient surgical and medical practices, prehistoric art, the primitive hunter-gatherers, and recent human interventions to the biosphere. The contributors are from many countries and almost all continents are represented in their contributions.

Here, I present the glimpses of the findings of various research contributions highlighting different perspectives contributing to the greater understanding of human origins and ongoing evolution. Considering the complexities of the contributions, I have tried to fix the order of the articles in a temporal perspective, starting from the earliest to the recent as they cover a geological period of late Mio-Pliocene through Pleistocene and Holocene of human prehistory. For the benefit of the general readers I have tried to bring in a normal flow between the individual contributions by inserting my own remarks and observations wherever needed and possible.

Let me begin with the first article from South China where the Kunming region has gained a very prominent place in Mio-Pliocene palaeoanthropology. Kunming has yielded several species of a late Miocene hominoid *Lufengpithecus*. The cranial finding discussed here by Ji XUEPING, Deng CHENGLONG & Yu TENGSONG from Zhaotong City, is of a juvenile hominoid dated to the ~7.1-8.2 million years ago. It is the second sub-adult cranium of the *Lufengpithecus*, and the notable Chinese palaeoanthropologists believe that the specimen ZT 299 is morphologically distinct from known species of *Lufengpithecus* and may be another species. The discovery of the cranium reaffirms biogeographical and ecological scenarios invoking endemism in Southwest China during the later Miocene.

My own visit to Kunming in July 2009 to organise a similar conference panel in the 16th IUAES, I got an opportunity to visit the Lufeng site with famous Professor Xinji Wu and his associates – Ji Xueping, Gao Feng, Lingxia Zhao, etc., and including French scholar Anne Dambricourt Malasse. I also inspected the hominoid's collections from China, which lead me to accept that the Kunming region was in fact an important refugium of the later Miocene hominoids. Their striking similarities with the Indian Siwalik hominoids *Ramapithecus* and *Sivapithecus,* with whom they were classified in 1980s, indicate them as Siwalik migrants at the terminal stage of the Miocene when savannahs expanded and colder conditions set in the sub-Himalayas. Kunming provided them the easy access and equable climatic and ecological refugium and also promoted their diversification and speciation such that they now look different from their antecedents. The late survival of the Siwalik giant ape, *Gigantopithecus bilaspurensis* in the Chinese Middle Pleistocene and its differentiation and coronation to a Chinese species, *Gigantopithecus blacki*, is a testimony to the fact that Kunming became a haven and refuge of the later hominoids between 8-6 million years ago. The South/East Asian hominoids are therefore quite attractive and may not be ruled out from the last common ancestry of the earliest hominids – the australopithecines, and the great apes.

A logical connection of the latest Miocene hominoids with the Plio-Pleistocene African australopithecines and of the later to the earliest Homo is surfacing in the next article *"Australopithecines shoulders: New remains for Old Debate"* by Jean-Luc VOISIN. The notable French scholar is a specialist on the hominid shoulder complex, and therefore competently handles the important academic debate. Of course, it is widely considered that the shoulder complex constituted by the clavicle, scapula, humerus and cervical vertebrae, is a key system for knowing the arboreal part of hominid locomotion. Jean-Luc conducted a comparative study on three well preserved *Australopithecus* skeletons, such as Malapa (*A. sediba*), Woranso-Mille (*A. afarensis*) and Dikika (juvenile *A. afarensis)*. Previous scholars had held divergent views on the locomotory repertoires of these hominids and so on their relative affinities to early humans (*Homo habilis*). But, he concludes that the australopithecines shoulder is closer to the first *Homo* species and the extant apes, but not similar to any of them. This way, Jean-Luc is strengthening the link between *Australopithecus* and the first *Homo* (*Homo habilis*).

Homo habilis is considered as the antecedent of *Homo erectus* – the immediate ancestor of our species, *Homo sapiens*. But, when and where it evolved to our species is still a matter of hot debate since a number of species have been recognised and the archaic types of our species are discovered in most of the regions of the Old World. I had assigned the difficult task to Professor Robin DENNELL – a renowned British palaeoanthropologist to overview the status of current researches on it. So, in the third article entitled "*Hominin palaeoanthropology in Asia comes of age*" Dr. Dennell has tried to find out changes in our perceptions on the status of select fossil important discoveries with special reference to Asia. He records his observations right from the early form of *Homo erectus* from Dmanisi, Georgia to the late Pleistocene hobbit, *H. floresiensis*, from Indonesia, including the Neanderthals and "Denisovans". He concludes that despite several major gaps, we are now beginning to glimpse the antiquity and complexity in human evolution.

Robin Dennell however, missed to take note of the recent hominin postcranial fossil findings from India since scholars in general consider the cranial findings for phylogenetic conclusions, and ignore the rest of the skeletal elements and leave them to the relevant specialists. Perhaps, it is left for the discoverer (A.R. SANKHYAN) to critically review the "*Pleistocene hominin fossil discoveries in India: implications for human evolution in South Asia*" in the next, 4th article.

Sankhyan critically reviewed the major features of the solitary partial cranial and five postcranial hominin fossils discovered so far from the Central Narmada valley of India in their faunal, archaeological and stratigraphic contexts. He concluded that the partial cranium and a new femur morphologically, culturally and biostratigraphically conforms to a robust Acheulian hominin who inhabited Central Narmada valley between ~250 to 150 Kya and was akin to *Homo heidelbergensis*. The other fossils – the two clavicles and 9th rib support the existence of another hitherto unknown 'short and stocky' "archaic" hominin who used Middle Palaeolithic Mode 3 technology between 150 to 70 Kya. The upper datum is supported by a recent humerus fossil. The author has distinguished the 'short and stocky' "archaic" *Homo sapiens* by a new name, *Homo "narmadensis"* as it continued to colonize most of the Central India until late Pleistocene. He considered that this species attained anatomical modernity as a separate lineage of the subsequent short-bodied ancient Indians including the Andaman pygmies, also supported by recent genomic evidences.

Hominin fossil discoveries from Narmada valley and the human evolution in South Asia can better be understood if we familiarise ourselves with the geomorphology and palaeoenvironmental aspects of the central Narmada basin. In the next (5th) article Satya DEV & A.R. SANKHYAN make a modest attempt to add fresh data and observations based on a brief field work on "*Geoarchaeological and environmental aspects of the central Narmada alluvium*". They describe some important archaeological and geo-environmental processes at work on the basis of palaeontological and archaeological findings recovered in stratigraphic contexts from the alluvial deposits of Hoshangabad and Sehore area in Central Narmada valley. The new findings corroborate with earlier interpretations of Sankhyan and associates on the emergence of *Homo sapiens* in South Asia.

The middle Pleistocene fossil evidences for hominins are important from any region of the world for understanding the origins and expansion of *Homo sapiens*. In the next (6th) article "*The role of Balkans in peopling of Europe: New evidence from Serbia*" Mirjana ROKSANDIC – a notable young palaeoanthropologist in Canada analyses the importance of the partial mandible discovered from Mala Balanica Cave, Sićevo, Serbia – the solitary fossil from the Balkan Peninsula and securely dated to the Middle Pleistocene. She highlights that the primitive morphology and lack of derived Neanderthal traits in the mandible point to an important role Balkans played in population movements during Pleistocene.

Past human diversity can better be understood if we have a total perspective on their habitats. For this Sally REYNOLDS – a competent young British palaeoanthropologist throws out a new perspective of early hominin habits in the next (7th) article "*Role of landscapes in shaping hominin habitats in Africa*". She follows the Tectonic Landscape Model of Bailey et al. (2011) applied to famous hominid site Sterkfontein in South Africa. She concludes that the landscapes, surface water and vegetation in a 'mosaic' habitat comprised of a lake or river setting with combination of forest or woodland with savannah grasslands in close or spatial proximity are critical to the characterization of hominin niches as adaptive strategies employed by our ancestors to past climatic changes.

Where fossil evidences are very meagre but the human remains contain genetic material, the status of the ancestral population can be known through the DNA evidence. Unknown to science a few years ago, a new recent prehistoric population called, Denisovan, has been discovered on the basis of ancient DNA isolated from a finger phalanx discovered in the Denisova Cave, in the Altai Mountains of Russia. This individual represents a population dating to roughly 50,000 years ago. A renowned American palaeoanthropologist, John HAWKS in his concise article (8th) entitled, "*The Denisova Genome: an unexpected window into the past*", has critically reviewed and emphasized that the Denisovan genome provides new evidence about the biology and dynamics of ancient populations, besides the functional insights about the immune system, pigmentation, and genes involved in brain development. John concludes that the Denisovans contributed five percent of genes to later populations in island Southeast Asia and Aboriginal Australians, New Guinea, the Philippines. He also indicates their possible relationship with Neanderthals and the subsequent population movements and interactions in East and Southeast Asia that may explain the present distribution of Denisovan ancestry.

The next article (9th) represents France in which Chérif Ousmane TOURÉ & Anne Dambricourt MALASSÉ present their "*Preliminary results on the first paleontological, anthropological and archaeological Pleistocene locality in Adrar, Mauritania*". The authors evaluate the Palaeolithic assemblages at El Beyyed identifying two stratified sites, especially Yeslem III which contained fossilised faunal and human bones discovered since 2000 field works initiated by Theodore Monod in 1997, a French naturalist and founder of the Mauritanian Institute for Scientific Research (I.M.R.S.). The preliminary analyses allow them to conclude a Pleistocene age without confusion with Neolithic occupations. The fossil fauna from the site attest a Sahelian environment having large permanent lakes, as observed in Zouerat – Hammami La défense in 1967 by Yves Coppens. The taxonomic identity of the human fibula is, however still unknown.

While the Central Narmada valley of India is well known for the Middle to Late Pleistocene human fossils, the western end of the valley has also come out with a startling human fossil discovery of a recent time which provides some linkages with the earlier findings says Dambricourt MALASSÉ, Rachna RAJ and S. SHAH in the next (10th) article entitled "*The Orsang Man: a robust Homo sapiens in Central India with Asian Homo erectus features*". An isolated cranium was discovered in situ at Orsang River valley, a tributary of the Narmada River in western India. IRSL dating of the host sediments provided an age ranging from 50 to 30 Kya. She considers that the skull is significant for understanding the links between South Asian *Homo erectus* and robust *Homo sapiens* or "Asian Cromagnoid" populations still few known in Continental Asia and presents evidence for endocrine activity and genetic continuity.

The next article (11th) is a detailed field report on the "*Geoarchaeology of the fluvial terraces of Tejo River in Central Portugal" by* a young geoarchaeologist of India, Satya DEV. The richness of data indicates that Satya has done intensive work in the Alto Ribatejo region of central Portugal (but the editor has to put lots of efforts to bring his data and numerous figures to a proper shape). Satya recognized six river terraces in the Tejo River basin right from the Lower Palaeolithic to the Bronze Age. He has emphasized the archaeological potential of the lower (T6) and the middle (T4) terraces and describes their lithology and archaeological findings in details. He contends that T6 may belong to a late phase of the Upper Pleistocene or marine isotope stage 3 (MIS), whereas the T4 is dated to marine oxygen isotope stage 9-6 (MIS) and its sedimentological analysis indicates temperate climate significant for prehistoric human occupation in Portugal. He also recovered Middle/Lower Palaeolithic and the Bronze Age artefacts in surface contexts.

Yannick KORPAL, a young French palaeoanthropologist from Reunion Island in Paris has done an excellent doctoral comparative work on the frontal bone of apes, early hominins and recent humans. In his next (12th) paper "*Morphometrics of the Frontal bone: a new method for measuring intracranial profiles*" he highlights the spheno-occipital flexion at the base of the skull as characteristic of the genus *Homo*. Yannick contends that it is not a rigid bone block and acts as a dynamic complex; its bone growth, constitution and position affects the entire skull and even the axial skeleton. There is an interesting evolutionary trend in the frontal conformation about the midsagittal frontal profile from a flat conformation of the great apes to a more straightened and rounded frontal of modern humans. He invented a new ellipse method that enables differentiation across taxa of apes and hominins by the endocranial profiles, which otherwise remain same. He considered that the method permits intergroup diagnosis of the profiles of man and therefore useful for specific diagnosis.

In the next article (13th) Ratna BHATTACHARYA from Kolkata takes us to the eastern Indian prehistory through her preliminary study on the "*Discovery of two prehistoric sites at Galudih in east Singbhum, Jharkhand: A study in typo-technology and geomorphology*". She describes the site geomorphology in a natural geographical setting along with the typo-technology of the findings and observes a continuous cultural development in this part of India. Her contention may be correct on the strength of field data need to be handled with proper documentation and presentation. The area, is however very promising and important to understand the eastward expansion and migrations of the later Middle to Upper Pleistocene hominins, also seen by me at Susunia in West Bengal (Sankhyan et al., 2009).

One of the most remarkable discoveries of the 21st century has been of the skeleton of a tiny woman (popular as Hobbits) assigned to a new species, *Homo floresiensis* by Peter brown and others in 2003 & 2004, discovered at the Liang Bua cave on Flores Island of Indonesia. I was also fortunate to have visited the cave site in July 2007 after deliberating (Sankhyan & Rao, 2007) in the *International Seminar on South East Asian Palaeoanthropology* held in Yogyakarta in which I was also entrusted to act as the Secretary of the conference panel on Hobbits. I noted that discovery has created great excitement and controversy and hype in international media and the anthropological community. The tiny bipedal woman, who lived as late as ~18,000 years ago, had a small-brain like that of a chimpanzee and survived in isolation but with also dwarf Stegodon elephants but giant dragons. The discovery has passed a decade of debate, scientific publication, and intense media scrutiny, but still remains debated as for its evolutionary models and dispersion, whether insular island dwarfed descendant of *H. erectus* or relative of the Plio - Pleistocene African hominins or the varied evolution of its body parts due to heterochrony (Van Heteren & Sankhyan 2009) or a descendant of a 'short and stocky' lineage represented in Narmada valley fossil record (Sankhyan & Rao, 2007).

Although, Peter Brown had contributed an abstract, but could not make it to send a full paper due to his heavy preoccupations and nothing new to add to his earlier publications on the finding. So, the opportunity was awarded to a noted young palaeontologist of The Netherlands, Anneke H. VAN HETEREN. She in her article, 14th in the order, *"Unbalanced endemic island faunas: are hominins the exception?"* partly reusing her earlier data (Van Heteren & Sankhyan 2009) highlights that evolution on islands differs fundamentally from evolution on mainland. Drawing parallelism with other island mammals in Flores Island she concludes that *Homo floresiensis* was not an exception and adapted to a 'low-gear locomotion' and heterochrony, and there are reversals to ancestral states. Therefore, she considers its evolution from *Homo erectus* in an island environment as the most parsimonious solution. Of course, this explanation would stand as long as we do not have direct fossil evidences to support a pygmy/tiny ancestral lineage.

It is a great challenge to demonstrate differences in brain development from the Oldowan and Acheulean stone tool technology practiced by *Homo habilis* and *Homo erectus*, respectively. In the next article (15th) *"Imaging Oldowan-Acheulian knappers: Scope & limitations"* Tanusree PANDIT & A.R. SANKHYAN have reviewed the importance of modern fMRI, PET, MRI scanning techniques used in such studies recently by Stout et al. (2011) and Faisal et al. (2010). They found differences in activation of brain parts in the two techniques by experimenting on modern 'naïve' (=Oldowan) and 'expert' (=Acheulean) knappers. It is quite interesting, but we cannot extrapolate the technological behaviour of our distant ancestors who lived between the last 200,000 and 50 Kya by simple experiments on modern humans. We need to have lots of controls since such behaviours were governed by a multitude of selection pressures, ecological settings, the availability/seasonality of diet, stone tool raw material – its quality, hardness and softness, hominins morphological idiosyncrasy – body form, sexual dimorphism, manual dexterity, etc. We need to extend the experiments on people at different levels of technological developments to arrive at meaningful conclusions.

However, early human behaviours and cognitive developments could be relatively better inferred from the archaeological evidences. Professor Robert BEDNARIK – a famous rock art archaeologist of Australia in his next (16th) article, *"Pleistocene beads and cognitive evolution"* has made a good attempt in this direction. He highlights the beads and pendants found in archaeological contexts as most instructive in understanding cognitive evolution of early hominins considering that such objects demand self-awareness and a theory of mind in their users as well as complex meanings as perfection was involved in their production. He also contended that advanced cognition was already in the oldest Lower Palaeolithic humans. So, restricting modern human behaviour to the Late Pleistocene is false according to him.

We can also better understand the technological and behavioural adaptations of our Stone Age ancestor from similar most ancient survivors, such as the Negrito pygmies. Therefore, in the next (17th) article, *"The Andaman pygmy: origins and new adaptations"* A.R. SANKHYAN & R. SAHANI have attempted to study them from field studies and hard fossil and archaeological evidences. The Negrito pygmy populations of the Andaman Islands are very fascinating and have long since been a centre of attraction and curiosity due to their tiny body, prehistoric life styles; the scholars debate the mystery of their entry into the Andaman archipelago as well. They have deeply probed into the mysteries of their origins by taking stock of the historical, archaeological, palaeoanthropological and genomic evidences, and the present-day bio-cultural adaptations of the four major populations of the Negrito pygmy – the Onge, Jarawa (Ang), Greater Andamanese and the Sentinelese. Their re-synthesis sheds new remarkable light on their origin in Indian heartland over 40 Kya and their common ancestry with the Rajbanshi, Pauri Bhuiya, Munda, Lodha and the Chenchu as early as 60 Kya. It is quite late that they got differentiated into the four major groups of the Andaman pygmy subsequently, which are at different levels of acculturation. Among them the Sentinelese are still in isolation, whereas the Jarawa are just recently in touch with the modern humanity but retaining their traditional forest life style of hunter-gatherers, which helps them maintain a good physique. The outside contact made them venereal to common diseases – viral fever, cough and cold, measles, malaria etc., which can be life threatening to them. The most acculturated population is that of the Great Andamanese, who were once the dominant inhabitants of the Andaman Archipelago before 1800 AD, but now their habitats are captured by the Jarawa and the mainland people. The Onge territory in the Little Andaman Island has also been occupied by mainland prisoner colonies. So, the Onge is also acculturated, and have shed off their traditional forager life style, and they receive ration from the Govt. agencies. But, this has made them to acquire morbid maladaptations of obesity and decreased their average physical activity level (PAL). They are greatly marginalized and vulnerable to extinction as their current number is less than a hundred.

The Neolithic and the Bronze Age period of human late prehistory and proto-history had developed great expertise in medical surgery performed on human brain reviewed in the next article (18th) *"Amazing Skills: The practice of Trepanation around the world"* by Alexandra COMSA & Anek R. SANKHYAN. It was an interesting ancient specialized practice of brain surgery wide spread since 6000 years ago and still in vogue among some very primitive tribes – like the Masai of eastern Africa. The authors observed that the surgical intervention upon human brain case is very risky and scaring even today with modern technology, but wonder that the spectacular Trepanation was followed globally as early as 5000-4000 years ago with awesome primitive stone and bronze instruments and archaic anaesthetics and painkillers! This amazing practice always stirred a great interest among anthropologists, archaeologists, the historians of medicine and religion, and the public at large. A brief overview presents its global history and continuity, the causes, the beliefs, the significance, the operational process and kinds of instruments and methods used, responses and complications of the operation, the symbolic and pseudo-trepanations, etc.

The 19th article is a very interesting illustrated account of ancient *"Decryption of the ethnic identity of the white mummies in Tarim Basin of China"* by Xinyan CHI & Jia WANG. The authors provide all together 32 beautiful images and briefly remark on the origins and significance the symbolic relics representing many types, unearthed at Xiaohe (Little River) in Sichuan Basin of China. Xiaohe cemetery of the dead White people. It was a great discovery of the 21st century, dated to between 2000-1500 BC. Located in the Tarim Basin of China, the landscape of the cemetery is extremely strange, mysterious and exaggerative. Focusing on its vegetation and digital phenomenon, the authors try to decrypt the secrets of Xiaohe people with the heavenly stems and earthly branches (HSEB), oracle-bone and bronze inscriptions, and historical documents.

The next article (20th) is a brief report of a very important *"Identification of a breast cancer BRCA1 mutation in West Bengal, India"* by Abhishikta GHOSH ROY and others as a part of her doctoral work. The inclusion of this paper is because of the fact that the breast cancer is becoming an increasing phenomenon in modern society. The survival of a species depends on feeding its young, thereby making the breast critical to the perpetuation of the human race and the development of the breast gland takes many years longer than any other organs, which allows evolution to tinker with the breast. The paper though small, but an important contribution to understand the molecular evolution of the breast cancer and isolating the BRCA1 gene, wide spread in West Bengal, eastern India. The study contemplates practical applications in early detection and prognosis of the disease.

The next article (21st) is included on *"Depleting biosphere reserves: traditional & modern concerns in India"* by Umesh KUMAR – a young upcoming human ecologist of India who has done extensive research on Indian biospheres. Understanding biodiversity is critical for sustainable human welfare and for adaptive change and our own survival. Umesh explores the environment-human nexus and the causes of biodiversity loss due to industrialization and urbanization. From Indian context he suggests that the ideal approach is involving people's participation – a 'bottom-up' approach as stakeholders in the management of natural resources and ecosystem conservation. It is so because the indigenous populations living within the BR have their traditional ecological knowledge and perceptions in terms of sacred species, sacred groves, sacred water bodies, sacred landscape, etc., which help preservation of the ecosystems.

In the 22nd article Somnath CHAKRAVERTY, a famous Indian rock art scholar, delineates the preliterate and prehistoric societies in its totality from the rock art inventory of India, focusing his observations on the world famous Bhimbetka, Adamgarh and Pachmarhi rock sites in the hilly confinements of Narmada valley, Central India. In his challenging work he traces a continuum from the prehistoric era to the living reminiscent cultures of the tribal egalitarian societies still living in partial isolation and maintaining their traditional cultures, value-system and life-activities.

In the last (23rd) very interesting article entitled *"Astronomical orientation of the trepanned Neolithic woman of Burzahom, Kashmir"* Iharka SZÜCS-CSILLIK, Alexandra COMŞA and Anek R. SANKHYAN have conducted the first archaeo-astronomical investigations on the Neolithic Burzahom skeleton of a 25-30 years female, dated to about 4000 BP. The woman's skull was multiply trepanned for a possible brain anomaly and head injury, but the ancient surgeon who performed the operation was a riddle. We observed that the skeleton No. 6 had a different astronomical orientation than the rest of the Burzahom skeletons. Sankhyan's recent study on it (now published in 2015) informed that it belonged to a male of 50-55 years of age, who was 'very robust' and medium tall with round head, broad nose and receding forehead, and he noted that the man was phenotypically different from the rest of the tall, narrow-nosed and long-headed Burzahom skulls; he noted one such comparable skull in the Neolithic Nagarjunakonda (Andhra Pradesh) in southern India. Hence our study claims that the skeleton No. 6 most likely belonged to the Shaman-cum-surgeon who treated the woman spiritually and surgically, and for that act he was accepted by the Burzahom pit-dwellers. and buried among them.

1.
A NEW JUVENILE CRANIUM FROM ZHAOTONG CITY, SOUTHWEST CHINA INDICATES COMPLEXITY OF HOMINOID EVOLUTION IN EASTERN ASIA

Ji XUEPING
Yunnan Institute of Cultural Relics and Archaeology, Kunming 650118, China
jxpchina@foxmail.com

Deng CHENGLONG
State Key Laboratory of Lithospheric Evolution, Institute of Geology and Geophysics, Chinese Academy of Sciences, Beijing 100029, China

Yu TENGSONG
Zhaotong Institute of Cultural Relics, Zhaotong, 657000, China

Abstract: *A newly discovered juvenile hominoid cranium dated to the Late Tertiary of Southwest China is only the second sub-adult found for the genus Lufengpithecus and one of the few found anywhere in the world. Specimen ZT 299 is morphologically distinct from known species of Lufengpithecus and plausibly represents a new species. The discovery of the cranium also reaffirms biogeographical and ecological scenarios invoking endemism in Southwest China during the Late-Tertiary as likely drivers of primate diversity.*

Keywords: *hominoid evolution, Lufengpithecus, latest Miocene, primate diversity, Zhaotong, China*

INTRODUCTION

More than 80 ape species have been recovered from Miocene deposits in Africa, Europe and Asia (Fleagle, 2013). Yet, the evolution of the hominoids remains poorly understood. Begun et al. (2012) have proposed that the Middle Miocene *Griphopithecus* might be a stem hominid, while the Middle Miocene *Kenyapithecus* and Late Miocene *Oreopithecus* could be stem great apes (hominoids). With respect to the living apes, except humans, only two fossils have been recovered that show clear links to any extant hominoid (McBrearty & Jablonski, 2005; Suwa et al. 2007). However, these are very incomplete to the point where a reliable species diagnosis is not yet possible despite showing clear affinities to chimpanzees (McBrearty & Jablonski, 2005) and gorillas (Suwa et al. 2007).

In East Asia, a number of fossil apes have historically been associated with the orangutan clade (*Pongo*). These include *Khoratpithecus* (Harrison, 2006), *Sivapithecus* (Kelley, 2002; Begun et al. 2012), and *Gigantopithecus* on account of its similarities to *Sivapithecus* (Fleagle, 2013). Yet, affinities to *Pongo* are far from straightforward with a lack of comparable anatomical elements across taxa, precluding proper comparison, as well as many important morphological and biomechanically relevant distinctions between them and living orangutans. From this, Fleagle (2013) has concluded that the best extinct candidate for membership of the Pongo clade is *Sivapithecus,* despite important locomotion differences between these genera, concluding that, "it seems reasonable to view this as evidence that the orangutan clade contained a greater diversity of apes in the past" (p. 337).

A similar situation applies to the lesser apes with various Miocene taxa (e.g. *Propliopithecus, Pliopithecus*, and *Micropithecus*) having historically been linked to gibbons on account of features, such as small body size, reduced snout length or orbit size (Fleagle, 2012). Yet, only the Late Miocene genus *Yuanmoupithecus* from Yunnan Province, Southwest China, seems to be a genuine candidate for a sister group to the living gibbons (Harrison, 2010).

Besides *Yuanmoupithecus*, Yunnan contains various localities that have provided evidence for the Miocene ape genus *Lufengpithecus. Lufengpithecus* is possibly the common ancestor of African apes and hominidae (Wu, 1987); a new species of *Ramapithecus* (Zhang et al., 1987), linked to the orang-utan clade (Schwartz, 1990); either a primitive hominid that represents the sister taxon of the Ponginae+Homininae or a primitive sister taxon to the Ponginae, but favour the second alternative; a side branch split from the basal stem of the hominidae (*Lufengpithecus hudienensis*) (Zheng, 2006); an early member of hominidae (Xu & Lu, 2008); either a stem hominid or possible basal member of the ape / human clade (Kelley & Gao, 2012); a possible primitive hominid (extant great ape and human) morphotype, with an overlay of a few autapomorphic features such as the exceptionally wide orbits resulting in a very wide mid-facial region (Ji et al., 2013: ZT299).

Four species have been described: *L. keiyuanensis*, recovered near Kaiyuan in the Xiaolongtan Basin and dated ~10-11 Mya (Woo, 1957; Zhang, 1987; Qiu & Qiu, 1995); *L. hudienensis*, found at multiple sites in the Yuanmou Basin and dated ~7.1-8.2 Mya (Zhang et al. 1987; Qi et al. 2006; Yue & Zhang, 2006); and *L. lufengensis* from Shihuiba in the Zhaotong Basin and dated ~6.13 or 6.24 Mya (Yue & Zhang, 2006; Xu et al. 1978; Wu et al. 1981), (Ji et al., 2013).

NEW HOMINOID FIND

New fieldwork at the locality of Shuitangba, Zhaotong, Northeast Yunnan, has provided a new, remarkably complete and undistorted juvenile ape cranium (ZT 299) belonging to the *L. cf. lufengensis*. Specimen ZT 299 was recovered from an open-pit lignite mine located in the Zhaotong Basin, a brickyard near Zhaotong Airport (Fig. 1.1, 1.2, 1.3 & 1.4). It comprises vertebrate fossils recovered previously from the Zhaotong Basin, thought to be of Pliocene to early Pleistocene in age (Zhou et al., 1962, Zhang et al., 1978).

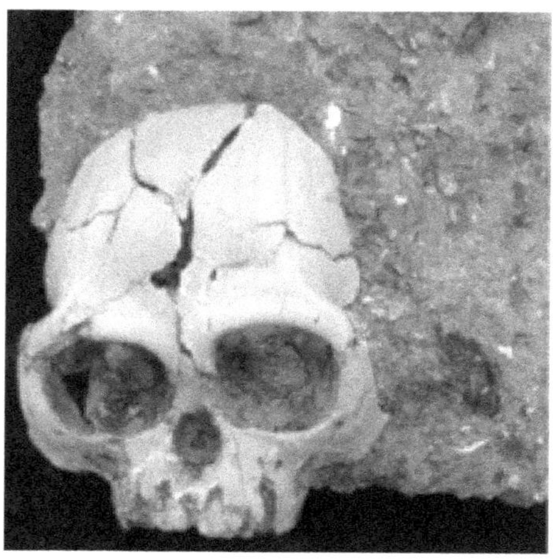

Fig. 1.1. Zhaotong juvenile hominoid skull

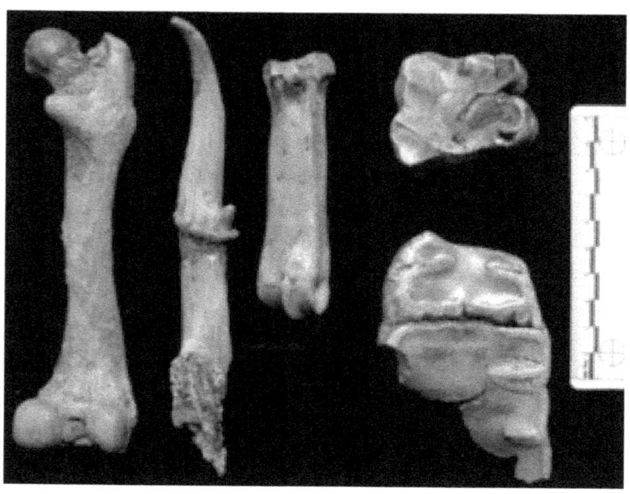

Fig. 1.2. Associated mammalian fossils

Fig. 1.3. Hominoid skull in situ in the bed

Fig. 1.4. Excavation for salvaging the fossil remains

However, the biochronological and geochronological investigations indicate a terminal Miocene age (Ji et al. 2013). In particular, the fauna from Shuitangba belongs to the Asian Baodean Stage/Age (Fig. 1.2), a standard terrestrial biochron for the late Miocene bracketed by dates in the range of 7.2-5.3 Mya (Zhu et al., 2008).

The Palaeomagnetic dating of ZT299 is estimated to be 6.13 Mya or 6.24 Mya. This indicates that ZT 299 is slightly younger than the youngest *L. lufengensis* fossils from Shihuiba (Ji et al., 2013) where the diagnostic characteristics of ZT299 are given.

Significance

The new ZT 299 hominoid is only the second relatively complete juvenile cranium belonging to *Lufengpithecus* found to date and one of the few fossil sub-adult ape crania found anywhere in the world. Thus, it offers new opportunities to understand the ontogeny of Miocene apes, especially *Lufengpithecus,* and to place it within a comparative framework of the well-characterized development of extant great apes. This will be vital for understanding the ontogenetic basis of specific differences among the hominoids and how evolution has acted to alter developmental trajectories to produce novel ape morphologies.

Many taxonomically relevant distinctions have been found between ZT 299 and the ontogenetically similarly aged YV0999, representing *L. hudienensis,* and reconstructed adult cranium of *L. lufengensis,* PA 644. Further comparison can be made with features employed by Kelley & Gao (2012) to compare YV0999 with extant apes. In doing so, it seems clear also that ZT 299 is distinct from both extant apes and *L. hudienensis*. This suggests that ZT 299 probably represents a new species, adding greatly to known diversity in the Miocene apes of East Asia, especially China. For the moment, ZT 299 is assigned to *Lufengpithecus cf. lufengensis*, pending further work.

This new and morphologically distinctive juvenile cranium also raises important questions concerning the broad processes involved in the generation of diversity during the later Neogene of Southwest China. Faunal endemism throughout the region is exemplified by the remarkable dissimilarity found in the composition of the primate faunas as a whole from the three principal hominoid-bearing sites in Yunnan (Ji et al., 2013). It seems probable that the apes not related to living species emerged during the Miocene and perhaps persisting in the Pleistocene in various part of Asia (Ciochon, 2009; Fleagle, 2013). The fossil record of Southwest China including Yunnan may hold a key to understand the primate diversity and ecological causes explicitly linked to the uplift of the Qinghai-Xizang (Tibet) Plateau and development of the East Asian monsoon which set the stage for the modern floral and faunal communities of the broad region (Qiu & Li, 2005).

Acknowledgements

This field and laboratory work of this project was supported by the Yunnan Natural Science Foundation (2010CC010), the Zhaotong Government, the National Basic Research Program of China and the National Natural Science Foundation of China (2012CB821900 and 40925012), National Science Foundation (BCS 1035897, BCS 1227838 and BCS 0321893), Bryn Mawr College. Thanks also to Dr. Darren Curnoe of the University of New South Wales, Australia for his beneficial discussions.

References

BEGUN, D.R., NARGOLWALLA, M.C. & KORDOS, L. 2012. European Miocene hominids and the origin of the African ape and human clade. *Evolutionary Anthropology* 21(1): 10-23.

CANDE, S.C. & KENT, D.V. 1995. Revised calibration of the geomagnetic polarity timescale for the Late Cretaceous and Cenozoic. *J Geophys Res* (Solid Earth), 100: 6093-6095.

CHOW, M. et al. 1962. Early Pleistocene Mammals of Chaotung, Yunnan, with notes on some Chinese Stegodonts. *Vertebrata Palasiatica* 6(2): 138-147.

CIOCHON, R.L. 2009. The mystery ape of Pleistocene Asia. *Nature* 459: 910-911.

CLARK, M.K., HOUSE, M.A., ROYDEN, L.H. et al. 2005. Late Cenozoic uplift of southeastern Tibet. *Geology* 33: 525-528.

FLEAGLE, J.G. 2013. Primate Adaptation and Evolution: 3rd Edn. Academic Press.

GRADSTEIN, F.M., OGG, J.G. & SMITH, A.G. 2004. Geologic Time Scale 2004. Cambridge: Cambridge University Press.

HARRISON, T, JI, X.P., DENISE, S.U. 2002. On the systematic status of the late Neogene hominoids from Yunnan Province, China. *Journal of Human Evolution* 43: 207-227.

HARRISON, T. 2010. Apes among the tangled branches of human origins. *Science* 327: 532-534.

JI, X.P., JABLONGSKI, N.G., et al. 2013. Juvenile hominoid cranium from the terminal Miocene of Yunnan, China. *Chin Sci. Bull.* 58: 3771-3779.

KELLEY, J. 2002. The Primate Fossil Record. In W. Hartwig (Ed.). Cambridge University Press, Cambridge, U.K., pp. 369-384.

KELLEY, J. & GAO, F. 2012. Juvenile hominoid cranium from the late Miocene of southern China and hominoid diversity in Asia. *Proceedings of the National Academy of Sciences* 109 (18): 6882-6885.

MCBREARTY, S. & JABLONSKI, N.G. 2005. First fossil chimpanzee. *Nature* 437: 105-108.

QIU, Z. & LI, C. 2005. Evolution of Chinese mammalian faunal regions and elevation of the Qinghai-Xizang (Tibet) Plateau. Science in China Series D: *Earth Sciences* 48(8): 1246-1258.

QIU, Z.X. & QIU, Z.D. 1995. Chronological sequence and subdivision of Chinese Neogene mammalian faunas. *Palaeogeogr. Palaeoclimatol. Palaeoecol.* 116: 41-70.

QI, G.Q., DONG, W., ZHENG, L. et al. 2006. Taxonomy, age and environment status of the Yuanmou hominoids. *Chin Sci. Bull.* 51: 704-712.

SCHWARTZ, J.H. 1990. Lufengpithecus and its potential relationship to an orangutan clade. *Journal of Human Evolution* 19 (6): 591-605.

SCHOENBOHM, L.M., BURCHFIEL, B.C. & CHEN, L.Z. 2006. Propagation of surface uplift, lower crustal flow, and Cenozoic tectonics of the southeast margin of the Tibetan Plateau. *Geology* 34: 813-816.

SUWA, G., KONO, R.T., KATOH, S., ASFAW, B. & BEYENE, Y. 2007. A new species of great ape from the late Miocene epoch in Ethiopia. *Nature* 448: 921-924.

YUE, L. P. & ZHANG, Y.X. 2006. Paleomagnetic dating of Lufengpithecus hudienensis localities (in Chinese). In: G.Q. Qi & W. Dong (eds), Lufengpithecus hudienensis Site. Beijing: Science Press, pp. 245-255.

WOO, J.K. 1957. Dryopithecus teeth from Keiyuan, Yunnan Province. Vertebrata Palasiatica 1: 25-32.

WU, R.K., HAN, D.F., XU, Q.H., et al. 1981. Ramapithecus skull found first time in the world (in Chinese). *Kexue Tongbao* 26: 1018-1021.

WU, R.K., XU, Q.H. & LU, Q.W. 1985. Morphological features of Ramapithecus and Sivapithecus and their phylogenetic relationships – Morphology and comparison of the teeth (in Chinese). *Acta Anthropol. Sin.* 4: 197-204.

YANG, Y., TIAN, K., HAO, J., PEI, S. & YANG, Y. 2004. Biodiversity and biodiversity conservation in Yunnan, China. Biodiversity & Conservation 13 (4): 813-826.

XU, Q.H., LU, Q.W., PAN, Y.Y., et al. 1978. Fossil mandible of the Lufeng Ramapithecus (in Chinese). *Kexue Tongbao* 9: 544-556.

XU, Q.H. & LU, Q.W. 2008. Lufengpithecus lufengensis—an early member of hominidae. Science Press, Beijing. 222-224.

ZHANG, Y.P., YOU, Y.Z., JI, H.X., et al. 1978. The Cenozoic deposits of the Yunnan Region (in Chinese). Prof Papers Strat. Paleontol. 7: 1-21.

ZHENG, L. 2006. Lufengpithecus hudienensis site. G.Q. Qi & W. Dong, (eds), Science press, pp. 101-108.

ZHANG, X.Y. 1987. New materials of Ramapithecus from Kaiyuan, Yunnan (in Chinese). *Acta Anthrop. Sin.* 6: 81-86.

ZHANG, X.Y., LIN, Y.P., JIANG, C., et al. 1987. A new species of Ramapithecus from Yuanmou, Yunnan (in Chinese). *J. Yunnan Univ* (Soc. Sci.) 3: 54-56.

ZHENG, L. 2006. New Lufengpithecus hudienensis fossils discovered within the framework of State Key Project of the 9th Five Year Plan (in Chinese). In: G.Q. Qi & W. Dong (eds), Lufengpithecus hudienensis Site. Beijing: Science Press, pp. 41-73.

2.
AUSTRALOPITHECINE SHOULDERS: NEW REMAINS FOR OLD DEBATE

Jean-Luc VOISIN

UMR 7268 *ADES*, Aix-Marseille Université/EFS/CNRS, Faculté de Médecine de Marseille, 51 Boulevard Pierre Dramard, 13344 Marseille, Cedex 15, France and UMR 7194 et USM 103, Département de Préhistoire, Muséum national d'Histoire naturelle. Institut de Paléontologie Humaine, 1 rue René Panhard 75013 Paris. France

jeanlucvoisin2004@yahoo.fr

Abstract: *Since the discovery of three well preserved australopithecine skeletons, viz., Malapa (*Australopithecus sediba*), Woranso-Mille (*Australopithecus afarensis*) and Dikika (juvenile* Australopithecus afarensis*) the debate about their putative arboreal locomotion is still not closed although their upper limbs, and in peculiar their shoulder girdles, are well preserved. Upper arms movements depend directly on shoulder architecture such that this complex is the key system for knowing the arboreal part of any primate behaviour. Notwithstanding within these three newcomers the shoulder remains are well preserved, the studies on them have lead to opposite conclusions. In present work I discuss why these three newcomers have not sold out the debate about australopithecine arboreal behaviour. There are a number of reasons for the divergent inferences, such as the conservation state as well as the hominoid species used for comparisons, being the major ones. The out shoulder morphology of these three newcomers lead me to draw a scheme of australopithecines shoulder which is close to the first* Homo *species and extant apes but not similar to any of them.*

Keywords: *Scapula; Clavicle; Humerus; Locomotion;* Australopithecus afarensis; Australopithecus sediba

INTRODUCTION

Shoulder is an important joint complex because it links the upper limb to the axial skeleton and allows, due to the clavicle, upper limb movements outside the parasagital plane. This allows climbing and manipulating habits within Theria – five orders still possess a complete clavicle, viz., Primates, Chiroptera, Insectivora (except Potamogale), Dermoptera, and Tubilidentata (Lessertisseur and Saban 1967). Although, shoulder complex is very important to understand extinct hominin behaviour, very little work has been done on it (Oxnard 1968a, b; Vallois 1976, 1977; Vrba 1979; Senut 1981; Voisin 2001a, 2004a, 2010) because shoulder remains are very scarce and fragmentary, especially for older species like australopithecines. Thereby, we knew very little about australopithecine shoulder before 2006. But, since 2006, three newcomers, Woranso-Milles (*Australopithecus afarensis*), Dikika (juvenil *Australopithecus afarensis*) and Malapa (*Australopithecus sediba*) have fundamentally enhanced our knowledge about australopithecines shoulder. However, studies on their shoulders lead to opposite conclusions, which apparently make no sense and compel a review.

This paper attempt a comparative analysis of the shoulder studies of these three newcomers (Alemseged et al. 2006; Berger et al. 2010; Haile-Selassie et al. 2010; Green and Alemseged 2012; Churchill et al. 2013) and try to explain the contradictions among conclusions arrived at by various studies. Nevertheless, this work would not classify publications from the worst to the best but just make a reflection about palaeoanthropological studies and give my point of view about australopithecines arboreal behaviour.

I will start with a summary of the characteristics of the shoulders of these three new australopithecines and I would try to highlight the weaknesses of these studies and complete, when possible, analysis made by authors. I will then draw an architectural scheme of the australopithecines shoulder.

AUSTRALOPITHECINES

Australopithecus sediba (Malapa)

The two Malapa fossils are not the most complete australopithecine remains, but they are exceptional for the shoulder study. In particular, MH2 individual displays a nearly complete clavicle and a very well preserved scapula and humerus. These three bones belong to the same side (Berger et al. 2010; Churchill et al. 2013).

The dorsal view morphology of the Malapa clavicles (UW88-38 & UW88-1) is described as closer to the chimpanzee than to modern human (Churchill et al. 2013) which is coherent with the Fig. 2 of their paper, especially for the complete clavicle UV88-38 (MH2 individual).

If *Australopithecus sediba* clavicles are morphologically closer to the chimpanzee than to modern humans thus Malapa clavicles would display two curvatures in dorsal view (Fig. 2.1), but picture from Churchill and colleagues does not allow to determinate any curvatures in dorsal view. These two curvatures are associated to a scapula

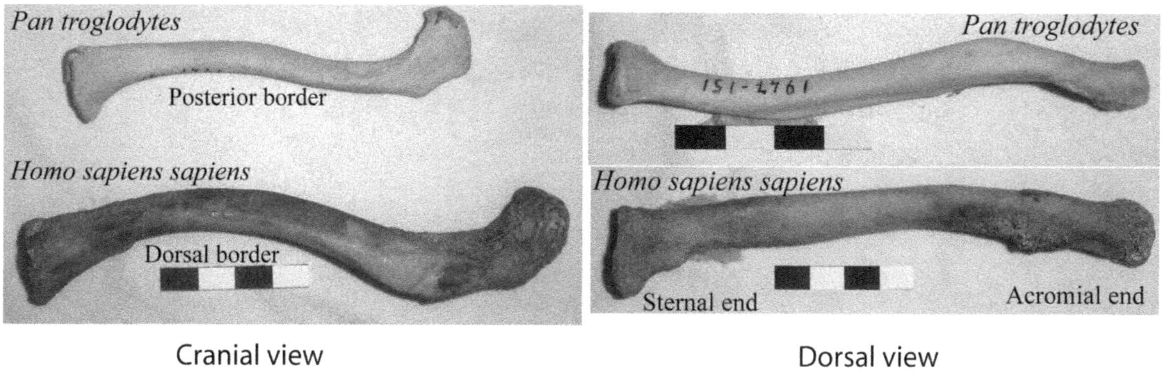

Fig. 2.1. Chimpanzee and human clavicles in superior (left) and in dorsal (right) view

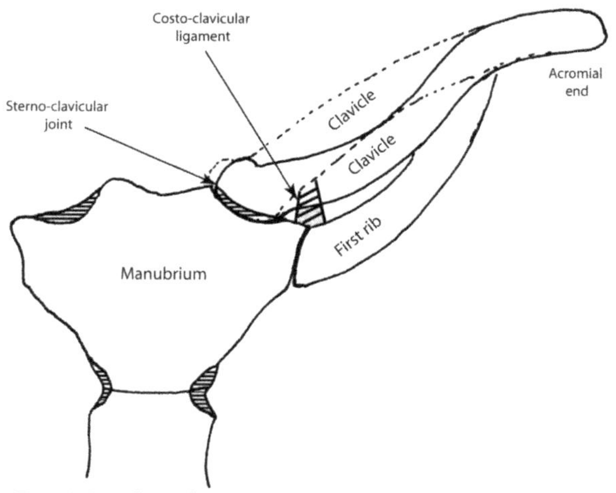

Fig. 2.2. Clavicles associated with a high scapula in regard to the thorax. Dotted line: clavicle with a unique inferior curvature (human morphology); full line: clavicle with two curvatures in dorsal view (great apes morphology). Note the high difference in regard to the manubrium between the two clavicular morphologies (after Voisin 2006)

which is higher in regard to the thorax than in modern human. The sternoclavicular joint is supported by the costoclavicular ligament (the Fig. 2 of Churchill and colleagues paper shows a clear costoclavicular imprint on the sternal end of the clavicle), which exists only in apes and humans (Cave 1961) and limits horizontal and vertical clavicle movements (Voisin, 2006). In this case, a costoclavicular ligament elongation increases the mobility and weakness of the sternoclavicular joint and involves greater muscular control, exerted by the subclavius muscle. In apes, an important superior curvature permits us to associate a high and dorsal scapula with respect to the thorax and a clavicle with its medial end nearly parallel to the manubrium (Fig. 2.2). This condition avoids the elongation of the costo-clavicular ligament (Voisin 2006). Hence, the scapula of *Australopithecus sediba* would be dorsal and higher in regard to the thorax than in modern human. Thereby, the shoulder architecture of this species might be closer to the great apes architecture than to modern human one, as Churchill and colleagues already proposed it.

Malapa clavicles are described as very short by Churchill et al. (2013) as they are shorter than other australopithecines clavicles (Tab. 2.1). However, there are only two australopithecine clavicles which are sufficiently well preserved to estimate correctly their absolute length: the Kadanuumuu one (but see below) and the AL333x-6/9 one. On the other hand, the absolute length of the Malapa clavicle fits with the lowest chimpanzee values (Tab. 2.1). To avoid size effect, the claviculo-humeral index (*clavicle maximum length/humerus maximum length * 100*) has been used. This ratio is the more appropriate method because there is a common isometric scaling relationship between clavicular and humeral length across all non human primates (Larson 2007, 2013). This ratio has not been calculated before Malapa discovery because no individual displayed complete clavicle and humerus from the same side. The value for *Australopithecus sediba* claviculo-humeral index is very low, outside the range of variation of modern human (Tab. 2.1), but inside the gibbon and chimpanzee range of variation (Voisin 2000). In other words, clavicles of Malapa australopithecine are relatively short and humerus is relatively long, which is an arboreal character.

The shortness of the Malapa clavicles implies a scapula, although dorsal, which is more lateral in regard to the thorax than in modern human, even more lateral than Kadanuumuu scapula (but see below). This more laterally position of the scapula corresponds to great apes shoulder architecture and close to what existed in oldest *Homo* species (Voisin 2010). Hence, the Malapa shoulder is characterized by a scapula which is higher and less dorsal in regard to the thorax than in modern human. This architecture is close to that of great apes (Schultz 1961; Sakka 1985; Voisin 2010) and hence might be associated to a funnel chest like in extant great apes.

Churchill and colleagues emphasize the great conoid tubercle which exists on the two clavicles remains UW88-38 (MH2) and UW88-1 (MH1). Following Vbra (1979) these authors consider that this character is associated to arboreal behavior. However, there is a great variation about the size of this tubercle, especially on australopithecines clavicles and its function is still unclear (Larson 2013).

Concerning the humerus, Churchill and his colleagues emphasize the humeral torsion, whereas they give numerous metrics data on the two humeral remains in

*Tab. 2.1. Absolute and relative clavicular length in apes, modern humans and some australopithecine clavicles. M: mean (in mm); SD: standard deviation; (L): left; (R): right; §: fossils and Primates not included in Haile-Selassie et al. (2010) and in Churchill et al. (2013); *: data calculated with the complete humeral length estimated. **: Data calculated with the humerus and clavicle length estimated; –: No data available (after table S8 of Haile-Selassie et al., 2010 and after table S1 of Churchill et al. 2013)*

Taxon	Sample	Clavicle Length	Claviculo-humeral index	Authors
§Hylobates	§22	§M: 93.9 SD: 7.7	§M: 41.2 SD: 4.6	Voisin (2000)
§Pongo pygmaeus	§24	§M: 163.4 SD: 22.3	§M: 49,1 SD: 2.35	Voisin (2000)
Pan troglodytes	25	M: 128.7 SD: 8.8	M: 43,8 SD: 2.6	Churchill et al. (2013)
§Pan paniscus	§19	§M: 104.2 SD: 2.9	§M: 37.1 SD: 1.6	Voisin (2000)
Gorilla gorilla	25-18	M: 151.2 SD: 20.6	M: 36.3 SD: 2.0	Haile-Selassie et al. (2010)
Homo sapiens	25-59	M: 148.7 SD: 11.3	M: 45.4 SD: 2.7	Haile-Selassie et al. (2010)
§West Neandertals	§5	§M: 157.9 SD: 16.9	–	Voisin (2011)
§Homo ergaster	§KNM-WT 15000 (R)	§129.0	§44.5 (41.7)*	Voisin (2000)
	§KNM-WT 15000 (L)	§127.0	–	Voisin (2000)
§Homo sp	§Narmada clavicle (R)	§ 90.0	–	Sankhyan (1997)
	§Narmada clavicle (L)	§~90.0	–	Sankhyan (2005)
§Homo habilis	§OH48 (L)	§129.5	–	Voisin (2000)
Australopithecus sediba	UW88-38 (R)	107.5	40.0	Churchill et al. (2013)
Australopithecus afarensis	§AL333x-6/9 (R)	97.9	–	Lovejoy et al. (1982)
	KSD-VP-1/1f (L)	156.4	§32.5**	Haile-Selassie et al. (2010)

their table S9. The humeral torsion of the two *Australopithecus sediba* humerus is very low but similar to other australopithecines values and inferior to extant apes as well as modern humans (Churchill et al., 2013, Table 2). This result confirms Larson (1996) affirmation suggesting that the similar high degree of torsion found in the latter two groups evolved independently and thus is derived trait.

Although, the Figure 1 from Churchill and colleagues displays an hominoid like overall morphology for the MH2 proximal humerus, they only give data about the antero-posterior diameter of the head which is close to that of modern human and chimpanzee. However, the humeral head morphology is not exactly similar between apes and modern humans. The former have a spherical humeral head or slightly broader than tall while the latter is characterized by an humeral head which is slightly taller than wide (Larson 2007, 2013). Thus a precise study of the proximal humerus has to be done. The morphology and size of the two tubercles are not similar between extant hominoids and are linked to function (Voisin 2000; Larson 2007). Hence, Malapa humerus would give us new insight about australopithecines locomotion pattern in the future with new studies. Indeed, Ciochon and Corrucini (1976) have shown the proximal part of the Sts7 humerus is peculiar in the size of the great tubercle and does not fit either with modern human morphology or of the apes.

Moreover, Churchill and colleagues do not take into account the relative position of the deltoid tuberosity even if they give the absolute position of this tuberosity in their table S9 (Table 2.3). If this position is calculated as a ratio (*deltoid position/Humeral maximum length * 100*) it might be possible to compare it between hominoids (Voisin 2004b). Voisin (2004b) has demonstrated that the lowest values are associated with the most arboreal apes such as Orangutan and Gibbon. The higher values are associated to knuckle walker and biped primates (Chimpanzee, Gorilla and human). As shown in table 2.3, *Australopithecus sediba* tuberosity position index is similar to that of *Pongo* and *Hylobates*. In other words, the insertion of the deltoid muscle on the humerus shaft of *Australopithecus sediba*, revealed by the deltoid tuberosity position, is characteristic of a well-developed arboreal behavior.

The scapula of the individual MH2 (UW88-56) is very well preserved (see figure 3 in Churchill and colleagues paper) and its overall morphology cluster it easily with great apes ones (Churchill et al. 2013). It possesses a cranially oriented glenoid fossa, a markedly cranially directed spine, a convex vertebral border, an acromion process (UW88-103[1]) which is long and curved with a large attachment area for supraspinatus muscle. All those traits are characteristic of apes (Vallois 1932) even if

[1] This remains comes from a left scapula.

*Tab. 2.2. Humeral torsion angles in extinct hominins and extant hominoids after table S8 of Churchill et al. (2013). M: Mean; SD: Standard deviation; n: Sample size; *: May be not* A. africanus *even not australopithecine (see Howell & Coppens (1976) and McHenry (1994); §: Fossils not included in Churchill et al. (2013) paper. According to picture S26 (Haile-Selassie et al., 2010) length of the humerus KSDP-VP-1/1f has been estimated to 235 mm. As this remains represent 60% of its real size, thus the real size is estimated to 480 mm*

Species	Individual	Adult	Subadult	Author
Australopithecus sediba	UW88-57, (MH2)	117.0°	–	Churchill et al. (2013)
	UW88-34, 42, 88 (MH1)	112.0°	–	Churchill et al. (2013)
Australopithecus afarensis	A.L.288-1 (Lucy)	124.0°	–	Churchill et al. (2013)
Australopithecus africanus	§Omo 119-73-2718*	130.0°	–	(Larson 1996)
	§Sts 7	127.0°	–	(Larson 1996)
Homo erectus	D4167	110.0°	–	Churchill et al. (2013)
	KNM-WT 15000	–	111.5°	Churchill et al. (2013)
	D2680	–	104.0°	Churchill et al. (2013)
Homo floresiensis	B1/50	115.0°	–	Churchill et al. (2013)
Homo sapiens		M: 165.0° SD: 6.9°, n: 27	M: 168.0° SD: 4.5°, n: 4	Churchill et al. (2013)
Gorilla gorilla		M: 159.9° SD: 5.1°, n: 17	M: 162.3° SD: 4.5°, n: 6	Churchill et al. (2013)
Pan troglodytes		M: 153.4° SD: 6.5°, n: 4	M: 155.8° SD: 2.0°, n: 2	Churchill et al. (2013)
Pongo pygmaeus/abelii		M: 135.0° SD: 6.5°, n: 5	M: 138.4° SD: 7.5°, n: 9	Churchill et al. (2013)
Hylobates lar		M: 116.6° SD: 8.1°, n: 11	M: 118.0° SD: 6.9°, n: 10	Churchill et al. (2013)

Tab. 2.3. Metric characteristics of A. sediba *proximal humerus (After table S9 of Churchill et al. 2013); §: Data not included in Churchill et al. (2013). M: Mean; SD: Standard deviation; n: Sample size*

Species	Humeral maximum length (1)	Humeral head antero-posterior diameter (2)	Deltoid tuberosity position (3)	Deltoid tuberosity position index (3)/(1)
Au. sediba	269 n: 1	32.3 n: 1	134 n: 1	§49.8 n: 1
australopithecines	M: 263.6 SD: 47.1; n: 3	M: 34.9 SD: 5.9; n: 7	117 n: 1	–
early/archaic *Homo*	M: 307.0 SD: 17.0; n: 2	–	151 n: 1	–
modern *Homo*	M: 305.7 SD: 21.3; n: 67	M: 37.9 SD: 3.7; n: 67	M: 155.0 SD: 12.8; n: 67	§M: 54.5 SD: 2.7; n: 25
Gorilla	M: 420.5 SD: 41.2; n: 20	M: 57.2 SD: 7.8; n: 20	M: 232.5 SD: 30.9; n: 20	§M: 58.8 SD: 3.1; n: 37
Pan troglodytes	M: 283.5 SD: 14.7; n: 23	M: 39.5 SD: 2.5; n: 23	M: 153.8 SD: 13.1; n: 23	§M: 55.3 SD: 3.3; n: 33
§*Pongo pygmaeus*	§M: 324.2 SD: 29.1; n: 20	§M: 36.8 SD: 5.3; n: 20	§M: 158.7 SD: 14.0; n: 20	§M: 49.0 SD: 2.4; n: 20
§*Hylobates sp.*	§M: 228,9 SD: 12.7; n: 17	§M: 15.7 SD: 1.4; n: 17	§M: 112.1 SD: 13.7; n: 14	§M: 49,5 SD: 4.6; n: 31

their functional role is not always clear. According to Churchill and colleagues, the glenoid fossa of UW88-56 (MH2) is narrow relative to its height, with proportions most similar to orangutans and gorillas. However, the value these authors give for the glenoid fossa index is lower than any extant Hominoid (Tab. 2.4). Thus, this value may be an typographic error. Moreover, between Gorilla, Chimpanzee and Human there are no significant differences concerning the glenoid fossa index (Vallois 1932; Senut 1981; Voisin 2000), Table 2.4). Anyway, data available for australopithecines are included in extant hominoid variation (Tab. 2.4) and there are no obvious differences between Hominoid, except Gibbon (Voisin 2000).

According to Churchill et al. (2013) principal component analysis (PCA) the MH2 scapula is morphologically more similar to *Pongo* scapula than other Hominoids (see

Tab. 2.4. Metric characteristics of MH2 scapula and other hominoids (After tables S3 & S4 in Churchill et al. 2013). M: Mean; SD: Standard deviation, n: sample size

	Glenoid-Axillary Angle		Glenoid Index		Supra and Infra Scapular Fossa Ratio			
Species	***Value***	***Authors***	***Value***	***Authors***	***SFB***	***IFB***	***(1)/(2)***	***Authors***
Au. Sediba (MH2)	123.4°	Churchill et al. (2013)	55.6	Churchill et al. (2013)	49.4	89.2	0.55	Churchill et al. (2013)
A. afarensis (AL. 288 1L)	132° 115° 130° 132.2°	Voisin (2000) Senut (1981) Stern & Susman (1983) Green & Alemseged (2012)	67.5 78.5	Senut (1981) Voisin (2000)	–	–	–	–
A. afarenis (DIK-1-1)	Right: 128.6° Left: 129.2°	Green & Alemseged (2012)	–	–	–	–	Right: 0.75 Left: 0.68	Green & Alemseged (2012)
A. africanus (Sts 7)	120° 116° 128°	Vrba (1979) Senut (1981) Green & Alemseged (2012)	60,6 60,6 51,4 58,5	Vrba (1979) Vallois (1977) Robinson et al. (1950) Senut (1981)	–	–	–	–
A. africanus (Stw 162)	124.3°	Green & Alemseged (2012)	–	–	–	–	–	–
Modern *Homo*	M: 137.8° SD: 4.6; n: 47	Churchill et al. 2013	68.3 5.2	Churchill et al. (2013)	7.2 6.4	108.3 8.24	M: 0.44 SD: 0.05; n: 47	Churchill et al. (2013)
Pan troglodytes	M: 119.0° SD: 4.9; n: 14	Churchill et al. (2013)	71.2 4.7	Churchill et al. (2013)	74.5 6.6	78.5 6.8	M: 0.96 SD: 0.12; n: 14	Churchill et al. (2013)
Gorilla	M: 122.4° SD: 4.3; n: 20	Churchill et al. (2013)	69.4 4.5	Churchill et al. (2013)	111.3 18.9	121.4 19.2	M: 0.91 SD: 0.08; n: 20	Churchill et al. (2013)
Pongo	M: 124.1° SD: 3.5; n: 8	Churchill et al. (2013)	62.1 4.3	Churchill et al. (2013)	39.1 9.6	102.3 7.4	M: 0.38 SD: 0.07; n: 8	Churchill et al. (2013)
Hylobates	M: 115.3° SD: 5.2; n: 33	Churchill et al. (2013)	77.8 7.5	Churchill et al. (2013)	44.6 3.6	20.2 3.2	M: 2.27 SD: 0.46; n: 33	Churchill et al. (2013)

SFB= Supraspinatus fossa breadth, IFB= infraspinatus fossa breadth

figure S2 in Churchill and colleagues paper). However, the inferior axillary border has a prominent flange at the origin of teres major, indicating hypertrophy of this adductor and medial rotator of the humerus like in most human (Vallois 1932). For some traits also, like the glenoid fossa orientation, Malapa values are closer to *Gorilla* than *Pongo* (Churchill et al. 2013). This latter remark is interesting because the overall morphology of Dikika scapula is closer to *Gorilla* one than to any other apes (see below). The differences could be explained by locomotion differences between this two species as well as age differences (see below). Moreover, *Australopithecus sediba* scapula shape, as measured by the scapular index, is most similar to that seen in *Homo* and *Pan* than in *Pongo* (Table 2.4). All these remarks explain why *Australopithecus sediba* scapula is also close to the human cloud in Churchill et al. (2013) PCA (see figure S2 in Churchill and colleagues paper). Therefore Churchill and colleagues works reach the first description of *Australopithecus africanus* scapula by Broom et al. (1950). These latter authors noticed that *Australopithecus africanus* scapula share most characters with *Pongo* and few with other Hominoids, especially humans.

However, Churchill et al. (2013) do not lent any attention to the coracoid process, maybe because it seems to very badly preserved according to their figure 3, but this process might be one of the australopithecines scapula part which is the best studied, at least for Sts 7 remains[2] (Broom et al. 1950; Oxnard 1968a; Ciochon and Corruccini 1976; Vallois 1976, 1977; Vrba 1979). Sts 7 coracoid process has been described as characteristic of brachiating habits but according to Vrba (1979) this may be due to a misinterpretation of the remains. Hence a description of a new remains would have helped to clear the situation because the coracoid process orientation is very peculiar in strictly bipedal primates (Voisin, 2000b).

Churchill et al. (2013) study about the shoulder shows an important arboreal behavior for australopithecines, at least for *Australopithecus sediba*. However, the UW88-56 (MH2 scapula) does not correspond exactly with any extant Hominoid species. This arboreal locomotion is associated to a usual bipedalism (Fleagle 1998; de Bonis, 1999; Cartmill and Smith 2009; Chevalier, 2013) which means when not climbing on trees, upper limb was in a pendant position like in humans. Hence, this mixed of arboreal and bipedalism behavior should explain the peculiar morphology of *Au. sediba* shoulder. In short, there are numerous traits in *Australopithecus sediba* shoulder morphology and some of them are relevant with

[2] Sts 7 Scapula might be one of australopithecine shoulder remains which have been the most studied.

arboreal displacement, especially: (i) the clavicle morphology in dorsal view; (ii) the orientation of the scapula spine as well as its glenoid fossa; (iii) the position of the deltoid tuberosity on humeral shaft and (iv) the humeral torsion. However, some traits are closer to modern human: (i) prominent flange at the origin of teres major and (ii) the dimension of the infra (and supra spinatus fossa.

Australopithecus afarensis (Kadanuumuu or Woranso-Mille)

The skeleton of Woranso-Milles is very well preserved (Haile-Selassie et al., 2010), even if this specimen is not beautiful as Malapa one concerning the shoulder girdle. Only the left clavicle (KSD-VP-1/1f) and the right scapula (KSD-VP-1/1g) are preserved. The right humerus (KSD-VP-1/1b) is also preserved but lacks the proximal epiphysis. Haile-Selassie and colleagues present the clavicle as being long without giving any value of the current length of the remains. However, authors estimate the real length of this clavicle to 156[3] mm which is very long, corresponding to the mean length of gorilla or western neandertal (Voisin 2011), which are known to be longer than modern human clavicles (Tab. 2.1). This value corresponds also to the upper part of modern human variations and to the lower part of the orangutan variations (Tab. 2.1). It is interesting to note that Haile-Selassie and colleagues do not have taken into account Asian apes and forgot also *Pan pansicus*. The close phylogenetic relationship between hominins and the African apes makes the latter the most relevant comparative group and the first we think to use. However, it is also useful to take into account Asian apes to examine the morphological diversity among all extant apes to identify traits that are primitive for hominoids and to understand the functional role of some of them (Larson 2013).

Haile-Selassie and colleagues use only relative lengths for comparing clavicles, but absolute length is also interesting by itself because it is correlated to the scapula position in regard to the thorax, the morphology of the chest and make evolutionary comparison available (Voisin 2006, 2010). AL333x-6/9 is the most complete australopithecine clavicle discovered before KSD-VP-1/1f and attributed to *Australopithecus afarensis*. AL333x-6/9 is nearly complete but, according to Lovejoy et al. (1982) its real length cannot be estimated because of the great variation of these traits within hominoid. However, according to its preserved morphology and comparison with chimpanzee clavicles, AL33x-6/9 clavicle would not exceed a length of 140 mm (Tab. 2.1). The *Homo habilis* one (OH 48), like AL333x-6/9, is nearly complete while bone loss might be greater than in AL333x-6/9 clavicle and its real length would not have exceed 140 mm too (Voisin 2000). Turkana boy skeleton (KNM-WT 15000) has two nearly complete clavicles which are also short with low bone loss at their extremities (Tab. 2.1). Bone loss might be explained by the young age of this individual, around 12 years old (Brown et al. 1985; Walker and Leakey 1993; Schwartz and Tattersall 2003) and the real length would not have exceed 140 mm too (Voisin, 2000). This length would not have change so much until adult age because juvenile stage would not have been as long as modern human (Schwartz and Tattersall 2003) and also because the maximum clavicle growth has already been done. Even the two Narmada clavicles, which are much younger, display a shorter length than the Woranso-Mille clavicle (Sankhyan, 1997, 2005).[4] In other words, the great clavicle length of Woranso-Mille clavicle is not consistent with all Australopithecines and old *Homo* species clavicle morphology. Concerning the relative length the more appropriate method is the claviculo-humeral index (see above). There are two specimens within early hominins where this ratio can be determined: Malapa skeleton (which is the first claviculo-humeral index calculated for australopithecines) and Turkana boy. Malapa claviculo-humeral index is very low (see above) and KNM-WT 15000 (which is more recent) get also a relatively low one (Tab. 2.1). Within extant hominoids, orangutan and modern human display the higher claviculo-humeral ratio which mean these two species posses a long clavicle in regard to their size contrary to other apes, *Homo ergaster* and *Australopithecus sediba*. In other words, the short clavicle is a primitive trait and it becomes longer during human evolution to reach its greatest relative and absolute length with *Homo antecessor*, anatomically modern human and neanderthal (Voisin 2010). Using the estimation length for the Woraso-Mille humerus and clavicle given by Haile-Selassie et al. (2010) it is possible to calculate the claviculo-humeral ratio which is extremely low, much lower than Malapa value (Table 1). In other words, the Woranso-Mille clavicle would be relatively short and its humerus very long. These remarks are not human like at all, and Woranso-Mille individual looks more as a "super" apes, at least for the shoulder contrary to Halie-Selassie et al. (2010) affirmation. As the humerus is not complete and does not come from the same clavicle side, these authors have used some other technique to estimate the relative clavicle length. However, this relative length may be not comparable between species because it might be no isometric scaling between the different bone lengths they used.

Australopithecines chest is usually described as "funnel-shaped", more or less similar to that of extant apes, which implies a scapula less dorsal than in modern human associated to a shorter clavicle (Voisin 2006, 2010). This kind of architecture with a less dorsal scapula is also found in the first *Homo* species (Larson 2007; Voisin 2010). Hence, the very long clavicle attributed to *Australopithecus afarensis* by Haile-Selassie and colleagues means a scapula located dorsally as in modern human associated to very broad thorax. It is thus difficult

[3] More precisely Haile-Selassie *et al.* (2010) propose a length comprises between 140 and 170 mm with the most probable value at 156 mm.

[4] The Narmada right clavicle is complete and 90 mm in length (Sankhyan, 2005) though on first reporting (Sankhyan, 1997) it was thought incomplete and estimated to be 100 mm long. But, the left clavicle is incomplete and appears to be of a similar short length.

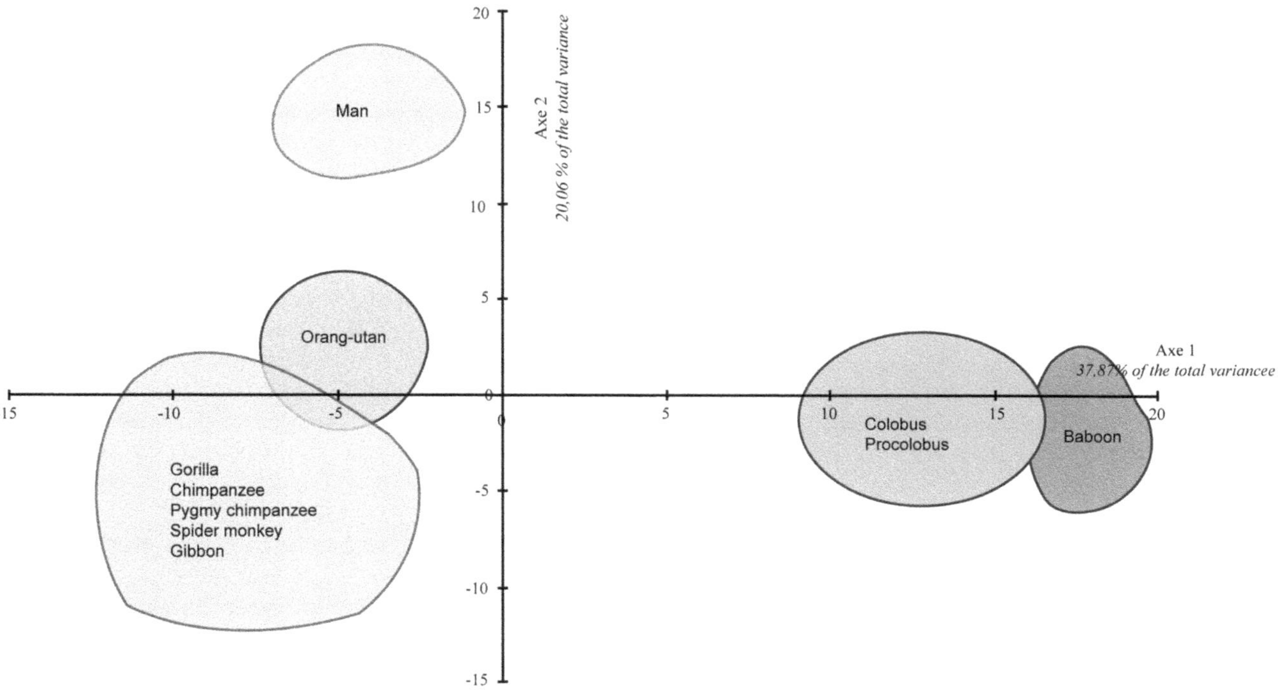

Fig. 2.3. PCA of shoulder data (clavicle, scapula and humerus) showing primates shoulder architecture (after Voisin 2000)

to conceive that, at least in *Australopithecus afarensis*, the clavicle becomes very long, associated to a broad thorax, to reduce in length after (associated again with a tight chest), with the appearance of genus *Homo*. Haile-Selassie and colleagues describe the chest morphology of Woranso-Mille skeleton as nearest to modern human than to other australopithecines.

In other words, this clavicle and the chest morphology associated to, do not fit with australopithecine morphology even if a great sexual morphology would have characterized *Australopithecus afarensis* (Cartmill and Smith 2009 for a review). Moreover, Larson (2013) in her review about australopithecine shoulder morphology does not take into account this clavicle while she includes the scapula, meaning thus the trouble it creates.

According to Haile-Selassie et al. (2010), only the right scapula (KSD-VP-1/1g) is preserved. This bone preserves a nearly complete infraspinous fossa, the entire spine (post-mortem bent) and a complete glenoid cavity but acromion and coracoid are lacking. A small portion of the supraspinous fossa, including the scapular notch, is preserved but not sufficiently to determine the size of this fossa. The supraspinous fossa lack does not allow estimating the proportion of the two fossa. The ratio of the two fossa is important for determining the kind of upper limbs use (see below about Dikika).

Moreover, Haile-Selassie et al. (2010) just give two pictures of this scapula. The two are oriented ventrally with one where the scapula is very little in size and the second is a radiologic one. Hence, no morphological information could be seen on the images.

Woranso-Mille scapula displays ape traits, like a glenoid fossa cranially oriented as well as traits bringing it closer to human scapula, like its overall morphology. This mixture of traits is usual in the australopithecine upper limbs and as lead to two ways of thinking. The first one considers the ape-like morphology as an indication of keeping the ability to climb trees (i.e. Stern and Susman 1983; Stern 2000) and the second one suggests that these features are simply retention of an archaic condition (i.e. Latimer 1991; Ward 2002). Haile-Selassie et al. (2010) show in their PCA that traits like glenoid fossa orientation are not a powerful variable for separating primate scapula. As most australopithecine scapulas remains are limited to the glenoid fossa with short part of the axillary border, they use the "bar-glenoid" angle (Stern and Susman 1983) to estimate the glenoid fossa orientation. To explain why this angle is not powerful to separate species they use a visual observation between the AL.288 scapula remain and a fragmentary modern human scapula (figure S21 in their paper). However, a careful study of this picture shows clear differences between the "bar-glenoid" angles of these two scapulas.

Haile-Selassie and colleagues prefer using geometric relationships within the scapula infrastructure that they consider more powerful to differentiate primates scapula. However, a trait could be not powerful enough to differentiate individuals inside a group, but have a great functional interest. Haile-Selassie and colleagues thus perform a PCA with their own data and the Woranso-Mille scapula fit closer to human cloud than any other species (see figure 4 in Haile-Selassie et al. (2010) paper). However, they use only African apes for comparison and, as we have seen above it is necessary to include all non human hominoid. Chucrchill et al. (2013) showed that the position of a specimen relatively to the hominid clade depended on the choice of a variable. Besides, it also depends on the species used for such comparison. This latter remark could be explained by the fact that Hominoid shoulder architecture is highly variable within hominoid and functional groups can be easily recognized contrary to Cercopithecidae and Colobidae (Voisin 2000) (Fig. 2.3).

Within hominoid shoulder three kinds of architecture could be define: a group including only Orangutan, an another one including only modern human and a group including *Gorilla*, *Pan Hylobates* and even *Ateles* (Voisin 2000) (Fig. 2.3).

***Australopithecus afarensis* (Dikika)**

Dikika is a presumed 3 years old female well preserved skeleton (Alemseged et al. 2006). Concerning the shoulder only the two scapulas are still present. However, the conservative stage is exceptional, especially for a young individual.

The glenoid fossa as well as the spine is oriented cranially, which corresponds to apes morphology and to other australopithecines remains (Tab. 2.4). The glenoid fossa orientation does not significantly change between childhood and adulthood in *Pongo*, *Pan* and *Gorilla* (Green & Alemseged, 2012) contrary to modern human where the glenoid fossa becomes more cranially, even if it still faces much more laterally than in apes (Green & Alemseged, 2012). Starting from DIK-1-1, a humanlike ontogenetic pattern would imply that adult *Australopithecus afarensis* individuals should have more cranially oriented shoulder joints than those displayed by other australopithecine remains. However, both juvenile and adult *Australopithecus afarensis* representatives have comparably oriented shoulder joints, suggesting that this trait remained relatively stable during ontogeny alike or very close to African apes growth trajectory.

The cranially oriented spine induces a greater supraspinatus fossa and a lesser infraspinatus fossa than in humans and similar to what it is found in the two other well preserved australopithecine scapulas (Malapa and Woranso-Mille) and extant apes (see above). These features are clearly seen from the figure 1 of Green & Alemseged (2012) where the two scapulas are shown from dorsal, ventral and lateral view. This spine orientation and the associated scapular fossa development provides a direct line of action for the supra spinatus muscle in preventing displacement of the humeral head during elevation of the humerus (Tuttle & Basmajian, 1978; Rouvière, 1982; Larson et al., 1986; Kapandji, 1994). Thus this muscle is more solicited in suspensory behaviors than in manipulating behaviors. Even if the infra spinatus acts also to prevent the elevation of the humeral head, its main action is to rotate laterally the humerus; which is less important during suspensory behaviors contrary to manipulating ones (Rouvière, 1982; Larson et al., 1986; Kapandji, 1994).

A first PCA analysis (Alemseged et al. 2006), and then confirmed by canonical variates analyses (CVA) performed by Green & Alemsegd (2012), shows that the overall morphology of the Dikika scapula is very close to that of juvenile *Gorilla*. As the multivariate analyzes not only use as much variables as possible but include also Asian apes which made the comparison more robust than the one made with Woranso-Mille scapula (see above). As seen above, Woranso-Mille scapula is close to human one according to Haile-Selassie et al. (2010), contrary to Dikika one although they belong to the same species. However, we have seen also that the human like shoulder of the Woranso-Mille specimen seem to be a misinterpretation, and that the *Australopithecus sediba* shoulder is closer to that of Dikika than to Woranso-Mille one.

As Dikika is a very young individual, a great part of the scapula is now lost because it was made of cartilage. The overall scapula morphology could change much when these cartilages are lost (see figure 1 of Hrdlička 1942). Moreover the overall morphology does change during juvenile growth in human (Hrdlička 1942) and especially among African apes (Green 2013). In other words, some conclusions must be taken with caution before another complete or near complete shoulder remains attributed to *Australopithecus afarensis* from different ages will be discovered. All these remarks could also explain differences between Dikika and Woranso-Mille differences. However, the adult shoulder of *Australopithecus sediba* is clearly adapted to arboreal locomotion and it seems strange that two very close species (*Australopithecus sediba* represented by Malapa and *Australopithecus afarensis* represented by Woranso-Mille) would have such important locomotion differences.

No clavicle and proximal humerus have been preserved with Dikika remains and thus conclusions have to be short concerning shoulder architecture concerning this individual. The clavicle, as shown above, allows to get much information as well concerning the shoulder architecture as the ability of arm rising. The humerus, especially the proximal part and the humeral torsion, give much information concerning the ability of arm rising too as well as brachiation capacity. In other words, clavicle and humerus would be useful not only to confirm the Dikika arborealism but also to precise it. Moreover, during growth there are some changes in scapula and clavicle morphology within modern humans and African great apes (Hrdlička 1942; Olivier and Capliez 1956; Green 2013) and thus some similar changes might also have existed in australopithecine shoulder during childhood.

DISCUSSION AND CONCLUSION

If we follow the line of thinking that the traits, like cranial orientation of the glenoid fossa, have no functional implications being merely archaic traits, then no relationship between bone morphology and function is implied. It is hard to consider so, otherwise most of the morphological studies on fossil remains would lose sense.

Tab. 2.5. Shoulder architectures within genus Homo. *Lateral: the scapula is more laterally placed than in modern human. Dorsal: the scapula is dorsally placed like in modern human (after Voisin, 2010)*

	Architecture 1	Architecture 2	Architecture 3
Clavicle curvatures (Dorsal view)	Two curvatures	Two curvatures	One curvature
Clavicle length	Short	Long	Long
Scapula position (in regard the thorax)	High	High	Low
	Lateral	Dorsal	Dorsal
Homo species	*Homo habilis, Homo ergaster*	Neanderthal *Homo antecessor*	Modern humans, Upper Paleolithic remains, Anatomically modern humans (i.e. Omo)

Anyway, Voisin (2010) has demonstrated there are three shoulder architectures within genus *Homo* (Table 2.5). The first architecture is characterized by a scapula which is high in regard to the thorax and less dorsal than in modern human and corresponds to the first *Homo* species (i.e. *Homo ergaster*). The second one is characterized by a scapula which is still high in regard to the thorax, but as dorsal as in modern human. This architecture corresponds to *Homo antecessor* and Neandertal. The last architecture is characterized by a scapula which is low in regard to the thorax and fully dorsal and corresponds to anatomically modern human (from Omo remains to living humans).

Furthermore, these architectures are linked to chest morphology. The shoulder architecture of the oldest *Homo* is associated to a "funnel shape" thorax, contrary to modern humans where the chest is more barrel like.

As Australopithecines *display* a short clavicle with two curvatures in dorsal view, at least for *Australopithecus sediba*, it is possible to conclude that their scapulas were more laterally placed and higher in regard to the thorax than in modern human. In this respect, their shoulder architecture was close to the oldest *Homo* species. However, they distinguished themselves by some arboreal traits like glenoid fossa orientation which do not exist in any *Homo* species. Australopithecine shoulder architecture is not identical to any apes too as Corrucini and Ciochon (1976) already propose it with their study on Sts 7 shoulder remains.

Some upper limbs movements would be possible or not (or not efficiency) depending shoulder architecture. Thereby, a more lateral scapula, as in australopithecines and old *Homo* species, would limit the gleno-humeral range of movement. With a more lateral scapula, flexion of the upper limbs will be easier than true abduction (Larson 2007). On the contrary, with a full dorsal scapula, the range of movements becomes larger and true abduction becomes possible (Larson, 2007).

Australopithecines shoulder is therefore more adapted to flexion than true abduction. This architecture associated to the scapula and humerus morphology, especially the humeral torsion, is characteristic of climbing ability.

Numerous factors explain the different conclusions obtained on australopithecines shoulder. First of all, interpretations of fossil remains depend on the preserved state of conservation, because it will impose traits we will use for studies. Moreover, to be sure to describe without any trouble any hominoid remains it is absolutely necessary to take into account all apes and not only the African ones. It is also important to have access to nearly all, if not all remains (at least good cast) of the species or relative species for comparison. It is also important to consider arboreal traits as functional ones and not as archaic features with any function, in peculiar for the shoulder. Indeed, bony structures of the scapula are not randomly arranged but depend on forces acting on the scapula (Roberts 1974; Preuschoft et al. 2012), which clearly shows there is a clear relationship between morphology and function for the shoulder. Moreover, the shoulder is a joint complex with more than 20 muscles (the exact muscles number depends on species). Such a complex system cannot function correctly in inappropriate way for a long time because of injuries (rotator cuff tear, tendinitis, etc.).

Australopithecine shoulder morphology is difficult to understand because it does fit exactly neither with the human one nor with the apes one. This is due to the fact that Australopithecines are totally biped when walking on the ground but still use also arboreal locomotion. These ubiquitous functions are also seen at least in *Australopithecus sediba* hand morphology (Kivell et al. 2011) as well as in foot morphology of *Australopithecus sediba* and *Australopithecus africanus* (Clarke & Tobias, 1995; Zipfel et al. 2011).

Acknowledgments

I would like to thank Professor A.R. Sankhyan for having invited me to participate to this book. This study has been first presented as a lecture at the Museum National d'Histoire Naturelle and I would thank Ms A. Dambricourt Malasse for having proposed me to do this lecture.

References

ALEMSEGED, Z., SPOOR, F., KIMBEL, W.H., BOBE, R., GERAADS, D., REED, D. & WYNN, J.G. 2006. A juvenile early hominin skeleton from Dikika, Ethiopia. *Nature* 443:296-301.

BERGER, L.R., DE RUITER, D.J., CHURCHILL, S.E., SCHMID, P., CARLSON, K.J, DIRKS PHGM, & KIBII, J.M. 2010. *Australopithecus sediba*: a new species of Homo-like Australopith from South Africa. *Science* 328:195-204.

BONIS (DE), L. 1999. La famille de l'homme. Des lémuriens à Homo sapiens. Paris: Bibliothèque Pour La Science, pp. 128.

BROOM, R., ROBINSON, J.T. & SCHEPERS, G.W.H. 1950. Sterkfontein Ape-Man *Plesianthropus*. *Transvaal Mus. Mem.* 4:11-117.

BROWN, F., HARRIS, J., LEAKEY, R. & WALKER, A. 1985. Early *Homo erectus* skeleton from west Lake Turkana, Kenya. *Nature* 316:788-792.

CARTMILL, M. & SMITH, F.H. 2009. The human lineage. Hoboken, New Jersey: Willey-Blackwell, pp. 609.

CAVE, A.J.E. 1961. Nature and morphology of the costoclavicular ligament. *J. Anat.* 95:170-179.

CHEVALIER, T. 2013. La diaphyse fémorale AL 333-61: preuve d'une bipédie comme unique mode locomoteur il y a 3,2 Ma en Ethiopie ? *Bull Mém Soc Anthropol Paris* 25:169-189.

CHURCHILL, S.E., HOLLIDAY, T.W., CARLSON, K.J., JASHASHVILI, T., MACIAS, M.E., MATHEWS, S., SPARLING, T.L., SCHMID, P., DE RUITER, D.J. & BERGER, L.R. 2013. The upper limb of *Australopithecus sediba*. *Science* 340 (Special section).

CIOCHON, R.L. & CORRUCCINI, R.S. 1976. Shoulder joint of sterkfontein *Australopithecus*. *S. Afr. j. Sci.* 72: 80-82.

CLARKE, R.J. & TOBIAS, P.V. 1995. Sterkfontein member 2 foot bones of the oldest South African hominid. *Science* 269:521-524.

FLEAGLE, J.G. 1998. Primate Adaptation and Evolution. New York: Academic Press, pp. 596.

GREEN, D.J. 2013. Ontogeny of the Hominoid scapula: the influence of locomotion on morphology. *Am. J. Phys. Anthropol.* 152:239-260.

GREEN, D.J. & ALEMSEGED, Z. 2012. *Australopithecus afarensis* scapular ontogeny, function, and the role of climbing in human evolution. *Science* 338:514-517.

HAILE-SELASSIE, Y., LATIMER, B.M., ALENE, M. et al. 2010. An early *Australopithecus afarensis* postcranium from Woranso-Mille, Ethiopia. *PNAS* (USA) 107:12121-12126.

HOWELL, F.C. & COPPENS, Y. 1976. An overview of Hominidae from the Omo succession, Ethiopia. In: Coppens, Y., Howell, F.C., Isaac, G.L. & Leakey, R.E.F. (Editors). Earliest Man and Environments in the Lake Rudolf Basin, Chicago: University of Chicago Press. p. 522-532.

HRDLIČKA, A. 1942. The juvenile scapula: further observations. *Am J. Phys Anthropol* 29:287-310.

KAPANDJI, I.A. 1994. Physiologie articulaire. 1. Membre supérieur. Paris: Edition Maloine. p. 296.

KIVELL, T.L., KIBII, J.B., CHURCHILL, S.E., SCHMID, P. & BERGER, L.R. 2011. *Australopithecus sediba* Hand demonstrates mosaic evolution of locomotor and manipulative abilities. *Science* 333:1411-1417.

LARSON, S.G. 1996. Estimating humeral torsion on incomplete fossil anthropoïd humeri. *J. Hum Evol* 31:239-257.

LARSON, S.G. 2007. Evolutionary transformation of the hominin shoulder. *Evol Anthropol* 16:172-187.

LARSON, S.G. 2013. Shoulder morphology in early Hominin evolution. In: Reed, K.E., Fleagle, J.G. & Leakey, R.E. (Editors). The paleobiology of Australopithecus. Dordrecht: Springer Science, pp. 247-261.

LARSON, S.G. & STERN, J.T.J. 1986. EMG of scapulohumeral muscles in the chimpanzee during reaching and "arboreal" locomotion. *Am J Anat* 176:171-190.

LATIMER, B. 1991. Locomotor adaptations in *Australopithecus afarensis*: the issue of arboreality. In: Coppens, C., & Senut, B. (Editors). Origine(s) de la bipédie chez les Hominidés. Paris: CNRS, pp. 169-176.

LESSERTISSEUR, J. & SABAN, R. 1967. Squelette appendiculaire. In: Grassé, P.-P. (Editor). Traité de zoologie anatomie, systématique, biologie, Tome 16 (Mammifères) Paris: Masson, p. 709-1078.

LOVEJOY, C.O., JOHANSON, D.C. & COPPENS, Y. 1982. Hominid upper limb bones recovered from the Hadar formation: 1974-1977 collections. *Am J Phys Anthropol* 57:637-649.

MCHENRY, H.M. 1994. Early hominid postcrania: phylogeny and function. In: Corruccini, R.S. & Ciochon, R.L. (Editors). Integrative Paths to the Past: Paleoanthropological Advances in Honor of F. Clark Howell, Englewood Cliffs: Prentice Hall. pp. 251-268.

OLIVIER, G. & CAPLIEZ, S. 1956. Anthropologie de la clavicule XI – La croissance. *Bull Mém Soc Anthropol Paris série 10, tome 8*:225-248.

OXNARD, C.E. 1968a. A note on the fragmentary Sterkfontein scapula. *Am J. Phys Anthropol* 28:213-218.

OXNARD, C.E. 1968b. A note on the Olduvai clavicular fragment. *Am J. Phys Anthropol* 29:429-432.

PREUSCHOFT, H., HOHN, B., SCHERF, H., SCHMIDT, M., KRAUSE, C. & WITZEL, U. 2012. Functional analysis of the primate shoulder. *Int. J. Primatol.* 31:301-320.

ROBERTS, D. 1974. Structure and function of the Primate scapula. In: Jenkins, F.A. (Editor). Primate locomotion. New York: Academic Press. p. 171-200.

SAKKA, M. 1985. Morphologie évolutive de la tête et du cou chez l'Homme et les Grands-Singes. Application aux Hominidés fossiles. Paris: CNRS éditions. p. 168.

SANKHYAN, A.R. 1997. Fossil clavicle of a middle Pleistocene hominid from the central Narmada valley, India. *J. Hum. Evol* 32:3-16.

SANKHYAN, A.R. 2005. New fossils of early stone age man from central Narmada valley. *Curr. Sci.* 88:704-706.

SCHULTZ, A.H. 1961. Vertebral column and thorax. *Primatologia* 4:1-66.

SCHWARTZ, J.H. & TATTERSALL, I. 2003. The human fossil record. Volume two: Craniodental morphology of genus *Homo* (Africa and Asia): Wiley-Liss. 603 p.

SENUT, B. 1981. L'humérus et ses articulations chez les Hominidés Plio-Pléistocènes. Paris: CNRS éditions. 141 p.

STERN, J.T.J. & SUSMAN, R.L. 1983. The locomotor anatomy of *Australopithecus afarensis*. *Am J. Phys Anthropol* 60:279-317.

ROUVIERE, H. 1982. Anatomie humaine descriptive, topographique et fonctionnelle: membres, système nerveux central. Paris: Masson. 679 p.

TUTTLE, R.H. & BASMAJIAN, J.V. 1978. Electromyography of pongid shoulder muscles. II. Deltoid, rhomboid and "rotator cuff". *Am J. Phys Anthropol* 49(1):47-56.

VALLOIS, H.-V. 1932. L'omoplate humaine. Etude anatomique et anthropologique (Chapitres VI, VII, VIII, IX et X). *Bull Mém Soc Anthropol Paris Série 8, tome* 3:3-153.

VALLOIS, H.-V. 1976. L'interprétation de l'omoplate de *Plesianthropus transvaalensis*. *L'Anthropologie* (Paris) 80:229-242.

VALLOIS, H.-V. 1977. Interpretation of the scapula of *Plesianthropus transvaalensis*. *J. Hum Evol* 6: 675-679.

VOISIN, J.-L. 2000. L'épaule des hominidés. Aspects architecturaux et fonctionnels, références particulières à la clavicule [Ph.D.]. Paris: Museum National d'Histoire Naturelle. p. 442.

VOISIN, J.-L. 2001a. Evolution de la morphologie claviculaire au sein du genre *Homo*, conséquence architecturale et fonctionnelle sur la ceinture scapulaire. *L'Anthropologie (*Paris) 105:449-468.

VOISIN, J.-L. 2001b. Réflexion sur le processus coracoïde des hominoïdes et des atèles; application à quelques hominidés fossiles. *C R Acad Sci, Ser 2, Earth Planet Sci* 332:299-306.

VOISIN, J.-L. 2004a. Clavicule: approche architecturale de l'épaule et réflexions sur le statut systématique des néandertaliens. *C R Palevol* 3:133-142.

VOISIN, J.-L. 2004b. Note sur la position de la tubérosité deltoïdienne (*tuberositas deltoidea*) et ses conséquences fonctionnelles. *L'Anthropologie* (Paris) 108:1-9.

VOISIN, J.-L. 2006. The clavicle, a neglected bone; morphology and relation to arm movements and shoulder architecture in Primates. *Anat Rec Part A* 288A:944-953.

VOISIN, J.-L. 2010. L'architecture de l'épaule au sein du genre *Homo*: nouvelles interprétations. *L'Anthropologie* (Paris) 114:354-369.

VOISIN, J.-L. 2011. A preliminary approach to the Neanderthal speciation by distance hypothesis: a view from the shoulder complex. In: Condemi, S., & Weniger, G.-C. (Eds). Continuity and discontinuity in the peopling of Europe 150 years of Neanderthal study. Dordrecht: Springer. p. 127-138.

VRBA, E.S. 1979. A new study of the scapula of *Australopithecus africanus* from Sterkfontein. *Am J. Phys Anthropol* 51:117-129.

WALKER, A. & LEAKEY, R. 1993. The post cranial bones. In: Walker, A., & Leakey, R. (Editors). The Nariokotome *Homo erectus* skeleton. Berlin: Springer-Verlag. p. 95-160.

WARD, C.V. 2002. Interpreting the posture and locomotion of *Australopithecus afarensis*: where do we stand? *Yearb Phys Anthropol* 45:185-215.

ZIPFEL, B., DESILVA, J.M, KIDD, R.S., CARLSON, K.J., CHURCHILL, S.E. & BERGER, L.R. 2011. The foot and ankle of *Australopithecus sediba*. *Science* 333:1417-1420.

3.
HOMININ PALAEOANTHROPOLOGY IN ASIA COMES OF AGE

Robin DENNELL
Department of Archaeology, University of Exeter EX4 4QE, U.K.
r.w.dennell@exeter.ac.uk

Abstract: *I review here the most important recent discoveries and changes in perception that are relevant to human evolutionary studies in Asia. The most important recent fossil discoveries are undoubtedly those of an early form of Homo erectus from Dmanisi, Georgia (1.75 Mya), and the late Pleistocene hobbit, H. floresiensis, from Flores, Indonesia. These apart, the most important changes have come about through the application of improved dating techniques to previous finds. These have considerably extended the time-depth of H. erectus in Asia, and in India, North China and SE Asia, it now extends back to at least 1.5-1.6 Mya. Major changes have also taken place in studies of recent human evolution in Asia. Although current archaeological, fossil human and genetic evidence indicates that our species dispersed across southern Asia between 40 and 60 ka, this process may have begun much earlier, and perhaps as early as the last interglacial. It is likely also considerably more complex that envisaged a few years ago because of genetic studies indicating that Neanderthals and H. sapiens in Eurasia (but not Africa) interbred, and thus all modern Eurasians contain some Neanderthal DNA. Additionally, analysis of the ancient DNA of a human phalange from Siberia indicates a sister population of "Denisovans" to Neanderthals that also contributed DNA to modern populations in Melanesia and SE Asia. Although several major gaps still exist in our understanding of human evolution in Asia, we are now beginning to glimpse its antiquity and complexity.*

Keyword: *hominin, palaeoanthropology, Asia, Dmanisi, Out of Africa 1, Out of Africa 2*

INTRODUCTION

Until the 1990's, Asia was the poor relation of Africa and Europe in studies of palaeoanthropology: of Africa because it contains the only Pliocene (pre-2.5 Mya) record of our earliest evolutionary stages, and also the earliest evidence for tool-making, our own genus, and later, our own species, *H. sapiens*; and of Europe because it contains by far the best record for Middle Pleistocene hominins, their technology and subsistence, and the most detailed record of our species behaviour during the Upper Palaeolithic from ca. 35-40 ka onwards. (Europe also contains the largest number of palaeoanthropologists and Palaeolithic archaeologists, and this factor cannot be ignored when considering why the European Peninsula figures so prominently in narratives of human evolution).

As a way of reminding ourselves how much has changed recently regarding our understanding of human evolution in Asia, a brief retrospective look at Asian palaeoanthropology in the 1980's is in order. At that time, 'Ubeidiya in Israel, then dated at a little over 1 Mya, was the earliest site outside Africa, and most researchers thought that *H. erectus* did not leave Africa until ca. 1 Mya (see e.g. Klein, 1989), for reasons never adequately explained. These early colonists left Africa with an Acheulean technology and used it in Southwest and South Asia (as well as Europe), but discarded it in favour of a simpler flake and core technology when they reached East and Southeast Asia. According to Movius (1948), these earliest inhabitants of China and SE Asia promptly lapsed into a long period of stagnation, whereas those in South and Southwest Asia were supposedly "progressive" because they made Acheulean handaxes, and later, Middle and Upper Palaeolithic technologies. At some point in the last glacial cycle, groups of *H. sapiens* appeared in Asia and were therefore interesting because they later reached Europe, Australia and the Americas; what they actually did in Asia was rarely given much attention. Regarding the East and Southeast Asian fossil record, there was a long, polarised and often acrimonious debate over whether *H. sapiens* developed indigenously or via immigration from Africa; this debate largely centred on a small number of poorly-dated finds from Java and China.

In the last 20 years, Asian palaeoanthropology has become far more exciting; indeed, we might be justified in saying that it has been revolutionised. This is partly through new discoveries, including that of two new species; and partly because researchers have moved beyond establishing when and where a hominin taxon is found to study early Asian hominins in relation to climatic change, and modelling dispersals and settlement trends. Genetic studies of modern populations have also played a major role, and the analysis of ancient DNA from human fossils has opened up research avenues that would have seemed impossible a few years ago. Dating has also played a major role in transforming our views of early Asia: some of the most important changes are summarised in Tab. 3.1.

To provide a structure to show how much has changed since 1990 in Asian palaeoanthropology, we can use the same framework that prevailed in the 1980's. This had three components: in chronological order, these were i) *Out of Africa 1* – the initial colonisation of Asia by the

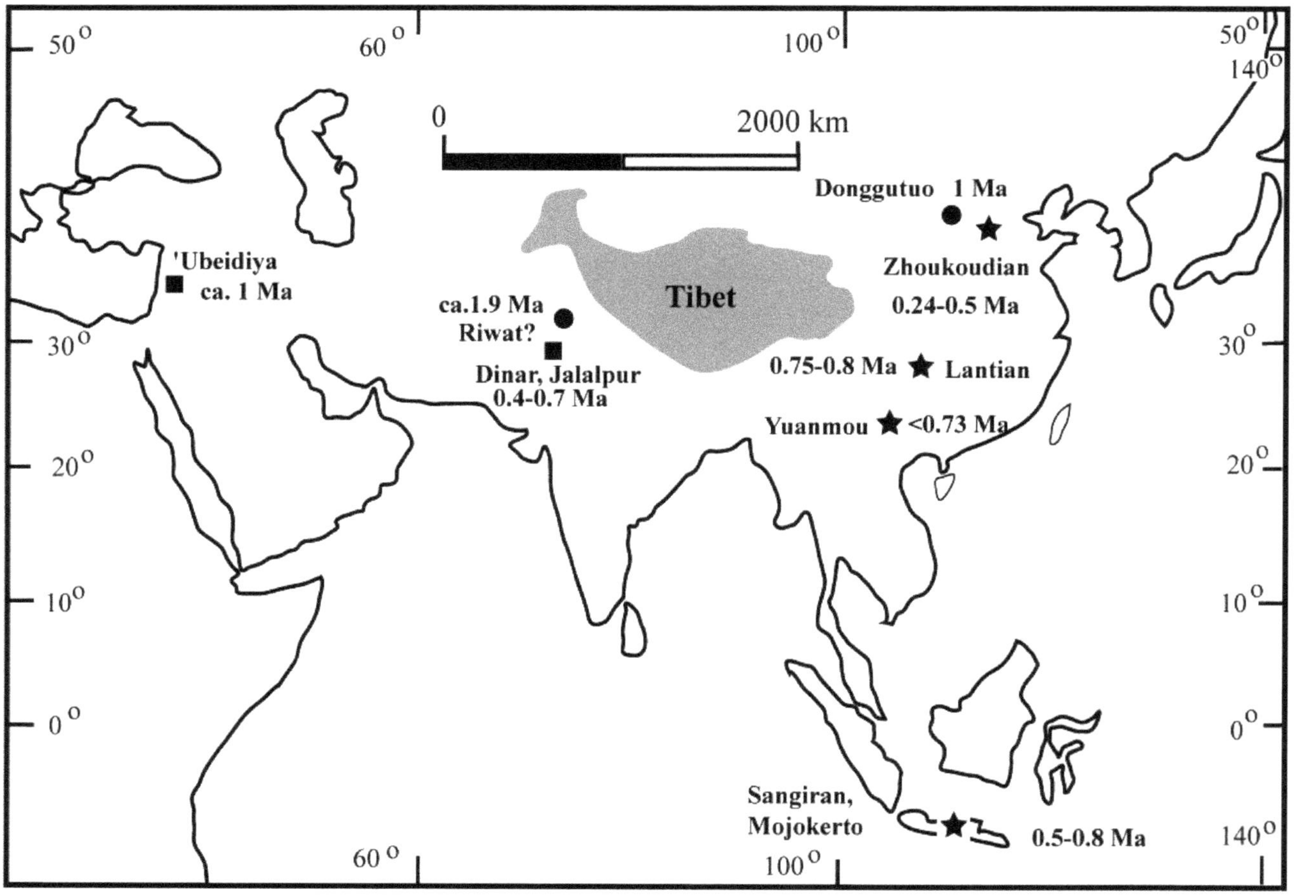

Fig. 3.1. Out of Africa, as envisaged in the 1980s. The map shows how limited the data base on the earliest appearance of hominins and the Acheulean was in the 1980s. The dating is from Klein (1989). Riwat is shown with a "?" because its estimated age was so obviously anomalous in relation to the dates from the rest of Asia, which appeared to indicate that it was not colonised until around 1 Mya – and seemingly much later in the case of Southeast Asia

Tab. 3.1. Major Changes in the dating of key Palaeolithic and hominin sites in Asia since the 1980's

Site	1980's dating	Later dating	Current dating
Out of Africa 1			
'Ubeidiya, Israel	1.0 Ma		1.0-1.4 Ma
Sangiran, Java	0.5-0.8 Ma	1.66 Ma (Swisher et al. 1994)	1.5-1.6 Ma (Larick et al. 2001)
Mojokerto, Java	0.5-0.8 Ma	1.8 Ma (Swisher et al. 1994)	<1.49 Ma (Morwood et al. 2003)
Ngandong, Java	250-300 ka	28-54 ka (Swisher et al. 1996)	146-546 ka (Indriati et al. 2011)
Yuanmou, China	<0.73 Ma		1.7 Ma (Zhu et al. 2008)
Nihewan Basin, China	1 Ma (Donggutuo)		1.66 Ma (Majuangou II) (Zhu et al. 2004)
Out of Africa 2			
Base of Tabun, Israel	74 ka	ca. 200-300 ka	ca. 400 ka

The 1980's datings are derived from Klein (1989). Note how the dominant trend was that the dating of the earliest sites was considerably underestimated, but that the estimated age of Ngandong for the most recent H. erectus specimens in Java was, by current standards, entirely reasonable. The other major change in Asia was the dating of the base Tabun sequence (and hence the Levantine Mousterian), which steadily increased in age from early in the last glacial cycle to over 400 ka for the basal, Acheulean levels F and G (for details, see Dennell, 2009, Table 8.2)

genus *Homo*; ii) a long period of occupation, during which its inhabitants were either "progressive" or "primitive", depending upon whether or not they used Acheulean bifaces; and iii) *Out of Africa 2*, when our own species appeared and eventually occupied the whole of Asia. We can begin with Out of Africa 1: who first colonised Eurasia, and when.

OUT OF AFRICA 1

Without doubt, the one site that totally transformed our understanding of the earliest hominins outside Africa was Dmanisi, Georgia. Back in 1991, I was fortunate enough to see the first viewing outside Georgia of the mandible D211 at a conference in Frankfurt; at the time it was thought to be a "normal" *H. erectus sensu lato,* and likely ca. 1 Mya old (Gabunia and Vekua, 1995). Move on to the present, and this remarkable site has now produced five hominin crania (Gabunia et al., 2000a; Lordkipanidze et al., 2005; Lordkipanidze et al., 2013) and large a mounts of post-cranial material (Lieberman, 2007; Lordkipanidze et al., 2007); substantial amounts of faunal remains (Gabunia et al., 2000b); and masses of very simple, unstandardized and unretouched stone artefacts (Lumley et al., 2005; Mgeladze et al., 2011). The hominins are now securely dated at ca. 1.77 Mya, just at the end of the Olduvai Subchron, and the earliest artefacts are slightly older, at ca. 1.85 Mya (Ferring et al., 2011). Although it has been suggested that these remains belong to a new taxon, *H. georgicus* (Lumley and Lordkipanidze, 2006), these hominins are probably best regarded as a very early form of *H. erectus* and may even predate the *H. erectus* finds from East Africa, if the earliest unambiguous ones are as late as 1.65 Mya (Gathago and Brown, 2006). They are remarkably small brained, with a cranial capacity of only ca. 600 cc3 for cranium D2700 (close to the average for *H. habilis*), and the largest, D2880 ca. 775 cc3, and show that the colonisation of Asia was not dependent upon brain size. Body size was also small, with an estimated height of ca. 145-66 cm and a weight of 40-50 kg. The simplicity of their lithic technology also indicates that was not a critical factor. They may not even have been the first hominins to colonise Asia as they appear primitive with respect to African specimens; indeed, Philip Rightmire and colleagues (Rightmire et al., 2006) have even suggested that following a very early dispersal of *Homo* from Africa, *Homo erectus* developed in SW Asia and in a back-migration, some of its members then re-entered Africa and developed into *H. ergaster* whilst others dispersed eastwards to China and Java. Asia might therefore have an important donor region to Africa, and not just a passive recipient.

Our understanding of when hominins first reached other parts of Asia has also been transformed since 1991. To take Java first: the consensus view in the early 1990's was that the earliest Javan specimens were either only 0.5-0.8 Mya in age (e.g. Klein, 1989, 194) or a little over 1.0 Ma or so in age (Itihara et al., 1994; see Tab. 3.1).

In one of the most seismic papers of the 1990's, Swisher and his team (Swisher et al., 1994) proposed that the earliest hominins at Sangiran were ca. 1.66 Mya, and the child's cranium from Mojokerto was no less than 1.88 Mya – even older than Dmanisi. However, these findings have not been maintained. Following sustained re-dating work at Sangiran, the earliest are likely slightly younger than Swisher's team estimated, and perhaps 1.5-1.6 Mya (Larick et al., 2001). Thanks to some fine and painstaking detective work by Hoffman and his team (Huffman et al., 2006), the original place of discovery of the Mojokerto find in 1936 has now been relocated, and is likely much less than 1.88 Mya in age; work by Mike Morwood and others indicates that it is <1.49 Mya, and perhaps near 1.0 Mya (Morwood et al., 2003) – in other words, its age estimate is now close to its estimated age of ca. 0.97 Mya (Hyodo et al., 1993) in the 1980s. Java provides a salutary reminder that dates, like house prices, can move down as well as up.

The dating of the earliest Chinese evidence has also been transformed in recent years. Donggutuo was the first site in the Nihewan Basin of North China to be dated palaeomagnetically to ca. 1.1 Mya (Jia and Wei, 1987). In 2001, a group led by Professor Ruxiang Zhu (Zhu et al., 2001) showed that Xiaochangliang was ca. 1.36 Ma, and then in 2004, the site of Majuangou III was dated to 1.66 Mya (Zhu et al., 2004), which may indicate a very rapid dispersal eastwards from Dmanisi after 1.75 Mya. This evidence shows that by 1.5 Mya, hominins – likely *H. erectus* – was distributed across much of Asia up to 40° N, at the latitude from Dmanisi to Nihewan, which appears to have remained the northern limit of the hominin range until the Upper Pleistocene. Whilst in North China, I have been collaborating with another Chinese group, led by Professor Zhaoyu Zhu from Guangzhou, on the re-dating of the *Homo erectus* cranium from Gong-wangling in Lantian County, Shaanxi. This was discovered in 1964 and dated palaeomagnetically in 1989 to ca. 1.15 Mya (An and Ho, 1989). This is now known to be a serious under-estimate of its probable age, which is 1.63 Ma (Zhu et al., 2015). Another site often mentioned in the context of the earliest dispersals across Asia is Yuanmou, where two incisors were found in 1965 that were attributed to *H. erectus* (Wu and Poirier, 1995). The dating of this evidence has been consistently problematic. On faunal grounds, an early Pleistocene age is likely (Zhu et al., 2003), and the layer from which the hominin incisors was much younger, and only ca. 0.8 Mya old. On faunal grounds, an early Pleistocene age is likely (Zhu et al., 2003), and the layer from which the hominin incisors reportedly came has recently been dated to the Olduvai Event, ca. 1.77-1.95 Ma (Potts and Teague, 2010), which would make them arguably the oldest hominin finds in Asia. However, Urabe et al., 2001 and Hyodo et al., 2002 have proposed that the Yuanmou hominin section was much younger, and only ca. 0.8 Ma old. On faunal grounds, the older dating seems more reasonable, but doubts must still surround the strati-graphic provenance of the incisors, and caution should still be exercised before accepting this evidence.

Although South Asia has yet to produce any Early Pleistocene fossil hominin evidence, my own work 30 years ago resulted in the discovery of a small assemblage of artefacts from Riwat, Pakistan, that was dated to a minimum of 1.9 Mya (Dennell et al., 1988; Rendell et al., 1987), which was totally unacceptable in the 1980s. As Tattersall (1997) remarked in 1997, their age is now "decreasingly anomalous" in the light of Dmanisi and the new dates from China and Java. Although I still stand by the stratigraphic context and age estimate for the Riwat assemblage and their status as artefacts rather than geofacts, the assemblage was very small, and that from Dmanisi rightly provides a better benchmark. The Pabbi Hills assemblage included artefacts that were found on fossil-bearing erosional surfaces 1.2-1.4 Mya and also on some that dated to or below the Olduvai Event (Dennell, 2004; Hurcombe, 2004). However, as none was

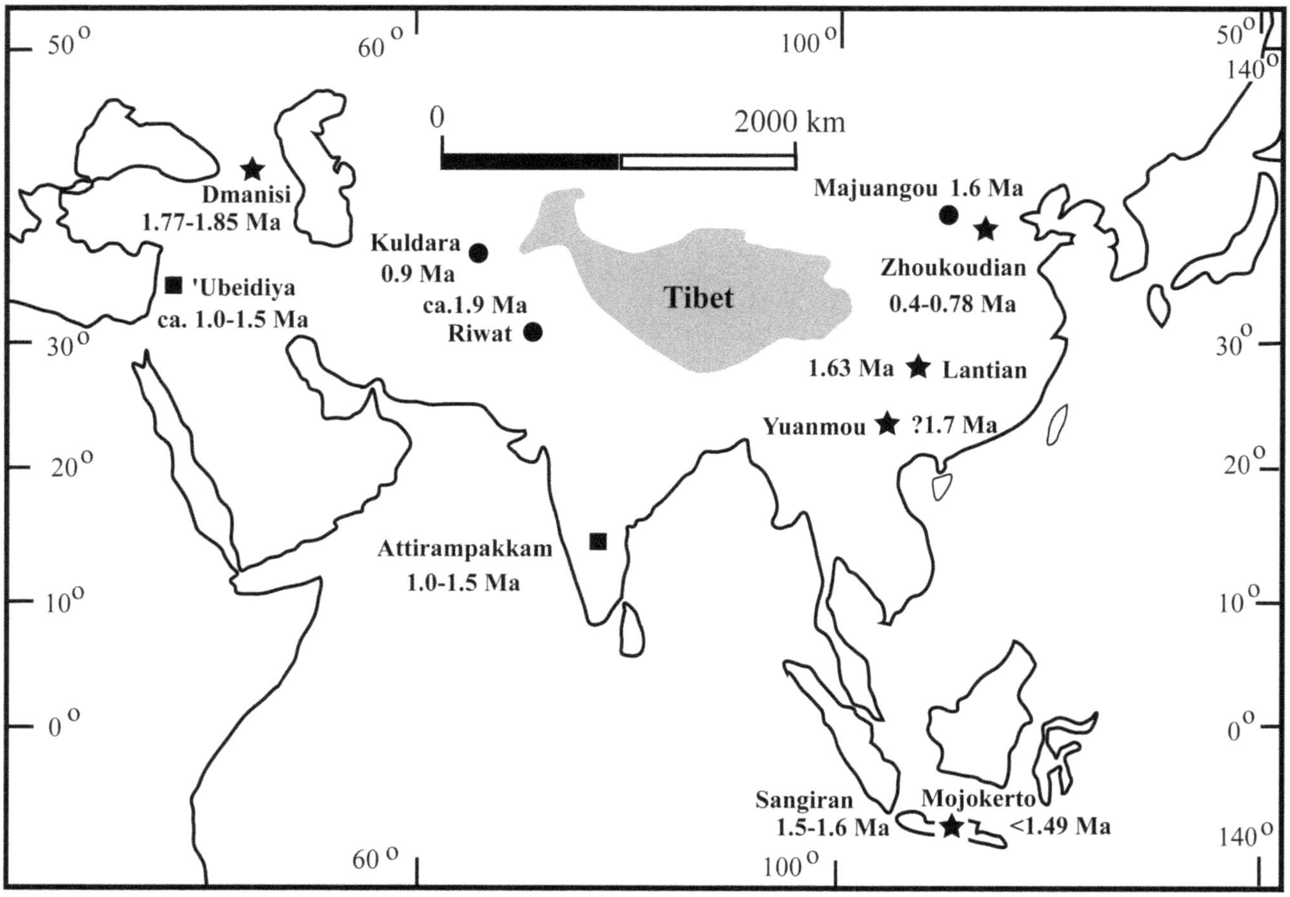

Fig. 3.2. Out of Africa 1, as envisaged in 2013. The obvious main change concerning Out of Africa 1 over the last 30 years is that the length of hominin occupancy in Asia has almost doubled. The dating is cited from Dennell (2009), with additional dates for the earliest Acheulean in South Asia at Attirampakkam from Pappu et al. (2011), Ferring et al. (2011) for Dmanisi, Shen et al. (2009) for Zhoukoudian, and Zhu et al. (2008) for Yuanmou, and Zhu et al. (2015) for Lantian. This site is shown here with a "?" because of doubts over the stratigraphic context of the two hominin incisors

excavated from a secure context, this provides only circumstantial evidence that hominins were present in the earliest part of the Pleistocene.

Between Out of Africa 1 and 2, and the Movius Line

Prior to the 1990's, there was little hard data on when Acheulean bifacial technologies were first used across Asia, what their users did apart from making handaxes, and who they were. Likewise, little information was available on the counterparts in East and Southeast Asia, where simple flake and core assemblages are found instead. Our understanding is still very imperfect, but is nevertheless far better than it was. For many years, the earliest indications of Acheulean bifaces in Asia was from Ubeidiya, Israel (Bar-Yosef and Goren-Inbar, 1993), with supporting evidence from the later, but poorly dated site of Latamne, Syria (Clark, 1969). A major addition to our knowledge was Gesher Benot Yaaqov, which is dated palaeomagnetically to ca. 0.7-0.8 Mya and has been investigated for many years with exemplary thoroughness by a team led by Naama Goren-Inbar (Goren-Inbar et al., 2000). The type of cleavers and bifaces at GBY are similar to sub-Saharan examples, and this may indicate a dispersal of hominins from Africa (Goren-Inbar and Saragusti, 1996), along with other types of mammals (Martínez-Navarro and Rabinovich, 2010), shortly before faunal connections appear to have been severed between the Levant and sub-Saharan Africa (O'Regan et al., 2005; Tchernov, 1992).

This point is highly relevant to considerations of hominin evolution in Asia after 1 Mya; see below. GBY has also provided convincing evidence that its inhabitants were hunting fallow deer (Rabinovich et al., 2008) and perhaps elephant (Goren-Inbar et al., 1994); repeatedly using fire (Goren-Inbar et al., 2004; Alperson-Afil and Goren-Inbar, 2010); shaping wood (Belitzky et al., 1991); and processsing plant foods such as nuts (Goren-Inbar et al., 2002). For many years, the earliest evidence for an Acheulean bifacial technology in South Asia were isolated handaxes found in stratigraphic contexts shortly above the Brunhes-Matuyama boundary at Dina and Jalalpur, Pakistan (Rendell and Dennell,1985), and thus dated to ca. 0.6-0.8 Mya. For this reason, the new evidence from Attirampakkam in South India is very welcome. Here Shanti Pappu and her team have recently dated Acheulean artefacts to at least 1 Ma and possibly 1.6 Mya (Pappu et al., 2011), and this radically transforms our understanding of when the Indian Peninsula was first colonised (Dennell, 2011). Even older dates for Acheulean artefacts have recently been obtained from Karakhach in Armenia, where artefacts were found below tuffs dated to >1.6 (Presnaykov et al., 2012). It will be interesting to see if these dates are confirmed, and replicated elsewhere in Southwest Asia. For the Middle Pleistocene, an outstanding long-term study was made of Acheulean land use in the Hunsgi-Baichbal Valleys of South India, where over 200 Acheulean sites were found (Paddayya, 2001).

The identity of the inhabitants of west and south Asia during the long period in which Acheulean bifacial technologies were used continues to be elusive. A mandible from Azyk Cave in the Caucasus dated to the mid-Middle Pleistocene has been compared to ones of *H. heidelbergensis* in western Europe (Howells, 1980). As this species in likely intrusive into Europe, its origin likely lies in Southwest Asia, which has been singled out as a centre of speciation (Bermudez de Castro and Martinon-Torres, 2013). However, a recent skull fragment from Koçibas, Turkey, has been identified as *H. erectus* s.s., i.e. like those from Zhoukoudian, China (Kappelman et al., 2008; Vialet et al., 2012). Some faunal evidence supports this suggestion: *Stegodon* is present at Evron Quarry and GBY in Israel, but is absent thereafter (Tchernov et al., 1994). As Stegodonts probably originate in Southeast Asia (Saegusa et al., 2004), there may also have been some hominin immigration from East Asia in the late Early Pleistocene. Nevertheless, this suggestion is at odds with the proposal that the inhabitants of GBY were African immigrants (see above) because of the type of cleavers used in the Levant and sub-Saharan Africa. An added uncertainty is *H. antecessor*, which is known only from Atapuerca, Spain, and dated to ca. 0.8 Mya (Carbonell et al., 1995): at present it remains unclear whether this was an isolated west European taxon or one that was distributed across southern Europe or even extended into Southwest Asia. (The mandible from Sima de Elefante, which is dated to ca. 1.2 Mya and currently the earliest specimen from Atapuerca, is so far identified only as *Homo* species indeterminate (Carbonell et al., 2008). Clearly, more specimens are needed to clarify to current confusion. Sadly, there are two specimens that remain unhelpful. The first is that partial cranium from Zuttiyeh, Israel, which is probably ca. 300 Kya and about which "virtually every opinion possible" (Sohn and Wolpoff, 1993) has been expressed. The other is the only Middle Pleistocene hominin cranial specimen from South Asia. This partial cranium from Hathnora in the Narmada Valley of Central India has been variously described as *Homo erectus* (Sonakia, 1985), *H. heidelbergensis* (Kennedy et al., 1991; Cameron et al., 2004), or archaic *H. sapiens* (Sonakia and de Lumley 2006), and prudently, as *Homo* species indeterminate (Athreya, 2007). The trajectory of human evolution in Southwest and South Asia remains unclear but is likely complex.

THE MOVIUS LINE

As originally defined by Movius (1948), the Middle Pleistocene inhabitants of Burma (Myanmar), mainland and island South East Asia, China and the Korean peninsula did not make Acheulean bifaces, and were therefore "primitive" and isolated. Leaving aside the prejudices that led Movius to equate backwardness with a technology that did not include handaxes (Dennell 2014a, 2014b), two features of Movius's synthesis continue to raise important issues regarding human evolution in East Asia. The first is the issue of isolation, as this has major implications for human evolution in East Asia. In particular, the type of multi-regional pattern of evolution envisaged by Weidenreich and later espoused by Wolpoff and others (see e.g. Wolpoff et al., 2000; Wolpoff and Caspari, 2007; Thorne and Wolpoff, 1992) envisaged continuous gene flow between populations from East Asia to Africa and Europe, which clearly could not have happened if populations were isolated from one another for lengthy periods of time. Yet isolation is likely to have become a major factor during the Middle Pleistocene, which was dominated by a small number of lengthy, high-amplitude climatic downturns during which much of Asia became colder and drier than at present. In particular, Asian deserts expanded and much of what I summarised as "Savannahstan" in the Early Pleistocene in Asia became "Aridistan" (Dennell 2009, chapter 7; 2013). This in turn makes it much less likely that *H. heidelbergensis* was distributed as far east as China during the Middle Pleistocene (contra Rightmire, 2001), or that *Homo sapiens* evolved in East Asia from its local background via gene exchange with populations further to the west.

Seemingly in conflict with this view is some evidence that bifaces were used in East Asia which were similar to Acheulean ones of western Asia. Examples are known from the Bose Basin, South China, dated to ca. 0.8 Mya (Hou et al., 2000; Huang et al., 2001), Dingcun and the Luonan Basin, north China, and Chong-nok-ni, South Korea (Norton et al., 2006). If these are regarded as "Acheulean" in a cultural sense, they imply contacts maintained across Asia, and thus a multi-regional model of later human evolution remains possible. Some have argued that the evidence for East Asian bifaces either totally invalidates the Movius Line (Yi and Clark, 1983), or that it should be modified into a "Movius Line sensu lato" (Lycett and Norton, 2010; Petraglia and Shipton, 2008). Three points can be raised about the biface evidence. First, although bifaces are known east of the Movius Line, they are nonetheless rare: there are, for example, sites such as Chirki in India with more bifaces than from most of eastern Asia. Secondly, although some East Asian bifaces look like Acheulean ones, many do not: those from the Bose Basin are considerably larger and thicker, and are often unifacial. Thirdly, although Acheulean assemblages contain bifaces, not all bifaces are Acheulean. Put another way, bifacial flaking is likely to end with some outcomes looking like Acheulean handaxes, irrespective of cultural affiliation and historical tradition.

OUT OF AFRICA 2

Few debates have raised more passion in palaeoanthropology than that over the origin of our own species, and the timing of when it left Africa. For a generation of students, this debate was largely a matter of deciding between two polarised positions: a multi-regional model, for which Milford Wolpoff (e.g. Wolpoff et al., 2000) was the main champion, and a replacement model, championed by in particular Chris Stringer (see Stringer and Andrews, 1988). For my part, I agree with a majority of researchers in seeing Africa as the most likely place where our species arose, as exemplified by specimens in East Africa from the Omo Valley (McDougall et al., 2005) and Herto, Ethiopia, dated to 160-190 Kya (White et al., 2003): there are no comparable examples from Asia of that age. Genetic evidence supports an African ancestry, although the current genetic diversity of modern Africans may also indicate that Africa had a larger population, and also expanded earlier in size than in Eurasia (Relethford, 1998; Eller et al., 2004). One key unresolved issue is when our species first left Africa. A commonly held view is that it remained in Africa until the last interglacial c. 125-115 Kya (see e.g. Frumkin et al., 2011), when some moved into the Levant, as shown by the skeletal evidence from Skhūl and Qafzeh Caves in Israel. These finds have often been portrayed as indicating a "failed dispersal" (Shea, 2008), in that it ended when Neanderthals re-colonised the Levant and there is little evidence that *Homo sapiens* spread beyond the Levant before 70 Kya. Yet two caveats are appropriate. First, our species might have left Africa before 125 Kya: it might, for example, have dispersed into the Levant during moist parts of MIS 6, shortly after it appeared in East Africa rather than 50-60 Kya later. The only relevant skeletal evidence from the Levant between 125 and 200 Kya is the Tabun C mandible which is normally attributed to a Neanderthal and dated to ca. 150 Kya, during MIS 6. This may indicate that Neanderthals were resident in the Levant during both MIS 6 and 7, or it may indicate a population that moved southwards during MIS 6, just as they did during MIS 4, when they displaced *H. sapiens* populations in what is now Israel. Secondly, we should bear in mind before we dismiss the Levantine record as indicating a "failed dispersal" the absence of any Pleistocene skeletal data from the Arabian Peninsula from between 125 and 70 ka, or from Iran and Iraq or South Asia, or even as far east as Southeast Asia – absence of evidence is not at all the same as evidence of absence (see Dennell, 2014c; Dennell and Petraglia, 2012).

For most scholars, the main expansion of our species across southern Asia dates from ca. 60-70 Kya. This view is largely derived from two sources. The first is the likely date of entry of humans into Australia as that serves as a *terminus ante quem* – humans obviously could not have left Africa after they had already reached Australia. Earliest estimates for entry, based on OSR dates (Roberts et al., 1994) and one controversial dating of the Mungo III skeleton (Thorne et al., 1999), are ca. 60 Kya, but other researchers prefer a more conservative dating of ca. 45-50 Kya, based on the earliest evidence from Niah Cave, Borneo (Barker et al., 2007), the earliest dates of ca. 47-49 Kya for entry into the New Guinea highlands (Summerhayes et al., 2010), and 14C dates from Australia itself (O'Connell and Allen, 1994). The second line of inferences comes from genetic studies of modern populations in South Asia, which suggest that modern lineages in this region have a likely time depth of ca. 60 ka (Quintana-Murci et al., 1999; Macaulay et al., 2005; Soares et al., 2009). This is, however, a consensus estimate, and recent mtDNA literature offers a wide range of estimates, from 45 to 85 Kya (see Boivin et al., 2013). There is also a small amount of archaeological evidence that is cited in favour of an African exit of ca. 60 ka. Mellars (2006) in particular has argued that the presence of microliths and incised shell in both the Howieson's Poort assemblages of South Africa, dated to ca. 65-60 Kya and those in India, dated at ca. 35-30 Kya, indicate an African exodus after 60 ka. The counter-argument is that the Indian microliths and incised shell are likely indigenous developments from within a long-standing Middle Stone Age tradition (see e.g. Clarkson et al., 2009; Petraglia et al., 2009). Furthermore, similarities in flaking techniques from before and after the great eruption of Toba in Sumatra at 74 Kya, the ash of which fell extensively over India, indicates that our species could have been in India before that time (Clarkson et al., 2011). What is clearly needed here is skeletal evidence from Arabia, southern Iran, Pakistan and India that is between 125 and 35 Kya. Sadly this is lacking at present, and the only fossil specimen from India – the cranial fragment from Hathnora – is not only taxonomically indeterminate, but likely to be late Middle Pleistocene in age, i.e. older than 125 Kya (see Athreya, 2007).

For those – like me – who think that our species could have left Africa for southern Asia before 60 ka, three recent pieces of evidence are encouraging. The first is the burial recently published by Demeter and his team from Tam Pa Ling in Laos, for which there is a direct dating of the cranium of ca. 63 Kya (Demeter et al. 2012); and the second is a human metatarsal from Callao Cave, Luzon, in the Philippines that may belong to *H. sapiens* (Mijares et al., 2010). (However, as it could equally well – in terms of size – be attributed to *H. erectus* and even *H. habilis*, this evidence on its own is inconclusive, although it is surely safe to assume that *H. habilis* never reached the Philippines). The third is a premolar associated with a Punung fauna dating from MIS 5 (80-125 Kya) that has been attributed to *H. sapiens* (Storm et al., 2005), although Barker et al. (2007) and Bacon et al. (2008) are sceptical over the dating and/or identification of this specimen.

One weakness of the genetic evidence is that it deals only with those past populations that have left living descendants: as example, there appears to be no genetic legacy from the earliest examples of *Homo sapiens* in the Levant ca. 100 Kya. There may therefore have been dispersals of *H. sapiens* before 60 Kya that have left no genetic imprint among extant populations. Secondly, estimates of the age of genetic lineages rests on a number of assumptions that are not always shared amongst geneticists. Scally and Durbin (2012), for example, have recently argued that the genetic clock can be revised, and in their view, the earliest exit of our species from Africa is likely to be ca. 125 Kya, not 60 Kya. The lesson from this is perhaps that genetic evidence is simply one source of information on past populations, but on its own, cannot provide a definitive answer. Genetic studies provide a valuable set of hypotheses that should be tested against archaeological and fossil skeletal data, but cannot and should not provide a definitive answer.

There remain here two subordinate issues relating to the earliest appearance of our species in Asia. The first is the date of extinction of *H. erectus* in Java and China. Java is the easier of the two to deal with. In one of the most controversial re-datings in recent times, Swisher and his team suggested that the likely age of the Ngangdong *Homo erectus* cranial remains from the Solo River, Java, were only 28-54 ka (Swisher et al., 1996), and this dating was largely confirmed by Yokohama et al. (2008). This dating was in direct conflict with faunal evidence that placed the Ngandong fauna – including the *H. erectus* remains – in the late Middle Pleistocene, and as preceding the last interglacial Punung Fauna, when animals such as orang-utan and sun bear made their appearance in Java (Storm, 2001). Recently, a thorough re-dating programme led by Indriati and including Swisher has indicated that the likely age of the Ngandong assemblage is between 146 and 546 Kya (Indriati et al., 2011). This not only confirms the faunal dating, but also implies that *Homo erectus* may have been extinct in Java before the appearance of *H. sapiens*. At present, the earliest unambiguous skeletal evidence of our species in island Southeast Asia is the cranium from Niah Cave, Borneo, dated by AMS procedures and U-series dating to 41-34 14C Kya BP (ca. 45-39 Kya cal. BP), some five millennia after the cave was first occupied.

The Chinese evidence is more problematic. The earliest definite evidence of our species comes from the recently excavated cave of Tianyuandong (Shang et al., 2010; Shang and Trinkhaus 2010) near Zhoukoudian in North China. Here, a partial skeleton (34 pieces) of a *H. sapiens* individual were found in level III, an unconsolidated breccia. Six bones were dated by AMS 14C to between 30,500 ± 370 and 39,430 ± 680 BP (uncalibrated). A sample from the human femur was dated to 34,430 ± 510 ka BP uncalibrated, or 40,328 ± 816 Kya BP calibrated, close in age to the cranium from Niah cave, Borneo (see above). Analysis of wear on the toe bones indicated that Tianyuan may also provide indirect evidence for footwear (Trinkhaus and Shang 2008).

Earlier evidence, particularly from South China, is harder to assess. Chinese (and some other) researchers have classified some Chinese specimens as "archaic *Homo sapiens*", and claim that some are dated in excess of 60 Kya, or even 100 Kya. These specimens are also attributed by most Chinese researchers to a local development from Middle Pleistocene *Homo erectus*. Examples here are Zhirendong (Liu et al., 2010a) and Liujiang from caves, and Maba and Dali from open air locations. There are four major problems with these specimens. The first is definitional: the term "archaic *Homo sapiens*" is already in use for describing early specimens from Africa, and comes with a completely different set of assumptions about ancestry from the Chinese specimens. Secondly, the stratigraphic provenance of several Chinese finds is uncertain, especially from caves where – as at Liujiiang – the remains were discovered by villagers digging for fertiliser (Shen et al., 2002, 2007). Thirdly and related to that is the dating. Maba, Xujiayao and Dali for example were dated to the Upper Pleistocene but may be considerably older and within the time span of *H. erectus*. The partial skeleton from Jinnuishan has a date of ca. 284 Kya and is described as "archaic *Homo sapiens*", but that age and identification is hard to reconcile with genetic estimates for the origins of *H. sapiens* in East Asia. Cave specimens have been particularly hard to date because of the problems of correlating flowstones that can be accurately dated to the hominin specimen found in the cave deposits. As example, the "*Homo sapiens*" mandible from Zhirendong in South China is published with an estimated age of ca. 100 ka, based on the age of the lowest of five flowstones (Liu et al., 2010a). Yet, as Kaifu and Fujita (2011) have pointed out, these five flowstones are very closely spaced within a band only 10 cm thick, and the mandible might quite feasibly have been associated with a flowstone only 50 ka in age and only 10 cm higher in the cave sequence. At Huanglong Cave, it is stated that it is not possible to determine whether the hominin teeth were above or below a flowstone dated to ca. 100 Kya (Liu et al., 2010b). Until specimens can be directly dated, these confusions will continue. Fourthly, there is a sampling problem. Many of the specimens attributed to "archaic *Homo sapiens*" are isolated teeth and ascribed to that taxon largely on the basis of size and then compared (usually) to much older, early Middle Pleistocene specimens from locality 1, Zhoukoudian. What is not known here is the degree of gracilisation that may have occurred during the later Middle Pleistocene. In the case of the mandible from Zhirendong, there are no late Middle Pleistocene mandibles from China for comparison, and the reference point is again the much older sample from Zhoukoudian (Dennell, 2010). What is needed with the isolated teeth are highly detailed micro-morphological studies of the kind carried out by Maria Martinon-Torres and her team at Atapuerca; recently, these techniques have been applied to the teeth from the late Middle Pleistocene site of Panxian Dadong, South China, and they find that they defy a simple, binary classification of either *H. erectus* or *H. sapiens* (Liu et al., 2013). The Chinese data is likely complex, and new evidence will hopefully bring clarity instead of perpetrating the current confusion.

Closely related to the appearance of when *Homo sapiens* dispersed across Asia is the thorny issue of "modernity": at which point did anatomically – or genetically – defined *H. sapiens* behave in a recognizably "modern" way like us, however we might care to define modernity and recognise it archaeologically. There is insufficient time here to explore this issue in depth, but a few general points can be usefully noted. The Palaeolithic world of Eurasia in the Upper Pleistocene before ca. 40 Kya had three main archaeological zones. In Europe, Southwest Asia, and Siberia, the Upper Palaeolithic marked a major technological break with the Mousterian, and Neanderthals became extinct. Given the evidence for art, music and skilled craftsman in the Upper Palaeolithic, there is little problem in regarding these as "modern" in their behaviour. However, what is questionable is whether the European evidence should be seen as the standard against which the rest of Eurasia (and Australasia) is judged. In the Levant, Arabia and India, the earliest *H. sapiens* communities operated within a Middle Palaeolithic or Middle Stone Age tradition; in Southeast Asia, South China and Australia, stone technologies were very simple, informal and unstandardized throughout the Upper Pleistocene. That is not to say that their owners were "simple", "primitive" or somehow "non-modern". As example, we know from recent work at Niah Cave, Borneo, that these early *H. sapiens* groups were doing some very complex and skilful things – for example, in hunting animals high in the tree canopy, or detoxifying plants to make them edible; or extracting and processing starch. "Modernity" takes different forms, but the key part is probably flexibility, and the ability to adapt to sudden environmental change, as happened throughout the last glacial cycle (see e.g. Rabett, 2012, chapter 8).

NEW SPECIES

Enough for the moment of *Homo erectus* and *Homo sapiens*, and let us turn to two of the most exciting discoveries in palaeoanthropology worldwide. These are the hobbit, *Homo floresiensis* and the island of Flores, Indonesia, and the Denisovans from Siberia.

Flores

The first major surprise from Flores was the discovery of stone tools at Mata Menge, dated at ca. 800 ka – the first time tools older than 100 ka had been found on a true island, i.e. one never joined to adjacent areas at times of low sea level (Morwood et al., 1998). Since then, even older tools, ca. 1 Ma old, have been found at Wolo Sege (Brumm et al., 2010). How and from where their owners arrived is still unclear. As boats seem unlikely at that time (contra Bednarik, 2003) – as otherwise we might have expected other islands, including Australia, to have been colonised much sooner, the most likely explanation is that they arrived by accident, by being, for example, swept out to sea on rafts of vegetation following a cyclone or tsunami (Smith, 2001). This explanation engenders its own set of problems, however. Although Flores lies near Lombok and then Java, the prevailing north-south flow of currents from the South China Sea to the Indian Ocean means that the accidental colonists likely arrived from the north; in Morwood and Oosterzee's (2007) view, from Sulawesi, on the Australian side of the Wallace Line. The problem here is that there is no evidence that hominins were on Sulawesi before the Upper Pleistocene, nor on neighbouring Kalimantan. So how did the source population of the colonists reach Sulawesi? Although *Stegodon* might have managed to swim the 40 km between Kalimantan and Sulawesi, it is unlikely that hominins could. One remote possibility is that they entered from South China via the Philippines, or via Hainan.

A second problem with Mata Menge and Wolo Sege is whether accidental colonisation following cyclones or tsunamis would result in the survivors growing to sufficient numbers to provide a permanent population. As Flores lies in a region regularly afflicted by tsuanmis (see Tsuji et al., 1995), and as tsunamis are far more likely to remove an island's inhabitants than transport them safely to a neighbouring island, one must assume that Flores too was regularly depleted of some inhabitants. I am not convinced that it would have been continuously occupied for a million years; perhaps a model of intermittent, short-lived occupation during the Early and Middle Pleistocene is more appropriate (Dennell et al., 2013).

Homo floresiensis

The discovery of the "hobbit" assemblage at Liang Bua cave, Flores, (Brown et al., 2004; Morwood et al., 2004) was one of the major fossil finds in the history of palaeoanthropology. As the remains of 9 individuals was found, dated from 95 to 17 Kya, the "hobbit" it is clearly part of a population and not just an aberration. Three views have formed on their significance. The first, that it is a pathological example of Laron's syndrome or simply a human pygmy (Herskovitch et al., 2007; Perry and Dominy, 2009; Richards, 2006), can be dismissed because neither is supported by detailed consideration of the cranial anatomy. The other two positions are more robust: i) that it represents a dwarfed population of *Homo erectus* (Kaifu and Fujita, 2010) or ii) that it represents a very primitive lineage with roots in the Earliest Pleistocene, and exhibiting some traits evident in early *Homo* or even gracile australopithecines (Aiello, 2010; Argue et al., 20006; Brown and Maeda, 2009; Falk et al., 2009; Jungers et al., 2009; Larson et al., 2009; Morwood and Jungers, 2009). The former view is more parsimonious and reduces *H. floresiensis* to a regional, SE Asian issue. The dwarfing is also consistent with the faunal evidence from Flores, which indicates – as would be expected on a small island with an impoverished fauna – some species dwarfed (e.g. *Stegodon*) whilst others became much larger (e.g. the rat and a flightless stork). If, however, the hobbit is as primitive as some believe, there is therefore a hidden lineage in southern Asia of a hominin that left Africa in the Earliest Pleistocene. Given the poverty of the fossil record in SW and mainland Southeast Asia, this is not impossible, although it is a little surprising that no trace has been found in the Upper Siwaliks of India and Pakistan, which have been investigated for over a century, or in Java, where *Homo erectus* is abundantly represented. Nevertheless, we should keep an open mind on this.

Denisova

The discovery of the "Denisovans" was as unexpected as that of *H. floresiensis*. Their discovery is all the more remarkable in that for the first time, an extinct type of hominin was identified from its ancient DNA instead of from its skeletal remains. This remarkable achievement was performed on a phalange and molar from Denisova Cave, Lake Baikal region, Siberia, dated to 30-50 Kya

(Krause et al., 2010; Reich et al., 2010). In terms of the DNA evidence, the Denisovans are likely a sister clade of Neanderthals. As with *H. floresiensis*, the discovery of the Denisovans throws up more problems than answers, particularly over their likely origins, distribution and genetic contribution to modern populations.

i) The origins of Denisovans

Krause et al. (2010) suggested that the Denisovans had an African ancestry, most likely from when *H. heidelbergensis* dispersed from Africa into Eurasia in the Early Middle Pleistocene. A different scenario was offered by Martinón-Torres et al. (2011). They pointed out the diverse range of opinions amongst physical anthropologists over whether *H. heidelbergensis* should be regarded as a European and Southwest Asian taxon; an African-European-Southwest Asian taxon; or even a taxon distributed across all of continental Eurasia as well as Africa. Several researchers would thus regard African specimens (e.g. Bodo, Kabwe) that some classify as *H. heidelbergensis* as instead *H. rhodesiensis* (McBrearty and Brooks, 2000). Additionally, the faunal evidence indicates that faunal connections between sub-Saharan Africa and SW Asia were severed in the Middle Pleistocene. They therefore suggest that an Asian origin is at least as likely.

ii) The distribution of Denisovans

By far the largest problem resulting from the identification of Denisovan DNA in a phalange and molar is that we have no idea what Denisovans looked like. More specifically, how similar were they in cranial and facial details to Neanderthals, and those specimens from China that are identified as either late *Homo erectus* or early (archaic) *Homo sapiens*? Until Denisovan DNA is extracted from a fossil specimens that has its face and cranial details preserved (which is at best a distant possibility at present), it is difficult to know whether some Asian fossil specimens are Denisovan, Neanderthal, late *H. erectus* or early *H. sapiens*.

iii) The genetic legacy of Denisovans

According to Reich et al. (2010), Denisovan DNA is present in Melanesians and New Guinea Highlanders, but very rare on the Asian side of the Wallace Line. This implies, they suggest, that the earliest colonists of Sahul had some genetic admixture with Denisovans, but later inhabitants of Sunda did not. It is difficult to see at present how this scenario can be satisfactorily integrated with archaeological evidence from island and mainland Southeast Asia.

Palaeodemographic studies

Until recently, the presentation of past Eurasian (and African) hominin taxa was largely shown either as points on a map, or as a generalised, maximum distribution; for example, the distribution of Neanderthals or *Homo erectus*. Implicit in many of these representations was the assumption that once established, populations were permanent and spatially continuous over the area of their distribution.

The fundamental shortcoming of those distribution maps was that across most of Eurasia (and probably much of Africa), hominin populations were most unlikely to have been continually resident in areas of profound climatic and environmental change, as happened repeatedly throughout the Pleistocene. Instead, as I (Dennell 2003) pointed out, in many areas they are likely to have occasional visitors, especially towards the northern limit of their range, which until the Upper Pleistocene, was roughly at 40° N., in continental Asia (from the Nihewan Basin to Kuldara in Central Asia to Dmanisi in Georgia). Even in more southerly regions, hominins are unlikely to have been continually resident in areas now desert, such as the Arabian Peninsula or the Thar Desert, and even equatorial regions, such as Southeast Asia, were subject to major climatic and environmental shifts. Attention should thus shift to modelling hominin populations at their minimal extent during those repeated periods of increased aridity and lower temperatures when conditions were at their most inhospitable. Under such conditions, hominin (and other taxa) would have been largely restricted to a small number of refugia, or core areas of settlement, where continuous residence was feasible at the prevailing level of technology and cognitive ability.

This type of study requires that hominin fossils and archaeological evidence can be related to their climatic and environmental context. Unfortunately this is rarely possible in Asia, although much greater progress has been made in Europe. There are still very few Palaeolithic sequences (and associated hominin fossils) in Asia that can be securely linked to specific types of climate, and the study by Ranov (1995) and Dodonov and Baiguzina (1995) is still outstanding as one of the few long-term Asian palaeolithic records that can be linked to climatic change (Fig. 3.3).

In many cases, hominin fossils and palaeolithic sites are too imprecisely dated to allow precise correlations with the highly detailed palaeoclimatic records now available from the ocean floor and loess-palaeosol sequences of Central Asia and especially the Loess Plateau of Central and northern China. However, it is reasonable to expect considerable improvement in the next few years as studies of palaeoclimate and palaeoenvironment are integrated into studies of Palaeolithic sites.

Recently, I, with Spanish colleagues, proposed how a "source-sink" model (see Hawks, 2009) could be used to model Middle Pleistocene populations in Europe (Dennell et al., 2011). In this model, source populations would have been restricted to a small number of refugia in southern Europe, most likely in the Iberian and Italian peninsulas and the Balkans. When conditions improved, daughter populations could disperse northwards. During this expansive phase, groups would also interact and interbreed; at their maximal extent, some would reach Britain, which is a classic example of an area where

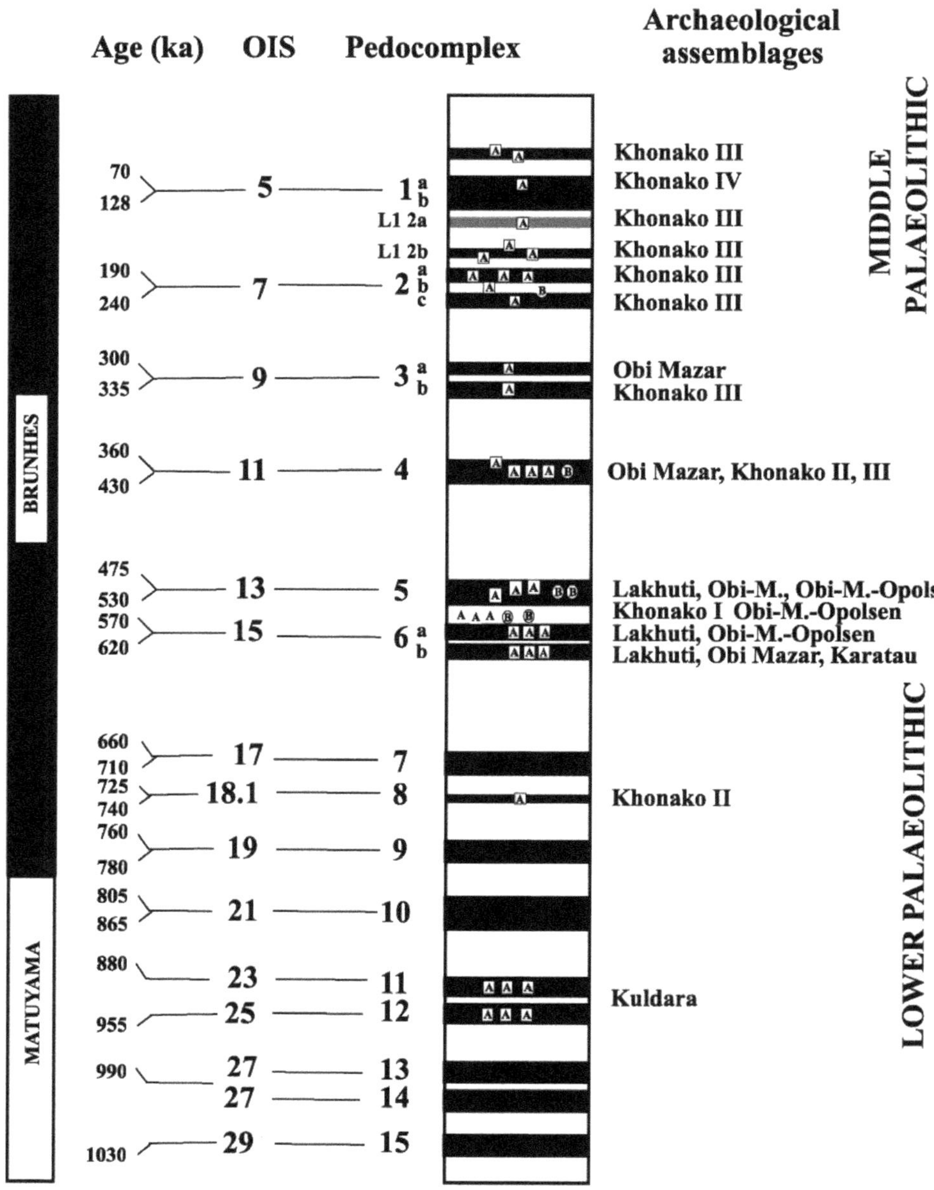

Fig. 3.3. Age and stratigraphic context of Early Palaeolithic sites in Tajikistan. Black bars denote interglacial pedocomplexes (palaeosols); ntervening white parts denote glacial loess: note the correlation of Palaeolithic sites with interglacial conditions. The study dates from the 1990's, and is still one of the best examples of an integrated archaeological and palaeoclimatic study that shows how hominin settlement in Asia was structured by climate Source: Dennell, 2009, Fig. 8.33

occupation was intermittent throughout the Pleistocene and largely restricted to interglacials and interstadials (see Stringer 2006). These daughter populations are regarded as "sinks" in that they probably required an influx of immigrants to remain viable, especially in their initial phases. When conditions worsened, many of these would have become extinct: we specifically proposed a model of repeated colonisation-extinction-recolonisation, as opposed to the type of "ebb and flow" model, whereby populations simply retreated into a refugia. This we feel would have been unlikely as the refugia would also have been under stress and its carrying capacity reduced during climatic downturns. Populations levels across Europe would thus have fluctuated considerably between glacial and interglacial periods. In a more recent study (McDonald et al., 2012), we tried to apply this type of model to the earliest settlement records from Britain and Iberia, which are among the most detailed settlement records from Europe. Even here, however, despite decades of intensive investigations, dating still remains a serious issue when trying to correlate settlement records, and what is probably required is a long period where the priority lies in building well-dated settlement records that can be linked to the Pleistocene climatic record.

Even where detailed palaeoclimatic data are lacking, simple modelling can be used to investigate likely palaeodemographic patterns. As example, a recent study (Dennell 2012) challenged the proposition that the Nihewan Basin in North China would have sustained a continuous hominin population throughout the Early and Middle Pleistocene. Because it lies in an area of strong seasonal contrasts, with cold dry winters and warm humid summers, any population resident there throughout the year would have needed sewn clothing, fire, shelter and probably the ability to store food in order to survive long periods of sub-freezing temperatures, which would have been even lower during much of the Pleistocene. As these skills were either lacking or at best very rudimentary, it is probable that hominins were in the Nihewan Basin during spring and summer, and likely only during interglacial conditions. The essential point here is that a long sequence of sites in an area is unlikely to indicate continuous occupation unless environmental conditions remained relatively constant. In areas near the northern limits of the hominin range – as in North China and Britain – occupation was likely highly episodic and limited to warm periods.

A recent useful step forward was taken by Stewart and Stringer (2012), who attempted to model later human evolution in Asia in terms of cold-period refugia and subsequent expansions from them when conditions improved. Although one can argue over specific aspects of this study – for example, the timing of when *Homo sapiens* left Africa – it represents a new direction in studies of Asian palaeodemography. Increasingly, these studies will utilise GIS data to model how hominin groups are likely to have responded to different environmental and climatic conditions.

Simulation studies

With the advent of GIS and powerful laptops, a number of researchers (e.g. Mithen and Reed, 2002; Nikitas and Nikita, 2005; Hughes et al., 2007; Holmes, 2007) have attempted to model the dispersal of hominins across Eurasia. The outcomes of these simulations are then tested against the current fossil and archaeological record. There are two problems here. The first is that plausible simulations require accurate environmental data across Eurasia, but at present, coverage is very uneven. Europe, for example, is abundantly documented by pollen profiles showing vegetational trends and patterns in great detail, whereas such data are lacking from most of continental Asia. Secondly, it is generally assumed that current data provides an accurate picture of past dispersals, whereas it is still an extremely imperfect record.

Hominins may have left Africa well before 1.8 Mya, and the earliest evidence for hominins in China, SE Asia and South and Central Asia may well under-represent their first appearance because the earliest evidence for colonisation is always later than than its actual date. Early studies (e.g. Mithen and Reed, 2002) found it problematic that Java appeared to have been colonised much earlier (ca. 1.8 Mya) than northern Europe (ca. 0.5 Mya): current dating for the earliest evidence from Java (ca. 1.6 Mya) and northern Europe (0.9 Mya; (Parfitt et al., 2005) considerably reduces this discrepancy from 1.3 Mya to 0.7 Mya. Indeed, simulation studies are valuable in showing what might have happened, where current archaeological and fossil hominin evidence is most likely to understate when a major dispersal event occurred, and which areas are most likely to contain very early evidence of hominins.

CONCLUSIONS

Almost all aspects of studies of human evolution in Asia have changed substantially since the 1990's. This applies particularly to understanding of when our genus first appeared outside Africa, and of the anatomy and likely behaviour of these early colonists. The length of occupation of Asia has now almost doubled, from just over one million years ago to almost two million years. The primitive nature of the Dmanisi hominins has led to speculation that Asia may have been colonised before 2 Mya, and perhaps shortly after stone tool making became routine ca. 2.6 Mya; indeed, we should even allow the possibility that australopithecines may have entered Southwest Asia (Dennell and Roebroeks, 2005). If *Homo floresiensis* is as primitive as many claim, there may even be a hidden lineage in southwest and southern Asia that dates from before 2 Mya. One of the main challenges of the next decades is to see if these speculations are true by locating well-dated skeletal and archaeological evidence in Asia that is older than the earliest artefacts at Dmanisi, i.e. >1.85 Mya.

Regarding the dispersal of our own species across Asia, one of the major debates to resolve is whether the presence of *H. sapiens* in the Levant between ca. 120 and 70 Kya represents a limited and therefore "failed" dispersal, or whether it was part of a much wider expansion across southern Asia. To phrase the question differently, did *H. sapiens* disperse eastwards to India and Southeast Asia before or after 60 Kya? Before taking up positions in this debate, we need to bear in mind the absence of skeletal evidence between 120 and 40 Kya in the Arabian Peninsula, South and Southeast Asia, and the currently inconclusive evidence for *H. sapiens* in China before 40 Kya: one single, well-dated discovery of *H. sapiens* within this enormous region could have far-reaching effects.

A welcome development within Asian palaeoanthropology is the way that palaeodemographic models are playing now a much wide role than before, with a major stimulus being the need to understand how Palaeolithic communities in Eurasia responded to the repeated and often profound climatic changes, particularly of the Middle and Upper Pleistocene. These studies require much greater integration than before of archaeological, human skeletal, environmental and climatic evidence, and much greater collaboration between specialists. Progress here has been encouraging, and the increasing sophistication (and decreasing cost) of GIS software and computers should result in regional case studies in Asia reaching the same standards as in western Europe. The unexpected addition of Denisovans and hobbits has also very forcefully indicated the complexity of human evolution in Asia by showing how little is known over vast areas of Asia; doubtless, further equally unexpected discoveries will be made in years to come. Far from being treated as a poor relation of Europe and Africa, Asian palaeoanthropology is now rightly accepted as containing a vital part of the story of our evolution.

References

AIELLO, L. 2010. Five years of *Homo floresiensis. American Journal of Physical Anthropology* 142: 167-179.

ALPERSON-AFIL, N. & GOREN-INBAR, N. 2010. *The Acheulian site of Gesher Benot Ya'aqov. Volume 2: ancient flames and controlled use of fire*. Springer Science + Business Media, Dordrecht, Heidelberg London & New York.

AN, Z. & HO, C.K. 1989. New magnetostratigraphic data of Lantian *Homo erectus. Quaternary Research* 32: 213-221.

ARGUE, D., DONLON, D., GROVES, C. & WRIGHT, R. 2006. *Homo floresiensis*: Microcephalic, pygmoid, *Australopithecus*, or *Homo*? *Journal of Human Evolution* 51: 360-374.

ATHREYA, S. 2007. Was *Homo heidelbergensis* in South Asia? A test using the Narmada fossil from central India. In *The Evolution and History of Human Populations in South Asia: Inter-disciplinary Studies in Archaeology, Biological Anthropology, Linguistics and Genetics*, ed. M.D. Petraglia and B. Allchin, pp. 137-170. Dordrecht: Springer.

BACON, A.-M., DEMETER, F., DURINGER, P., HELM, C., BANO, M., VU THE LONG, NGUYEN THI KIM THUY, ANTOINE, P.-O., BUI THI MAI, NGUYEN THI MAI HUONG, DODO, Y., CHABAUX, F. & RIHS, S. 2008. The Late Pleistocene Duoi U'Oi cave in northern Vietnam: palaeontology, sedimentology, taphonomy and palaeoenvironments. *Quaternary Science Reviews* 27, 1627-1654.

BARKER, G.W.W., BARTON, H., BIRD, M., DALY, P. et al. 2007. The 'human revolution' in lowland tropical Southeast Asia: the antiquity and behavior of anatomically modern humans at Niah Cave (Sarawak, Borneo). *Journal of Human Evolution* 52: 243-261.

BAR-YOSEF, O. & GOREN-INBAR, N. 1993. *The Lithic Assemblages of 'Ubeidiya: A Lower Palaeolithic Site in the Jordan Valley, Jerusalem*. Jerusalem: The Hebrew University of Jerusalem.

BEDNARIK, R. 2003. Seafaring in the Pleistocene. *Cambridge Archaeological Journal* 13, 41-66.

BELITZKY, S., GOREN-INBAR, N. & WERKER, E. 1991. A Middle Pleistocene wooden plank with man-made polish. *Journal of Human Evolution* 20: 349-353.

BERMUDEZ DE CASTRO, J.M. & MARTINÓN-TORRES, M. 2013. A new model for the evolution of the human Pleistocene populations of Europe. *Quaternary International* 295, 102-112.

BOIVIN, N., FULLER, D.Q., DENNELL, R. ALLABY, R. & PETRAGLIA, M.D. 2013. Human dispersal across diverse environments of Asia during the Upper Pleistocene. *Quaternary International* 300, 32-47. doi:10.1016/j.quaint.2013.01.008.

BROWN, P. & MAEDA, T. 2009. Liang Bua *Homo floresiensis* mandibles and mandibular teeth: a contribution to the comparative morphology of a new hominin species. *Journal of Human Evolution* 57: 571-596.

BROWN, P., SUTKINA, T., MORWOOD, M.J., SOEJONO, R.P. et al. 2004. A new small-bodied hominin from the Late Pleistocene of Flores, Indonesia. *Nature* 431: 1055-1068.

BRUMM, A., JENSEN, G.M., VAN DEN BERGH, G.D., MORWOOD, M.J. et al. 2010. Hominins on Flores, Indonesia, by one million years ago. *Nature* 464: 748-752.

CAMERON, D., PATNAIK, R. & SAHNI, A. 2004. The phylogenetic significance of the Middle Pleistocene Narmada cranium from Central India. *International Journal of Osteoarchaeology* 14: 419-447.

CARBONELL, E., BERMUDEZ DE CASTRO, J.M., ARSUAGA, J.L. et al. 1995. Lower Pleistocene hominids and artifacts from Atapuerca-TD6 (Spain). *Science* 269: 826-829.

CARBONELL, E., BERMUDEZ DE CASTRO, J.M., PARES, J.M. et al. 2008. The first hominin of Europe. *Nature* 452: 465-469.

CLARK, J.D. 1969. The Middle Acheulean occupation site at Latamne, Northern Syria (second paper). *Quaternaria* 10: 1-60.

CLARKSON C., PETRAGLIA, M., KORISETTAR, R., HASLAM, et al. 2009. The oldest and longest enduring microlithic sequence in India: 35,000 years of modern human occupation and change at the Jwalapuram Locality 9 rockshelter. *Antiquity* 83: 326-348.

CLARKSON, C., JONES, S.A. & HARRIS, C. 2011. Continuity and change in the lithic industries of the Jurreru Valley, India, before and after the Toba eruption. *Quaternary International* 258, 165-179.

DEMETER, F., SHACKELFORD, L.L., BACON, A.M., DURINGER, P., WESTAWAY, K., SAYAVONGKHAMDY, T. & KARPOFF, A.M. 2012. Anatomically modern human in Southeast Asia (Laos) by 46 ka. *Proceedings of the National Academy of Sciences, 109*(36), 14375-14380.

DENNELL, R.W. 2003. Dispersal and colonisation, long and short chronologies: How continuous is the Early Pleistocene record for hominids outside East Africa? *Journal of Human Evolution* 45: 421-440.

DENNELL, R.W. 2004. Early hominin landscapes in northern Pakistan: Investigations in the Pabbi Hills. *British Archaeological Reports (International Series)*, 1265: 1-454.

DENNELL, R.W. 2009. *The Palaeolithic Settlement of Asia*. Cambridge: Cambridge University Press.

DENNELL, R.W. 2010. Early *Homo sapiens* in China. *Nature* 468, 512-513.

DENNELL, R.W. 2011. An earlier Acheulian arrival in South Asia. *Science* 331: 1532-1533.

DENNELL, R.W. 2012. The Nihewan Basin of North China in the Early Pleistocene: Continuous and flourishing, or discontinuous, infrequent and ephemeral occupation? *Quaternary International* 30: 1-14.

DENNELL, R.W. 2013. Hominins, deserts, and the colonisation and settlement of continental Asia. *Quaternary International* 300, 13-21.

DENNELL, R.W. 2014a. East Asia and human evolution: from cradle of mankind to cul-de-sac. In *East of Africa: Southern Asia, Australia and Modern Human Origins*, ed. R.W. Dennell and M. Porr, pp. 8-20. Cambridge: Cambridge University Press.

DENNELL, R.W. 2014b. Hallam Movius, Helmut de Terra, and the Line that never was: Burma, 1938. In *Living in the Landscape: Essays in Honour of Graeme Barker*, ed. K. Boyle, R.J. Rabett and C. Hunt, pp. 11-34. Cambridge: McDonald Institute for Archaeological Research.

DENNELL, R.W. 2014c. 'Smoke and mirrors': the fossil record for Homo sapiens between Arabia and Australia. In *East of Africa: Southern Asia, Australia and Modern Human Origins*, ed. R.W. Dennell and M. Porr, pp. 33-50. Cambridge: Cambridge University Press.

DENNELL, R.W., LOUYS, J.L., O'REGAN, H.J. & WILKINSON, D.M. 2013. The origins and persistence of *Homo floresiensis* on Flores: biogeographical and ecological perspectives. *Quaternary Science Reviews* 96, 98-107.

DENNELL, R.W., MARTINÓN-TORRES, M. & BERMUDEZ DE CASTRO, J.M. 2011. Hominin variability, climatic instability and population demography in Middle Pleistocene Europe. *Quaternary Science Reviews* 30: 1511-1524.

DENNELL, R.W. & PETRAGLIA, M.D. 2012. The dispersal of Homo sapiens across southern Asia: how early, how often, how complex? *Quaternary Sciences Reviews* 47, 15-22.

DENNELL, R.W., RENDELL, H.R. & HAILWOOD, E. 1988. Early tool-making in Asia: two-million-year-old artefacts in Pakistan. *Antiquity* 62: 98-106.

DENNELL, R.W. & ROEBROEKS, W. 2005. An Asian perspective on early human dispersal from Africa. *Nature* 438: 1099-1104.

DODONOV, A.E. & BAIGUZINA, L.L. 1995. Loess stratigraphy of Central Asia: palaeoclimatic and palaeoenvionmental aspects. *Quaternary Science Reviews* 14: 707-720.

ELLER, E., HAWKS, J. & RELETHFORD, R. 2004. Local extinction size and recolonization, species effective population size, and modern human origins. *Human Biology* 76 (5): 689-709.

FALK, D., HILDEBROT, C., SMITH, K., MORWOOD, M.J. et al. 2009. LB1's virtual endocast, microcephaly, and hominin brain evolution. *Journal of Human Evolution* 57: 597-607.

FERRING, R., OMS, O., AGUSTI, J., BERNA, F., NIORADZE, M. *et al.* 2011. Earliest human occupations at Dmanisi (Georgian Caucasus) dated to 1.85-1.78 Ma. *Proceedings of the National Academy of Sciences* USA, 108(26), 10432-10436.

FRUMKIN, A., BAR-YOSEF, O. & SCHWARCZ, H.P. 2011. Possible paleohydrologic and paleoclimatic effects on hominin migration and occupation of the Levantine Middle Paleolithic. *Journal of Human Evolution* 60, 437-451.

GABUNIA, L. & VEKUA, A. 1995. A Plio-Pleistocene hominid from Dmanisi, East Georgia, Caucasus. *Nature* 375: 509-512.

GABUNIA, L., VEKUA, A., LORDKIPANIDZE, D. et al. 2000a. Earliest Pleistocene hominid cranial remains from Dmanisi, Republic of Georgia: Taxonomy, geological setting, and age. *Science* 288: 1019-1025.

GABUNIA, L., VEKUA, A., & LORDKIPANIDZE, D. 2000b. The environmental contexts of early human occupation of Georgia (Transcaucasia). *Journal of Human Evolution* 38: 785-802.

GATHAGO, P.N. & BROWN, F.H. 2006. Revised stratigraphy of Area 123, Koobi Fora, Kenya, and new age estimates of its fossil mammals, including hominins. *Journal of Human Evolution* 51 (5): 471-479.

GOREN-INBAR, N. & SARAGUSTI, I. 1996. An Acheulean biface assemblage from Gesher Benot Ya'aqov, Israel: indications of African affinities. *Journal of Field Archaeology* 23: 15-30.

GOREN-INBAR, N., LISTER, A., WERKER, E. & CHECH, M. 1994. A butchered elephant skull and associated artifacts from the Acheulean site of Gesher Benot Ya'aqov, Israel. *Paléorient* 20 (1), 99-112.

GOREN-INBAR, N., FEIBEL, C.S., VERSOUB, K.L., MELAMED, Y. et al. 2000. Pleistocene milestones on the Out-of-Africa corridor at Gesher Ya'aqov, Israel. *Science* 289: 944-947.

GOREN-INBAR, N., SHARON, G., MELAMED, Y. & KISLEV, M. 2002. Nuts, nut-cracking, and pitted stones at Gesher Benot Ya'aqov, Israel. *Proceedings of the National Academy of Sciences USA* 99 (4): 2455-2460.

GOREN-INBAR, N., ALPERSON, N., KISLEV, M.E., SIMCHONI, O. et al. 2004. Evidence of hominin control of fire at Gesher Benot Ya'aqov, Israel. *Science* 30: 725-727.

HAWKS, J. 2009. Local adaptation in population sinks. http://johnhawks.net/taxonomy/term/260.

HERSHKOVITZ, I., KOMREICH, L. & LARON, Z. 2007. Comparative skeletal features between *Homo floresiensis* and patients with primary growth hormone insensitivity (Laron Syndrome). *American Journal of Physical Anthropology* 134: 198-208.

HOLMES, K.M. 2007. Using Pliocene palaeoclimatic data to postulate dispersal pathways of early hominins. *Palaeogeography, Palaeoclimatology, Palaeoecology* 248(1), 96-108.

HOU, Y., POTTS, R., BAOYIN, Y., ZHENGTANG, G. et al. 2000. Mid-Pleistocene Acheulean-like stone technology of the Bose Basin, South China. *Science* 287:1622-6.

HOWELLS, W.W. 1980. *Homo erectus* – Who, when and where: A survey. *Yearbook of Physical Anthropology* 23: 1-23.

HUANG WEIWEN, NE NAIHAN & SAGAWA MASATOSHI 2001. *Comparative Studies on Handaxes Found at the Bose Sites in Guangxi, China II.* Tohoku Gakuin University, Sendai: Sagawa Matatoshi.

HUFFMAN, O.F., ZAIM, Y., KAPPELMAN, J. et al. 2006. Relocation of the 1936 Mojokerto skull discovery site near Perning, East Java. *Journal of Human Evolution* 50(4): 431-451.

HUGHES, J.K., HAYWOOD, A., MITHEN, S.J., SELLWOOD, B.W. & VALDES, P.J. 2007. Investigating early hominin dispersal patterns: developing a framework for climate data integration. *Journal of Human Evolution* 53: 465-474.

HURCOMBE, L.M. 2004. The stone artefacts from the Pabbi Hills. In *Early Hominin Landscapes in Northern Pakistan: Investigations in the Pabbi Hills* by R.W. Dennell, pp. 222-292. British Archaeological Reports (International Series) 1265.

HYODO, M., WATANABE, N., SUNATA, W., SUSANTO, E.E., & WAHYONO, H. 1993. Magnetostratigraphy of hominid fossil bearing formations in Sangiran and Mojokerto, Java. *Anthropological Science* 101(2): 157-186.

HYODO, M., NAKAYA, H., URABE, A., SAEGUSA, H. et al. 2002. Paleomagnetic dates of hominid remains from Yuanmou, China, and other Asian sites. *Journal of Human Evolution* 43: 27-41.

INDRIATI, E., SWISHER, C.C., LEPRE, C., QUINN, R.L. et al. 2011. The age of the 20 meter Solo River Terrace, Java, Indonesia and the survival of *Homo erectus* in Asia. PlOS 6: 1-10.

ITIHARA, M., WATANABE, N., KADAR, D. & KUMAI, H. 1994. Quaternary stratigraphy of the hominid fossil bearing formations in the Sangiran area, Central Java. *Courier Forschungsinstitut Senckenberg* 171; 123-128.

JIA LANPO & WEI QI 1987. Artefacts lithiques provenant du site pléistocène ancien de Donggutuo près de Nihewan (Nihowan), province d'Hebei, Chine. *L'Anthropologie* 91(3): 727-732.

JUNGERS, W.L., LARSON, S.G., HARCOURT-SMITH, W., MORWOOD, M.J. et al. 2009. Descriptions of the lower limb skeleton of Homo floresiensis. *Journal of Human Evolution* 57: 548-554.

KAIFU, Y. & FUJITA, M. 2012. Fossil record of early modern humans in East Asia. *Quaternary International* 248: 2-11.

KAPPELMAN, J., ALÇIÇEK, M.C., KAZANCI, N., SCHULTZ, M. et al. 2008. Brief communication: First *Homo erectus* from Turkey and implications for migrations into temperate Europe. *American Journal of Physical Anthropology* 135: 110-116.

KENNEDY, K.A., SONAKIA, A., CHIMENT, J. & VERMA, K.K. 1991. Is the Narmada hominid an Indian *Homo erectus*? *American Journal of Physical Anthropology* 86(4), 475-496.

KLEIN, R.G. 1989. *The Human Career.* Chicago: University of Chicago Press.

KRAUSE, J., FU, Q., GOOD, J.M., VIOLA, B. *et al.* 2010. The complete mitochondrial DNA genome of an unknown hominin from southern Siberia. *Nature* 464, 894-897. doi:10.1038/nature 08976.

LARICK, R., CIOCHON, R.L., ZAIM, Y. et al. 2001. Early Pleistocene 40Ar/39Ar ages for Bapang Formation hominins, Central Jawa, Indonesia. *Proceedings of the National Academy of Sciences of the USA* 98(9): 4866-4871.

LARSSON, S.G., JUNGERS, W.L., TOCHERI, M.W. et al. 2009. Descriptions of the upper limb skeleton of Homo floresiensis. *Journal of Human Evolution* 57: 555-570.

LIEBERMAN, D.E. 2007. Homing in on early *Homo*. *Nature* 449: 291-292.

LIU, W., JIN, C., ZHANG, Y., CAI, Y., XING, S., WU, J. et al. 2010a. Human remains from Zhirendong, South China, and modern human emergence in East Asia. *Proceedings of the National Academy of Sciences USA* 107, 19201-19206.

LIU, W., WU, X., PEI, S., WU, X.-J. & NORTON, C.J. 2010b. Huanglong Cave: A late Pleistocene human fossil site in Hubei Province, China. *Quaternary International* 211: 29-41.

LIU, W., SCHEPARTZ, L.A., SONG XING, MILLER-ANTONIO, S. *et al.* 2013. Late Middle Pleistocene hominin teeth from Panxian Dadong, South China. *Journal of Human Evolution* 64(5), 337-355.

LORDKIPANIDZE, D., DE LEÓN, M.S.P., MARGVELASHVILI, A., RAK, Y., RIGHTMIRE, G.P., VEKUA, A., & ZOLLIKOFER, C.P. 2013. A complete skull from Dmanisi, Georgia, and the evolutionary biology of early Homo. *Science*, 342, 326-331.

LORDKIPANIDZE, D., JASHASHVILI, T., VEKUA, A., PONCE DE L. et al. 2007. Postcranial evidence from early *Homo* from Dmanisi, Georgia. *Nature* 449: 305-310.

LORDKIPANIDZE, D., VEKUA, A., FERRING, R., RIGHTMIRE, G.P. et al. 2005. The earliest toothless hominin skull. *Nature* 434: 717-718.

LUMLEY, H. DE, NIORADZÉ, M., BARSKY, D., CAUCHE, D. et al. 2005. Les industries lithiques préoldowayennes du début du Pléistocène inférieur du site de Dmanissi en Géorgie. *L'Anthropologie* 109 (1): 1-182.

LUMLEY, DE M.-A. & LORDKIPANIDZE, D. 2006. L'homme de Dmanissi (*Homo georgicus*), il y a 1,810,000 ans. *Comptes Rendues Palévolution* 5: 273-281.

LYCETT, S. & BAE, C. 2010. The Movius Line controversy: the state of the debate. *World Archaeology* 42, 521-544.

MACAULAY, V., HILL, C., ACHILLI, A., RENGO, C. et al. 2005. Single, rapid coastal settlement of Asia revealed by analysis of complete mitochondrial genomes. *Science* 308: 1034-1036.

MARTÍNEZ-NAVARRO, B. & RABINOVICH, R. 2010. The fossil Bovidae (Artiodactyla, Mammalia) from Gesher Benot Ya'aqov, Israel: Out of Africa during the Early Middle Pleistocene transition. *Journal of Human Evolution* 30:1-10.

MARTINÓN-TORRRES, M., DENNELL, R.W. & BERMÚDEZ DE CASTRO, J.M. 2011. The Denisova hominin need not be an out of Africa story. *Journal of Human Evolution* 60: 251-255.

MCBREARTY, S. & BROOKS, A.S. 2000. The revolution that wasn't: a new interpretation of the origin of modern human behaviour. *Journal of Human Evolution* 39 (5): 453-563.

MCDONALD, K., MARTINÓN-TORRES, M., DENNELL, R.W. & BERMUDEZ DE CASTRO, J.M. 2012. Discontinuity in the record for hominin occupation in south-western Europe: Implications for occupation of the middle latitudes of Europe. *Quaternary International* 271, 84-97. doi:10.1016/j.quaint.2011.10.009.

MCDOUGALL, I., BROWN, F.H. & FLEAGLE, J.G. 2005. Stratigraphic placement and age of modern humans from Kibish, Ethiopia. *Nature* 433: 733-736.

MELLARS, P. 2006. Going east: new genetic and archaeological perspectives on the modern human colonization of Eurasia. *Science* 313: 796-800.

MGELADZE, A., LORDKIPANIDZE, D., MONCEL, M.H., DESPRIEE, J., CHAGELISHVILI, R., NIORADZE, M. & NIORADZE, G. 2011. Hominin occupations at the Dmanisi site, Georgia, Southern Caucasus: Raw materials and technical behaviours of Europe's first hominins. *Journal of Human Evolution, 60*(5), 571-596.

MIJARES, A.S., DÉTRIOT, F., PIPER, P., GRÜN, R. et al. 2010. New evidence for a 67,000-year-old human presence at Callao Cave, Luzon, Philippines. *Journal of Human Evolution* 59, 123-132.

MITHEN, S. & REED, M. 2002. Stepping out: a computer simulation of hominid dispersal from Africa. *Journal of Human Evolution* 43: 433-462.

MORWOOD, M. & OOSTERZEE, P. VAN 2007. *The Discovery of the Hobbit.* Sydney: Random House Australia.

MORWOOD, M.J. & JUNGERS, W.L. 2009. Conclusions: implications of the Liang Bua excavations for hominin evolution and biogeography. *Journal of Human Evolution* 57: 640-648.

MORWOOD, M.J., O'SULLIVAN, P.B., AZIZ, F. & RAZA, A. 1998. Fission-track ages of stone tools and fossils on the east Indonesian island of Flores. *Nature* 392: 173-176.

MORWOOD, M.J., O'SULLIVAN, P., SUSANTO, E.E. & AZIZ, F. 2003. Revised age for Mojokerto 1, an early *Homo erectus* cranium from East Java, Indonesia. *Australian Archaeology* 57: 1-4.

MORWOOD, M.J., SOEJONO, R.P., ROBERTS, R.G., SUTNIKA, T. et al. 2004. Archaeology and age of a new hominin from Flores in eastern Indonesia. *Nature* 431, 1087-1091.

MOVIUS, H. 1948. The lower Palaeolithic cultures of southern and eastern Asia. *Transactions of the American Philosophical Society* 38 (4): 329-420.

NIKITAS, P. & NIKITA, E. 2005. A study of hominin dispersal out of Africa using computer simulations. *Journal of Human Evolution* 49, 602-617.

NORTON, C.J., KIDONG B., HARRIS, J.W.K. & LEE, H. 2006. Middle Pleistocene handaxes from the Korean Peninsula. *Journal of Human Evolution* 51(5): 527-536.

O'CONNELL, J.F. & ALLEN, J. 1998. When did humans first arrive in Greater Australia and why is it important to know? *Evolutionary Anthropology* 132-146.

O'REGAN, H.J., BISHOP, L.C., LAMB, A., ELTON, S. & TURNER, A. 2005. Large mammal turnover in Africa and the Levant between 1.0 and 0.5 Ma. In *Early-Middle Pleistocene Transitions: The Land-Ocean Evidence*, ed. M.J. Head and P.L. Gibbard, pp. 231-249. Geological Society of London Special Publications 247.

PADDAYYA, K. 2001. The Acheulean Culture Project of the Hunsgi and Baichbal Valleys, peninsular India. In *Human Roots: Africa and Asia in the Middle Pleistocene* (ed. L. Barham and K. Robson-Brown), pp. 235-258. Bristol: Western Academic and Specialist Press Ltd.

PAPPU, S., GUNNELL, Y., AKLILESH, K., BRAUCHER, R. et al. 2011. Early Pleistocene presence of Acheulian hominins in South India. *Science* 331, 1596-1599.

PARFITT, S.A., BARENDREGT, R.W., BREDA, M., CANDY, I. et al. 2005. The earliest record of human activity in northern Europe. *Nature* 438: 1008-1012.

PERRY, G.H. & DOMINI, N.J. 2009. Evolution of the pygmy human phenotype. *Trends in Ecology and Evolution* 24: 218-225.

PETRAGLIA, M.D. & SHIPTON, C. 2008. Large cutting tool variation west and east of the Movius Line. *Journal of Human Evolution* 55: 962-966.

PETRAGLIA, M., CLARKSON, C., BOIVIN, N., HASLAM, M. et al. 2009. Population increase and environmental deterioration correspond with microlithic innovations in South Asia ca. 35,000 years ago. *Proceedings of the National Academy of Sciences* USA 106: 12261-12267.

POTTS, R. & TEAGUE, R. 2010. Behavioral and environmental background to 'Out-of-Africa-I' and the arrival of *Homo erectus* in East Asia. In *Out of Africa I: the First Hominin Colonization of Eurasia*, ed. J.G. Fleagle, J.J. Shea, F.E. Grine, A.L. Baden and R.E. Leakey, pp. 67-85. Springer Science + Business Media B.V.

PRESNYAKOV, S.L., BELYAEVA, E.V., LYUBIN, V.P., RODIONOV, N.V. et al. 2012. Age of the earliest

Paleolithic sites in the northern part of the Armenian Highland by SHRIMP-II U–Pb geochronology of zircons from volcanic ashes. *Gondwana Research* 21: 928-938.

QUINTANA-MURCI, L., SEMINO, O., BANDELT, H.-J., PASSARINO, G. et al. 1999. Genetic evidence of an early exit of *Homo sapiens* from Africa through eastern Africa. *Nature Genetics* 23: 437-441.

RABETT, R. 2012. *Human adaptation in the Asia Palaeolithic: hominin dispersal and behaviour during the late Quaternary*. Cambridge: Cambridge University Press.

RABINOVICH, R., GAUDZINSKI, S. & GOREN-INBAR, N. 2008. Systematic butchering of fallow deer (Dama) at the early Middle Pleistocene Acheulian site of Gesher Benot Ya'aqov, (Israel). *Journal of Human Evolution* 54: 134-149.

RANOV, V.A. 1995. The 'Loessic Palaeolithic' in South Tadjikistan, Central Asia: its industries, chronology and correlation. *Quaternary Science Reviews* 14: 731-745.

REICH, D., GREEN, R.E., KIRCHER, M., KRAUSE, J. et al. 2010. Genetic history of an archaic hominin group from Denisova Cave in Siberia. *Nature* 468, 1053-1060.

RELETHFORD, J.H. 1998. Mitochondrial DNA and ancient population growth. *American Journal of Physical Anthropology* 105: 1-7.

RENDELL, H.M. & DENNELL, R.W. 1985. Dated lower Palaeolithic artefacts from northern Pakistan. *Current Anthropology* 26(5): 393.

RENDELL, H.M., HAILWOOD, E.A. & DENNELL, R.W. 1987. Magnetic polarity stratigraphy of Upper Siwalik Sub-Group, Soan Valley, Pakistan: implications for early human occupance of Asia. *Earth and Planetary Science Letters* 85: 488-496.

RICHARDS, G.D. 2006. Genetic, physiologic and ecogeographic factors contributing to variation in *Homo sapiens*: *Homo floresiensis* reconsidered. *Journal of Evolutionary Biology 19:* 1744-1766.

RIGHTMIRE, G.P. 2001. Comparison of Middle Pleistocene hominids from Africa and Asia. In *Human Roots: Africa and Asia in the Middle Pleistocene*, ed. L. Barham and K. Robson-Brown, pp. 123-33. Bristol: Western Academic and Specialist Press Ltd.

RIGHTMIRE, G.P., LORDKIPANIDZE, D. & VEKUA, A. 2006. Anatomical descriptions, comparative studies and evolutionary significance of the hominin skulls from Dmanisi, Republic of Georgia. *Journal of Human Evolution* 50 (2): 115-141.

ROBERTS, R.G., JONES, R., SPOONER, N.A., HEAD, M.J., MURRAY, A.S. & SMITH, M.-A. 1994. The human colonisation of Australia: optical dates of 53,000 and 60,000 years bracket human arrival at Deaf Adder Gorge, Northern Territory. *Quaternary Geochronology (Quaternary Science Reviews)* 13, 575-583.

SAEGUSA, H., THASOD, Y. & RATANASTHIEN, B. 2004. Notes on Asian stegodontids. *Quaternary International* 126-8: 31-48.

SCALLY, A. & DURBIN, R. 2012. Revising the human mutation rate: implications for understanding human evolution. *Nature Reviews Genetics*, 13(10), 745-753. doi:10.1038/nrg3295.

SHANG, H., TONG, H., ZHANG, S., CHEN, F. & TRINKHAUS, E. 2010. An early modern human from Tianyuan Cave, Zhoukoudian, China. *Proceedings of the National Academy of Sciences USA* 104: 6573-6578.

SHANG, H. & TRINKHAUS, E. 2010. *The early modern human from Tianyuan Cave*. Texas A and M University Press.

SHEA, J.J. 2008. Transitions or turnovers? Climatically-forced extinctions of *Homo sapiens* and Neanderthals in the east Mediterranean Levant. *Quaternary Science Reviews* 27: 2253-2270.

SHEN, G., WANG, W., WANG, Q., ZHAO, J. et al. 2002. U-series dating of Liujiang hominid site in Guangxi, southern China. *Journal of Human Evolution* 43: 817-829.

SHEN, G., WANG, W., CHENG, H. & EDWARDS, R.L. 2007. Mass spectrometric U-series dating of Laibin hominid site in Guangxi, southern China. *Journal of Archaeological Science* 34: 2109-2114.

SHEN, G., GAO, X., GAO, B. & GRANGER, D.E. 2009. Age of Zhoukoudian Homo erectus determined with 26Al/10Be burial dating. *Nature* 458: 198-200.

SMITH, J.M.B. 2001. Did early hominids cross sea gaps on natural rafts? In *Faunal and Floral Migrations and Evolution in SE Asia-Australasia*, ed. I. Metcalfe, J.M.B. Smith, M. Morwood, and I. Davidson, pp. 409-16. Lisse: Balkema.

SOARES, P., ERMINI, L., THOMSON, N., MORMINA, M. et al. 2009. Correcting for purifying selection: an improved human mitochondrial molecular clock. *American Journal of Human Genetics* 84: 740-759.

SOHN, S. & WOLPOFF, M.H. 1993. Zuttiyeh face: A view from the East. *American Journal of Physical Anthropology* 91: 325-347.

SONAKIA, A. 1985. Early *Homo* from the Narmada Valley, India. In *Ancestors: The Hard Evidence*, ed. E. Delson, pp. 334-338. New York: Alan Liss Inc.

SONAKIA, A. & LUMLEY, H. DE 2006. Narmada *Homo erectus* – A possible ancestor of the modern Indian. *Comptes Rendues de Palévolution* 5: 353-357.

STEWART, J.R. & STRINGER, C.B. 2012. Human evolution out of Africa: The role of refugia and climate change. *Science* 335: 1317-1321.

STORM, P. 2001. An environmental approach to the fate of *Homo erectus* in Australasia. In: L. Barham, L., Robson-Brown, K. (Eds.), *Human Roots: Africa and Asia in the Middle Pleistocene*. Western Academic and Specialist Press Ltd., Bristol, pp. 203-15.

STORM, P., AZIZ, F., VOS, J. DE, KOSASIH, D. et al. 2005. Late Pleistocene *Homo sapiens* in a tropical rainforest fauna in East Java. *Journal of Human Evolution* 49: 536-545.

STRINGER, C. 2006. *Homo Britannicus: The Incredible Story of Human Life in Britain*. London, Allen Lane.

STRINGER C. & ANDREWS, P. 1988. Genetic and fossil evidence for the origin of modern humans. *Science* 239: 1263-1268.

SUMMERHAYES, G.R., LEAVESLEY, M., FAIRBAIRN, A., MANDUI, H. et al. 2010. Human adaptation and plant use in Highland New Guinea 49,000 to 44,000 years ago. *Science* 330: 78-81.

SWISHER, C.C. III, RINK, W.J., ANTÓN, S.C., SCHWARCZ, H.P. et al. 1996. Latest *Homo erectus* of Java: Potential contemporaneity with *Homo sapiens* in Southeast Asia. *Science* 274: 1870-1874.

SWISHER, C.C. III, CURTIS, G.H., JACOB T., GETTY, A.G. et al. 1994. Age of the earliest known hominids in Java, Indonesia. *Science* 263: 1118-1121.

TATTERSALL, I. 1997. Out of Africa again...and again? *Scientific American* 276 (4): 46-53.

TCHERNOV, E. 1992. Eurasian-African biotic exchanges through the Levantine corridor during the Neogene and Quaternary. In *Mammalian Migration and Dispersal Events in the European Quaternary*, ed. W. von Koenigswald and Lars Werdelin, pp. 103-123. *Courier Forschungsinstitut Senckenberg* 153.

TCHERNOV, E., HORWITZ, K., RONEN, A. & LISTER, A.I. 1994. The faunal remains from Evron Quarry in relation to other lower paleolithic hominid sites in the southern Levant. *Quaternary Research* 42: 328-339.

THORNE, A., GRÜN, R., MORTIMER, G., SPOONER, N.A. et al. 1999. Australia's oldest human remains: age of the Lake Mungo 3 skeleton. *Journal of Human Evolution* 36: 591-612.

THORNE A.G. & WOLPOFF M.H. 1992. The multiregional evolution of humans. *Scientific American* 266(4): 76-83.

TRINKHAUS, E. & SHANG, H. 2008. Anatomical evidence for the antiquity of human footwear: Tianyuan and Sunghir. *Journal of Archaeological Science* 35; 192 8-1933.

TSUJI, Y., MATSTOMI, H., IMAMURA, F, TAKOE, M. et al. 1995. Damage to coastal villages due to the 1992 Flores Island earthquake tsunami. *Pure and Applied Geophysics* 44: 481-524.

URABE, A., HIDEO NAKAYA, TETSUJI MUTO, et al. 2001. Lithostratigraphy and depositional history of the Late Cenozoic hominid-bearing successions in the Yuanmou Basin, southwest China. *Quaternary Science Reviews* 20: 1671-1681.

VIALET, A., GUIPERT, G. & ALÇIÇEK, M.C. 2012. Homo erectus found still further west: Reconstruction of the Kocabaş cranium (Denizli, Turkey). *Comptes Rendues Palévolution* 11, 89-95.

WHITE, T.D., ASFAW, B., DEGUSTA, D., TILBERT, H. et al. 2003. Pleistocene *Homo sapiens* from Middle Awash, Ethiopia. Nature 423: 742-747.

WOLPOFF, M. & CASPARI, R. 1997. *Race and human evolution: a fatal attraction.* New York: Simon and Schuster.

WOLPOFF, M., HAWKS, J. & CASPARI, R. 2000. Multiregional, not multiple origins. *American Journal of Physical Anthropology* 112: 129-136.

WU, XINXHI & POIRIER, F.E. 1995. *Human Evolution in China: A Metric Description of the Fossils and a Review of the Sites.* New York/Oxford: Oxford University Press.

YI, S. & CLARK, G. 1983. Observations on the Lower Palaeolithic of Northeast Asia. *Current Anthropology* 24: 181-202.

YOKOYAMA, Y. et al. 2008. Gamma-ray spectrometric dating of Late *Homo erectus* skulls from Ngandong and Sambungmacan, Central Java, Indonesia. *Journal of Human Evolution* 55: 274-277.

ZHU, R., ZHINSHENGAN, POTTS, R. & HOFFMAN, K.A. 2003. Magnetostratigraphy of early humans in China. *Earth-Science Reviews* 61: 341-359.

ZHU, R.X., HOFFMAN, K.A., POTTS, R., DENG, C.L. et al. 2001. Earliest presence of humans in northeast Asia. *Nature* 413: 413-417.

ZHU, R.X., POTTS, R., XIE, F., HOFFMAN, K.A. et al. 2004. New evidence on the earliest human presence at high northern latitudes in northeast Asia. *Nature* 431: 559-562.

ZHU, R.X., POTTS, R., PAN, Y.X., YAO, H.T. et al. 2008. Early evidence of the genus *Homo* in East Asia. *Journal of Human Evolution 55:* 1075-1085.

ZHU, Z.-Y., DENNELL, R., HUANG, W.-W., WU, YI, RAO, Z.-G., QIU, S.-F., XIE, J.-B., LIU, W., FU, S.-Q., HAN, J.-W., ZHOU, H.-Y., OU YANG, T.-P. & LI, H.-M. 2015. New dating of the *Homo erectus* cranium from Lantian (Gongwangling), China. *Journal of Human Evolution* 78, 144-157.

4.
PLEISTOCENE HOMININ FOSSIL DISCOVERIES IN INDIA: IMPLICATIONS FOR HUMAN EVOLUTION IN SOUTH ASIA

Anek R. SANKHYAN
Ex-Sr. Anthropologist (P), Anthropological Survey of India, Kolkata.
President, 'Palaeo Research Society, Ghumarwin-174021 (H.P.), India
arsankhyan@gmail.com

Abstract: *Since 1830s the Central Narmada valley of India is known for the Quaternary mega-terrestrial fauna and later for their unique co-occurrences with the Lower Palaeolithic implements in stratified contexts. But, the anticipated human fossil discoveries could only be made after over a century and half. First came a partial calvarium, followed by the discovery of two clavicles and a 9th rib from two layers of the Surajkund Formation exposed at Hathnora. More recently, an excavation yielded a femur and a humerus from another site, Netankheri – 3 km upstream Hathnora. The femur came from the calvarium-yielding bio-stratigraphic horizon whereas the humerus just above the clavicles-yielding Surajkund Formation. A large number of mega-terrestrial mammals and Palaeolithic implements were recovered from the hominin beds, which allow sorting out of the hominin fossils into three types. The earliest type represented by the calvarium and the femur and datable ~250 Kya is a robust 'large-bodied' Mode 2 (Acheulian) archaic man closer to Homo heidelbergensis. This was succeeded by a 'small-bodied' Mode 3 archaic type represented by two clavicles and the 9th rib emerged ~150 Kya, provisionally named as Homo narmadensis as it was succeeded by a similar 'small-bodied' 'late archaic' or 'early modern' third type represented by the humerus associated with late Mode 3 or early Mode 4 stone and bone artifacts documented at around 80-70 Kya. Overall evidences, including the 60 Kya genomic signatures, indicate continuity of the 'short-bodied' type leading to the similar 'small-bodied' indigenous populations of Central India/South Asia.*

Keywords: *cranial, postcranial, hominin, fossils, Acheulian, bone artifacts, Homo heidelbergensis, Homo narmadensis*

INTRODUCTION

Since 1830s the Siwalik Hills of the Indian sub-continent as well as the Central Narmada valley (Fig. 4.1) were hunting grounds for the palaeoanthropologists seeking the fossil remains of human ancestors. The result was discovery of numerous fossils of the later Miocene 'large hominoid apes' and "ape-men" in the Siwaliks and Middle Pleistocene mammals in Narmada valley followed later of the Palaeolithic implements (see Sankhyan, 2009 & 2010 for detailed review). Since the Central Narmada valley yielded the mammalian fossils and the Palaeolithic implements in the stratified contexts, speculations continued with optimism that one day the fossils of the Stone Age man would also be recovered from there. But, the speculations came true only after a long wait of a century and a half in the 1980s. First came the discovery of partial skullcap (calvarium) from Hathnora reported by Sonakia (1984), followed by the discovery of a right clavicle (Sankhyan, 1997a, b) and later of the partial left clavicle and the 9th rib (Sankhyan, 2005). Recent rigorous explorations coupled with trial excavations conducted by the author and his team from the Anthropological Survey of during 2005 to 1010 yielded large assemblages of mammalian fossils and Palaeolithic implements which included two more hominin bones systematically described by Sankhyan (2010) and briefly reported by Sankhyan et al. (2012a, b).

The postcranial hominin fossils have not received any alternate interpretation than what is interpreted by the reporter, but the cranial specimen remained hotly debated and subjected to various interpretations ever since it was reported.

THE CALVARIUM

The Narmada skullcap (calvarium) eluded different scholars differently. It was considered as "evolved" *H. erectus* by Lumley and Sonakia (1985, and the view is recently championed by Mallasse (2009). However, it was conceived as "archaic" *H. sapiens* by Kennedy et al. (1991, Kennedy, 2000, 2007; Badam et al., 1986; Sankhyan 1999, 2007), Dennell & Roebroeks (2005) and Cameron et al. (2005). The latter, noted a special cladistic affinity with the Steinheim man, whereas (Athreya, 2007) recognised it a separate species, *H. heidelbergensis*, a view also shared by Sankhyan, 2010 and Sankhyan et al., 2012a, b).

It may be worthwhile to list the salient features of the Narmada calvarium which unite it to *Homo erectus* and *Homo sapiens* including its own idiosyncrasies. To start with Kennedy et al. (1991), the *H. erectus* traits itemized by Rightmire (1988, 1990, and 1998), the traits present in Narmada are as follows:

a) ***The H. erectus traits:*** Only 43 percent of them occurred viz.: (i) small mastoid process, (ii) narrow post-orbital constriction, (iii) the maximum breadth of the cranial vault is across the mastoid portions of the temporal bone, (iv) a prominent bilateral bony structure called a torus angularis. Overall, the

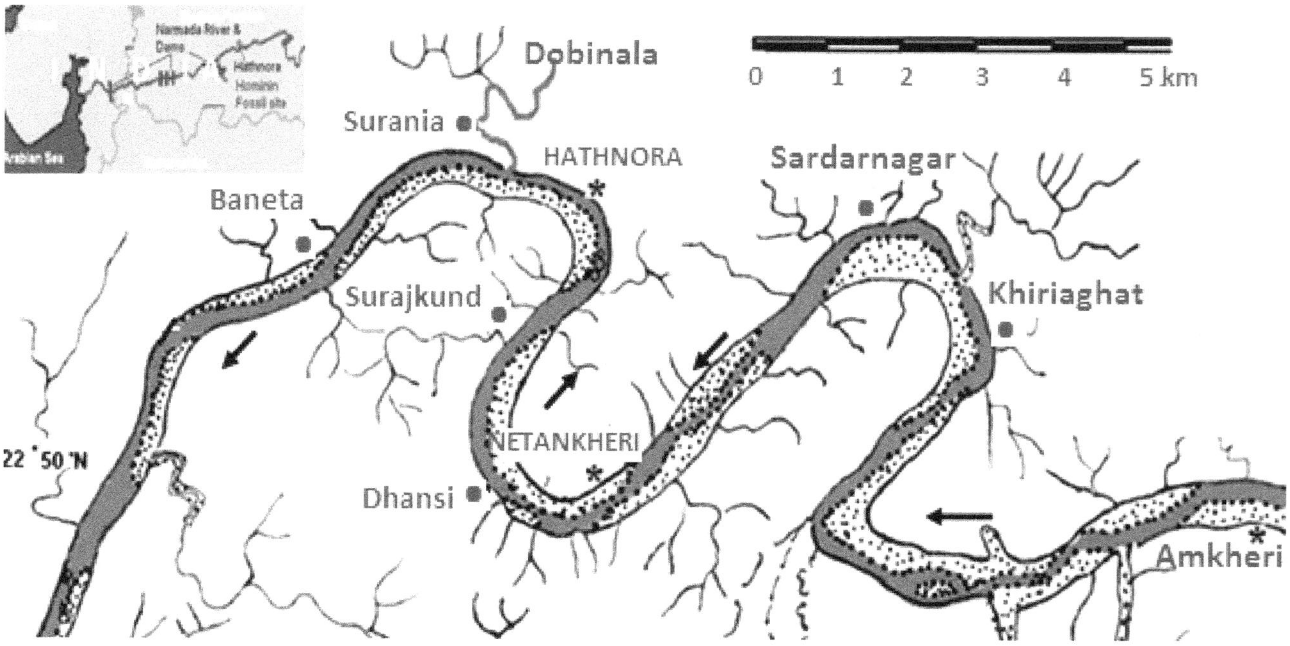

Fig. 4.1. Map of a portion of the Central Narmada valley around Hathnora showing hominin-yielding sites marked with star (modified from Sankhyan et al., 2012b)

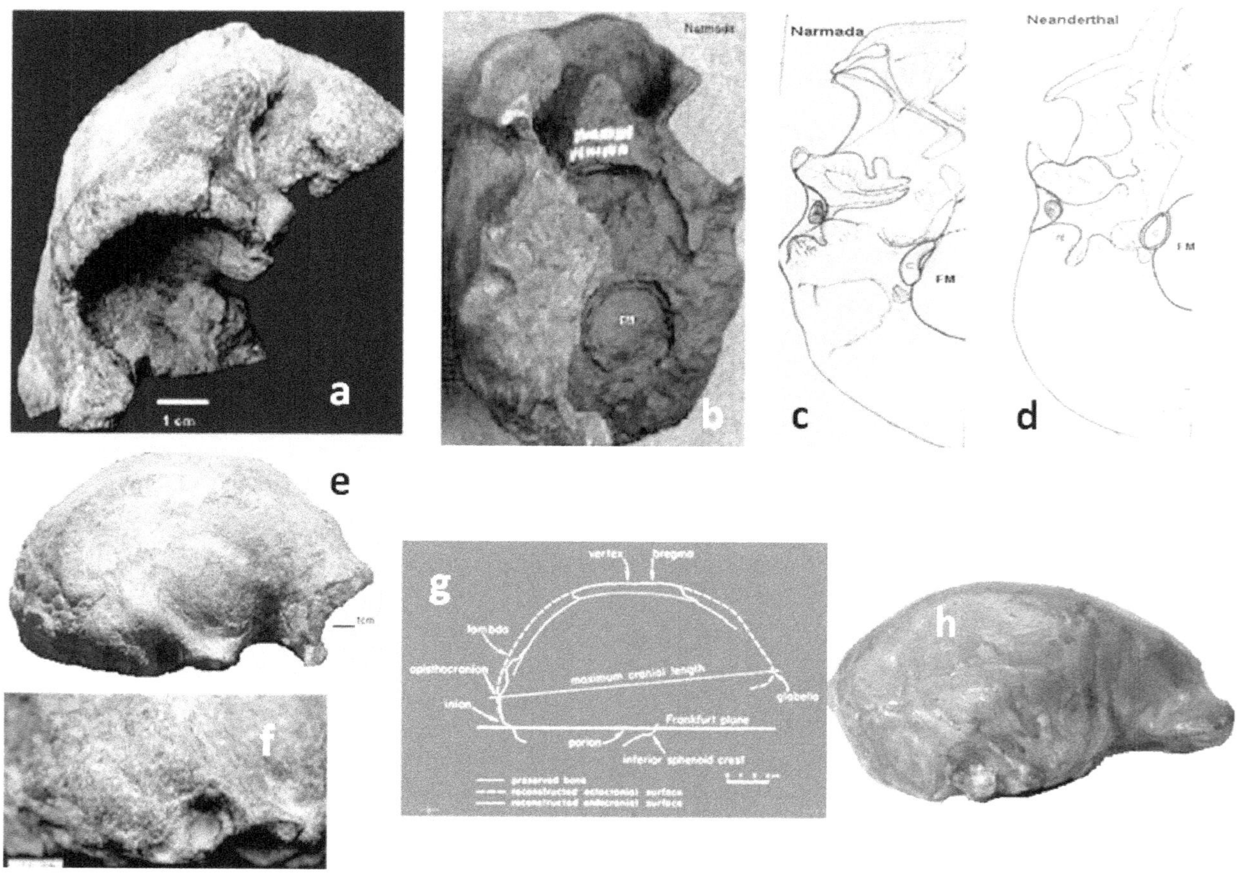

Fig. 4.2. Narmada calvarium in different views and compared

temporal and sagittal regions in particular suggest Narmada as *H. erectus* status (Fig. 4.2f).

b) ***The H. sapiens traits:*** (i) a relatively high elevation of the cranial vault, (ii) the landmarks bregma and vertex are not coincident, (iii) the most posterior point in the instrumental calibration of maximum cranial length falls superior to the landmark inion (where it lies in *H. erectus* skulls), and (iv) the estimated higher cranial capacity between 1155 and 1421 cubic

Petralona =11 (40.74%)-Eu
La Ferrassie=10 (37.04%)-Eu
La Chapelle Aux Saint=8 (29.63%)-Eu
Kabwe =7 (25.93)-Af
Dali = 7 (25.93)-EA
Zkd-10 =7 (25.93)-EA
Ngadong-11 =7 (25.93)-SEA
Sangiran 17 =7 (25.93)-SEA
Ceprano =6 (22.22%)-Eu
Steinheim =6 (22.22%)-Eu
Arago =5 (18.52%)-Eu
Saccapastor =5 (18.52%)-Eu
Amud=5 (18.52%)-Eu
Sambrumenchan =5 (18.52%)-SEA
Shanidar-1 =4 (14.81%)-Eura
Mt Carmel =4 (14.81%) Eura
ER3733=3 (11.11%)-Af
OH-9=3 (11.11%)-Af
Zkd-11, 12=3 (11.11%)-EA
Swascombe=3(11.11%)-EU
Ehrndorf=3(11.11%)-EU
Skhul=2 (7.41%)-Eura
Cro-Magnon=1 (3.0%)-Eu

a

A. Traits Narmada shares with *H. erectus*:
- Small mastoid process
- Narrow post-orbital constriction
- Maximum breadth of the cranial vault across the mastoid
- A prominent torus angularis

B. Traits Narmada shares with *H. sapiens*:
- A high cranial vault
- Non-coincident bregma and vertex
- Non-coincident opisthocranion and inion
- High cranial capacity (~1155-1421 cc)

C. Unique Traits of Narmada
- The furrowing of the sagittal ridge
- Large external auditory meatus
- Unusually long temporal bone

b

Box 4.1. Degree of affinities Narmada hominin calvarium with the Pleistocene hominins: a. Metric affinities, b. non-metric affinities

centimetres, which in *H. erectus* averages at about 1000 cubic centimetres. All these morphological traits indicate that Narmada fits most nearly to the phenotypic pattern of an anatomically "archaic" or early *H. sapiens* (Fig. 4.2g).

c) ***The unique traits:*** which are infrequent or absent in *erectus* and modern *sapiens,* namely: (1) the furrowing of the sagittal ridge along the top of the cranial vault, (2) a large external auditory meatus (ear hole), and (3) an unusually long temporal bone (Fig. 4.2f).

Based on these features, they opined that Narmada hominin also qualifies to be distinguished at the trinomial level, '*narmadensis'*, but both bivariate and multivariate analyses were concordant with a conclusion that Narmada is a late Middle Pleistocene anatomically "archaic" *H. sapiens,* though viewed as *H. heidelbergensis* by the followers of cladistics who remove Neanderthals from *H. sapiens* and re-designate it as *H. neanderthalensis.* The problem with the cladistic methodology is that it ultimately leads us to distinguish only a few shared derived morphologies (synapomorphies), leaving out the major chunk as primitive retentions, i.e. symplesiomorphies, besides a few, autapomorphies as well. A closer scrutiny of the data source used by Cameron et al. (2004) and a comparative metric study by Sankhyan (2007, 2010) reveal no such specific affinity of Narmada with the Steinheim. Rather, Narmada presented closer affinity with other European Middle to Late Pleistocene hominins, e.g., *H. neanderthalensis*, Atapuerca, *H. heidelbergensis* and *H. antecessor* as well and lesser with the Asian though in a graded way (Box 4.1). In fact, both metric and non-metric comparisons show that Narmada calvarium present a generalised *mosaic* of primitive, shared and unique morphology which anticipate the earlier conclusions of an "evolved" *H. erectus* or "archaic" *H. sapiens*, but cladistically we may attribute Narmada to the Old World species, *H. heidelbergensis.* Athreya (2007) has tested different Models and reached the same conclusion. Such mosaic morphology could be due to Narmada's 'hybridization' or 'assimilation' with the immigrant hominins from Europe, Africa and East and Southeast Asia, but "replacement" is not implied.

THE CLAVICLES

Narmada clavicle fossils have gained greater significance for reconstructing hominin body dimensions, viz., (i) stature, (ii) shoulder width= biacromial diameter, and (iii) hominin adaptations and affinities.

The stature estimates range from 115 to 130 cm which tally closely with the stature of the shortest Andaman female Onge pygmy (135 cm). Interestingly, there is no size parallel for the Narmada clavicles in the known

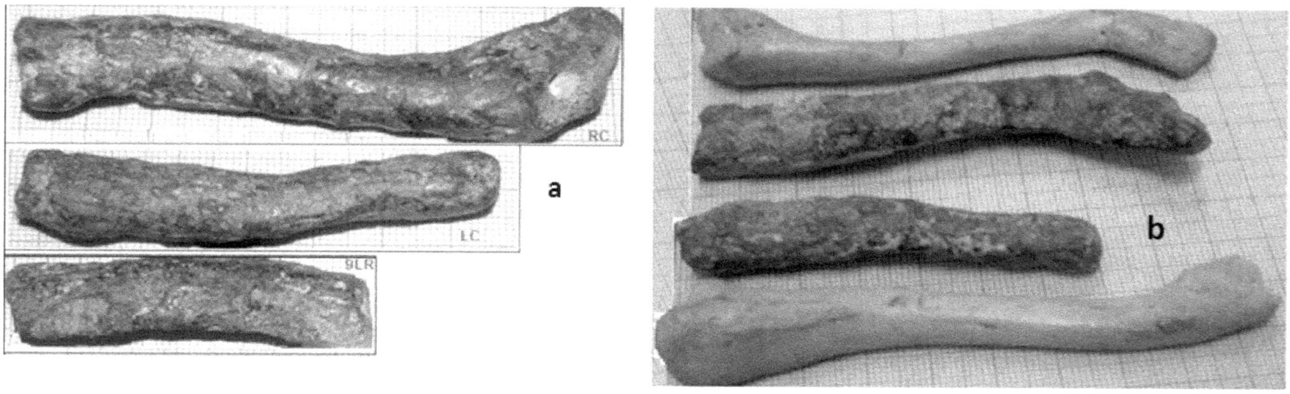

Fig. 4.3. a. Narmada right and left clavicles and 9th rib; b. Clavicles compared with those of the Onge

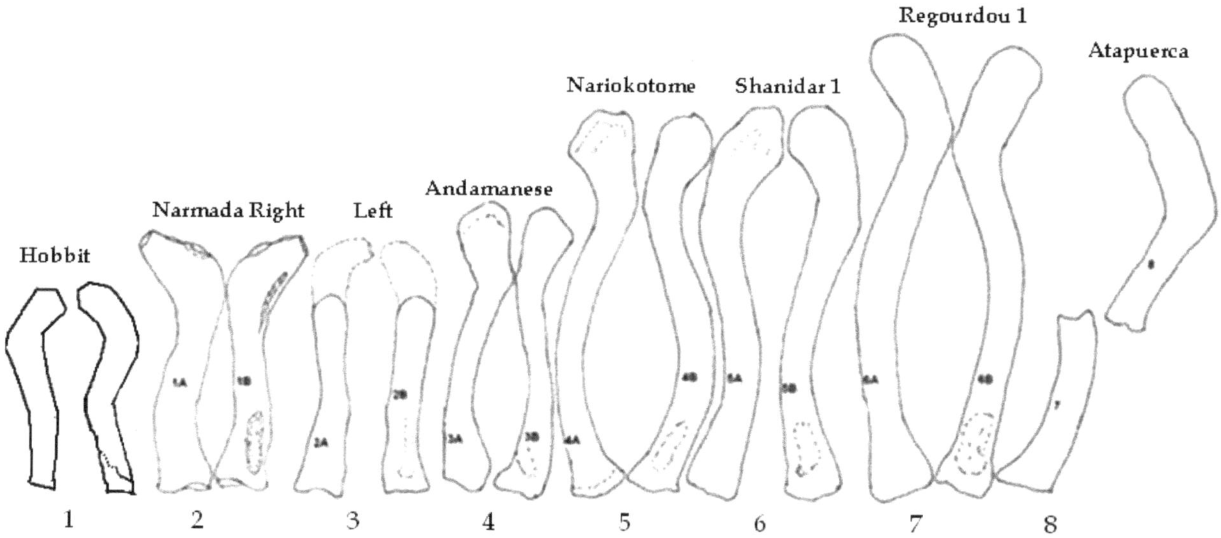

Fig. 4.4. Narmada right and left clavicles compared with the Andaman Pygmy and other Middle Pleistocene hominin clavicles in same scale

Pleistocene hominin fossil record, except that the Hobbits (*H. floresiensis*) clavicles are exceptionally small and smaller than the Narmada clavicles as well (Fig. 4.3).

The shoulder width estimate from a 90 mm right and 88 mm left clavicle averages at 30.6 cm, which tallies with the biacromial diameter (Chatterjee, 1956) of the Onge female population [mean: 31.2±1.1 cm: range 29-33.5 cm: N=29] and also approaches the Sakai pygmy (33.5 cm). Another study (Sahani, 2001) personal communication) on 16 adult Onge females also endorse the results revealing 30.14 cm mean acromial diameter with a narrow range (28-33 cm). Among other parameters, the midshaft index presents the most striking affinity of Narmada (80.0/78.6) with the Kenyan *H. habilis* (83.3) touching the lower threshold (80) of modern Andaman Pygmy and mainland Indians [Table 1] reflecting a progressive trend in the diaphysial rounding in Narmada hominin over other Lower and Middle Pleistocene hominins. The conoid index reveals a moderate chest depth of the Narmada hominin like modern humans. These are important clues suggesting *H. sapiens* status of the Narmada 'short-bodied' hominin progressing towards modernity.

THE 9TH RIB

It is less curved compared to modern human 9th rib and its curvature corresponds to that of the 4th or 5th rib, and therefore it conforms to a shorter thoracic frame at par with those of Pygmy thorax since. Like the clavicles, its robust character indicates an archaic character suggesting strong chest and back musculature, especially the lateral adjacent to the angle (Fig. 4.5) compared here with modern human corresponding rib.

THE HUMERUS

The preserved portion of the humerus is fully mineralized and typically cylindrical proximally, widening and turning prismatic distally (Fig. 4.6a). Compared well with the average size of the modern Indian humeri (N=29), in particular with 4 humeri of the small-bodied Chaurite Nicobarese, notably the CH-27 humerus, indicating the existence of small-bodied late archaic or early modern *H.*

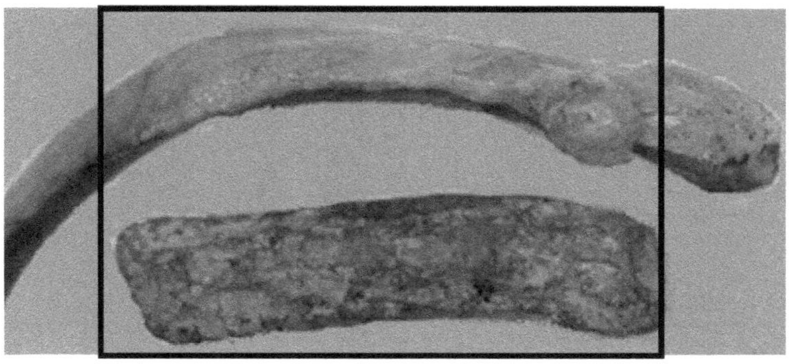

Fig. 4.5. Narmada and modern human left 9th ribs

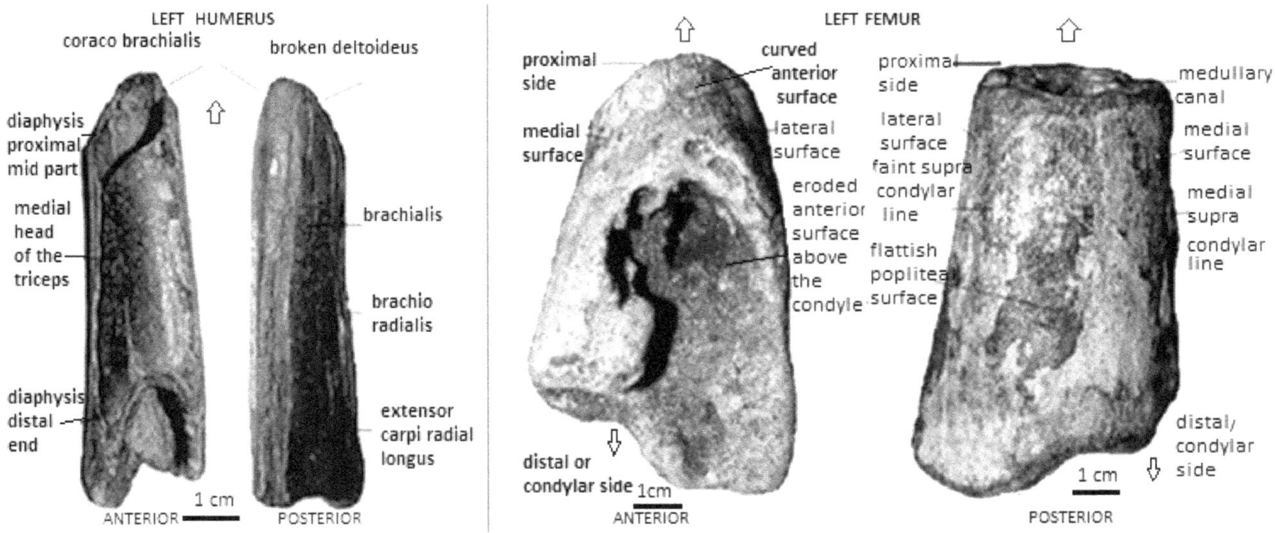

a b

Fig. 4.6. a. fossilised left humerus; b. left femur shaft portions from Netankheri, Central Narmada valley

sapiens since humerus does not provide much differentiation between the archaic and the modern morphology (Todd & Churchill, 2006). Nevertheless, fossilization and stout morphology may indicate a "late archaic" character of the NTK fossil humerus. Its 84 mm preserved fragment yields an estimated total length. There could be two stature estimates from the humerus fossil. If it belonged to a female the stature estimate comes to 138.59 cm, and 147.46 cm if comes from a male. Using the same formulae, the mean stature estimate for the five Chaurites female comes to 154.42 cm and for the males 162.32 cm. The stature estimates for the 33 humeri mixed sample comes to 152.36 cm for females and 160.38 cm for males (Tab. 4.2, bottom rows).

Interestingly, the NTK population was much shorter and stockier than the Chaurite Nicobari population including the mixed mainland Indians, and much more from other fossil populations, namely the Omo Kibish and Cro-Magnon 1 (Carretero, 2009; Bermudez de Castro et al., 2012, personal communication). Overall the Netankheri hominin (138-147.5 cm) stands more close to the Hathnora 'short and stocky' archaic hominin for which the maximum estimated stature from the right clavicle does not exceed 135 cm. The Netankheri humerus therefore, may be from a "late archaic" or "early modern" *Homo sapiens*. It suggests that the Central Narmada Valley was continuously occupied by the 'short and stocky' hominins all through the Middle to Late Pleistocene and that the Narmada hominins might have included the ancestors of the 'short-bodied' ancient South Asians, including the Andaman Pygmy.

Thus, Narmada valley documents a clear existence of a very 'short and stocky' archaic hominin during later Middle Pleistocene, and a moderately short and stocky one during early Late Pleistocene indicating a continuity of such progressive hominins leading to anatomical modernity.

THE DISTAL FEMUR

The specimen is fully mineralized, 81 mm distal-most shaft portion of the left femur, detached from the condyles (Fig. 4.6b). The popliteal surface is well preserved while articular surface of the patella intercondylar fossa is eroded.

Comparative morphometrics of femur specimens are presented in Tab. 4.2. Although limited by preservation,

Tab. 4.1. The Midshaft Index of Narmada and other hominins/humans (adapted from Sankhyan, 2007)

Species	N	Midshaft Index
Narmada *Homo*	2	80.0/78.6
Australopithecus afarensis (AL. 333 x -6/9)	1	68.6
African *Homo erectus / H. ergaster* (WT-15000)	2	55.8/61.7
Olduvai *Homo* (OH-48)	1	70.3
Kenyan *Homo habilis* (E.R. – 1808)	1	83.3
Chinese *Homo erectus* (Zhoukoudian)	1	76.3
Neanderthal Man *(Homo neanderthalensis)*: Mix sample Regourdou-1 (France) Atapuerca (Spain) Shanindar -1 left (Iraq)	 17 1 4 1	 68.4±9.1 60.0 66.8 63.0
Modern Humans *(am Homo sapiens)*: Coimbra, Portugal Andamanese Modern Indian Pygmy + Modern Indian	 160 16 14 30	 89.0+18.0 80 (91.3) 109.8 80 (94.7) 110.1 80 (92.2) 110.1

Tentative proposed classification for Midshaft Index =
Midshaft Supero-Inferior Diameter x 100 Midshaft Transverse Diameter

Hyperplatycleidic	Platycleidic	Mesocleidic	Eurycleidic	Hypereurycleidic
?-54.9	55-69.9	70-84.9	85-99.9	100-?

a b

Fig. 4.7. a. Excavations conducted by the author and his team at U1 level of Hathnora hominin calvarium site yielded in situ mandible of Elephas namadicus and other mega fauna including over a hundred 'Large Flake Acheulian' handaxes, cleavers and chopping tools for the first time (Sankhyan et al., 2012b); b. Author's revisit to Netankheri hominin locality in October 2012 yielded more Palaeolithic findings recovered in situ from the humerus bed shown in Figure 4B (bottom left)

the most notable feature of the NTK femur is its inconspicuous medial and lateral lips resulting in a sub-circular/ovoid shape of the diaphysis in cross section. It shows robust, rounded "archaic" morphology at the distal/condylar end compared to the medio-laterally expanded/flattened end in *H. sapiens*. In this respect Narmada femur is placed between Java *H. erectus* and *H. sapiens,* though relatively closer to *H. Neanderthalensis* than to *H. erectus* considering its diaphyseal robusticity comparable to the Neanderthals (Trinkaus & Ruff, 1999; Trinkaus, 2007; Churchill, 2007; Weaver, 2009). Though there is paucity of more comparative samples for the fossil hominins, but it interestingly goes well with the robust archaic calvarium, and therefore likely derived from the same hominin population.

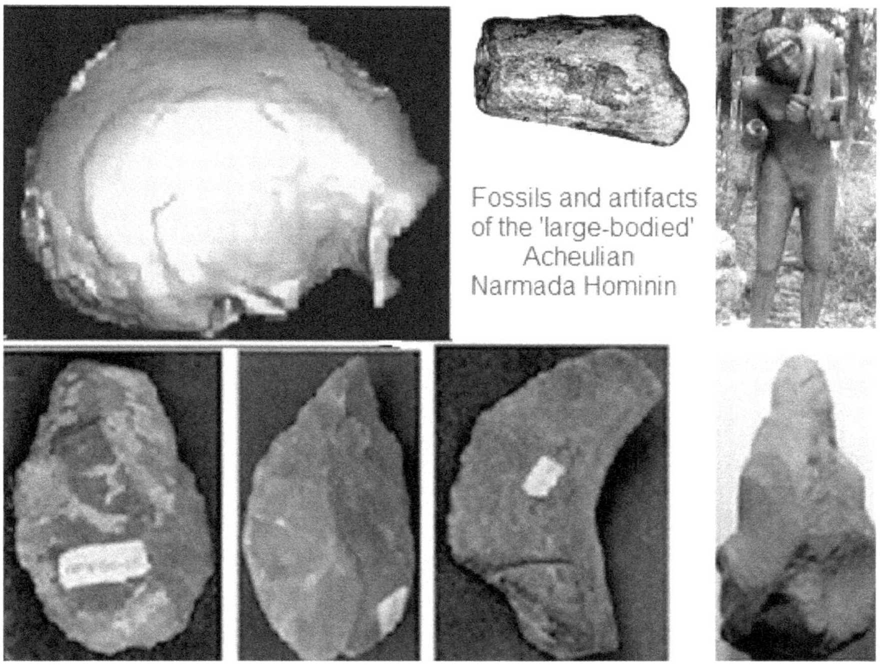

a

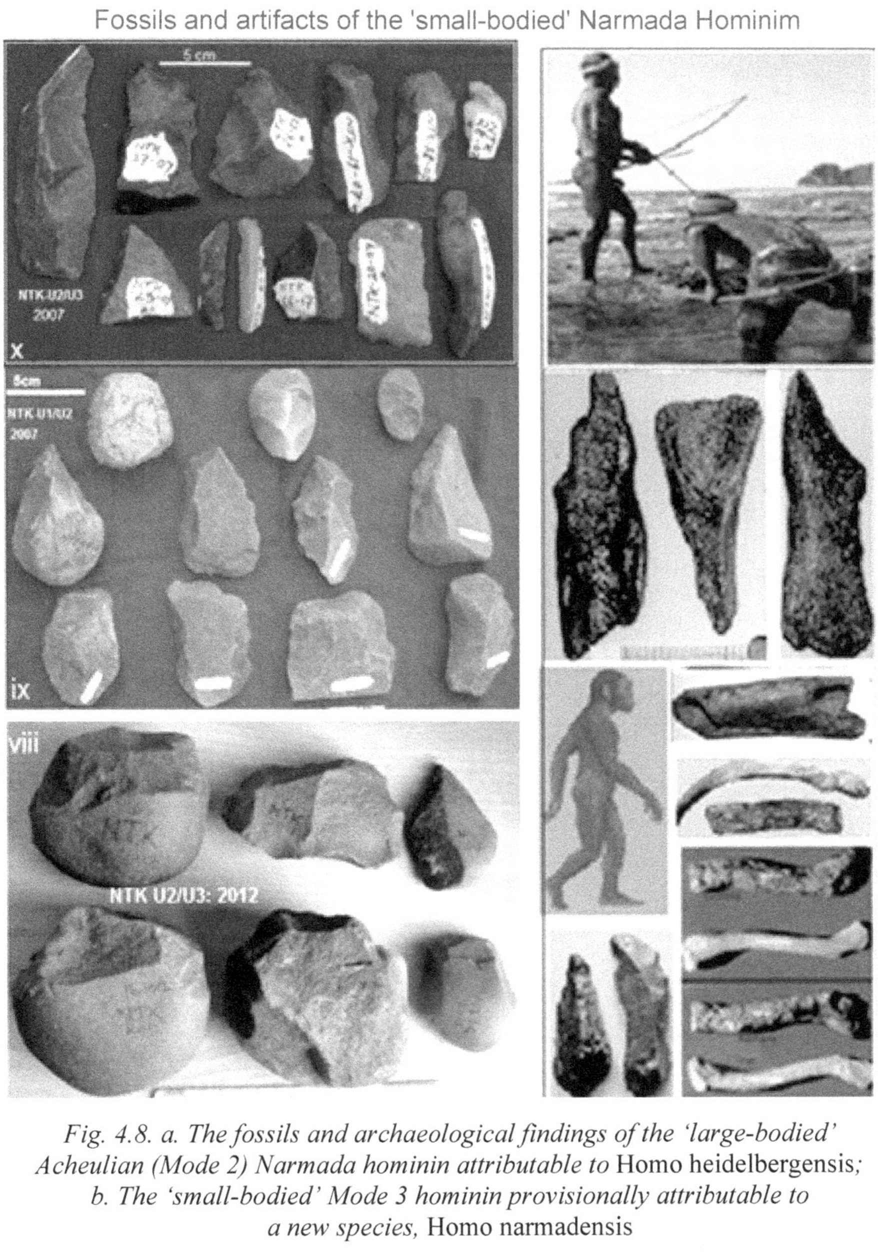

b

Fig. 4.8. a. The fossils and archaeological findings of the 'large-bodied' Acheulian (Mode 2) Narmada hominin attributable to Homo heidelbergensis*; b. The 'small-bodied' Mode 3 hominin provisionally attributable to a new species,* Homo narmadensis

Tab. 4.2. Metric comparison of the Narmada fossilized humerus and femur with modern mainland Indians and Chaurite Nicobari of Andaman-Nicobar Islands, and other Pleistocene hominins. M-L = medio-lateral diameter or lateral pillar thickness; A-P = antero-posterior diameter or middle pillar thickness; all measurements in millimeter

Humerus Mid-Shaft Measurement					
Skeletal Population	**N & Value**	**M-L/LPT**	**A-P/MPT**	**Girth**	**Length**
Netankheri (NTK-F- 02-07)	1	18.3	18.7	67.5	84=240 est
Mainland Indians + Chaurite Islanders	33	16.02	16.56	60.02	284.70
	SD	2.65	2.43	5.82	24.19
	SE	0.46	0.42	1.01	4.21
Chaurite Islanders	5	19.21	19.08	65.95	291.4
	SD	3.15	1.20	4.75	13.43
	SE	1.41	0.54	2.46	6.00
KNM-WT 15000F	1	17.3	11.8		
Gombore IB-7594	1	15.6	11.61		
ATD6-148	1	14.0	7.0		
BOD-VP-1/2b	1	18.0	9.0		
Kabwe	1	18.4	11.9		
Skhul IV	1	18.6	14.2		
Omo Kibish I-r (KHS-1-30)	1	19.9	11.5		
Omo Kibish I-l (KHS-1-31)	1	20.6	12.0		
Cro-Magnon 1	1	20.7	14.0		
Dolní Věstonice	4	16.8±1.7	9.1±2.8		
Sima de los Huesos	6/9	15.7±2.0	8.6±1.3		
Neandertals	21/23	15±2.2	7.7±1.8		
Femur Distal Shaft Measurement					
NTK-F-07-05 (L)	1	44.85	38.05	125.5	
Neanderthal (R) 1	1	47.3	40.7	133.5	
Homo erectus (L) (Java) 1	1	37.65	34.15	114.5	
Homo erectus (Tautavel) 2	1	36.9	32.55	–	
Modern Indians (Homo sapiens) 3	18	34.87	27.68	102.85	
	SD	34.87	28.11	104.65	
	SE	3.643	2.93	10.1	

Stature from NTK humerus: the 84 mm preserved length gives maximum reconstructed length of 240 mm: (HL 24) = 3.08 x 24 + 64.67 = 138.59 (female) & 2.89 x 24 + 78.10 = 147.46 (male). Stature from humerus of 5-Chaurite: (HL 29.14) = 3.08 x 29.1 + 64.67= 154.42 cm (female) & 2.89 x 29.14 + 78.10 = 162.32 (male); Stature from humerus of 33-Indian+5 Chaurite (HL 28.47) = 3.08 x 28.47 + 64.67 = 152.36 cm (female) & 2.89 x 28.47 + 78.10 = 160.38 (male); Stature from Femur (male) = 2.32 x 24 + 65.53; Stature from Femur (female) = 2.47 x + 54.10. est estimated, 1measurements on replicas; 2on original at Tautavel, France taken in July 2009, access courtesy Professor Henry de Lumley; and 3on original in Anthropological Survey of India, Palaeoanthropology Laboratory, Kolkata.

BIO-STRATIGRAPHIC AND ARCHAEOLOGICAL ASSOCIATIONS

The calvarium and the femur though come from two different localities but derived from the same U1 (Khan & Sonakia, 1992) stratigraphic level of the Pebble Conglomerate bed attributed to the Middle Pleistocene Surajkund Formation (Tiwari & Bhai, 1997) and are therefore synchronic. The fauna recovered in situ or excavated contexts include several complete crania and mandibles of the typical Middle Pleistocene mammals, viz., *Elephas namadicus, Equus namadicus, Bos namadicus, Bubalus palaeindicus,* and *Hexaprotodon namadicus.* More importantly, unlike previous findings of small-sized Late Acheulian (Lumley & Sonakia, 1985), our search and excavations at Hathnora partly submerged U1 bottom yielded a large assemblage of massive Acheulian handaxes, cleavers and chopping tools. The fauna and implements are concordant with stratigraphy and the *H. heidelbergensis* status. The two clavicles and the left 9th rib were collected from the from the U1/U2 juncture (Sankhyan, 1997a, b; 2005) which yielded a few light duty refined Late Acheulian, but more Mode 3 Middle Palaeolithic implements (Sankhyan et al., 2009). In other words, we notice Late Acheulian (Lower Palaeolithic) / Middle Palaeolithic and the Middle / Late Pleistocene juncture interface at Hathnora (Sankhyan, 2009b, 2010; Sankhyan & Talwar, 2009).

1. One hand-held large bone dagger with handle-shaped rounded and elongated butt softened by grinding for a very comfortable grip; the two edges are tapering and sharpened by secondary chipping along the converging sides.
2. One awl apparently for piercing and stitching animal hides.
3. One caprid antler, likely used as a borer or drill having tapering round and blunt end.
4. One burin made by step cut-flaking at the elongated working edge with smooth butt.
5. Two spatulas-cum end-scrapers with secondary retouch at the broadened end, and tail-like butt, likely used for scooping bone marrow/ fat/blood of the pray.
6. One backed side-scraper or knife with one edge sharp and serrated by retouch.
7. A bone-split with working edge formed by chipping out a few flakes.
8. One animal rib fragment with both sides tapering and very sharp to function as a blade.
9. A few bones fragments showed evidence of dental pitting and gnawing marks.

Box 4.2. Bone artefacts recovered from the humerus-yielding beds

Bio-stratigraphically and archaeologically the calvarium of the 'large-bodied' *H. heidelbergensis* from the U1 could be datable to ~250 Kya (Patnaik et al., 2009), whereas the clavicles should be younger to that, ~150 Kya? The palaeontological and archaeological associations of Netankheri humerus at U2/U3 juncture are also interesting. They occur within the Surajkund Formation (Tiwari & Bhai, 1997), but below the Baneta Formation stratigraphic boundary. Therefore, the humerus is quite likely older to or around the YTA datum of ~75 Kya (Rose & Chesner 1990; Chesner et al., 1991). It is so if we accept the conclusions of Acharyya & Basu (1993) that the YTA has not fallen during the Surajkund but during the overlying Baneta and Hirdepur sedimentations in Narmada valley.

The humerus was also associated with a few Upper Group fauna, like Axis *axix, Equus asinus, Cervus* sp. and certain species of turtles. In addition, the artifacts found at the U2/U3 interface include blades and flakes made on igneous rocks, but also included hammerstones as well as unique bone tools of characteristics of the late Mode 3 and early Mode 4 typo-technology described in the Box 2. They indicate special adaptations of the evolving 'short-bodied' hominin to the changing sedimentary regime.

DISCUSSION

Overall, the Pleistocene Narmada fossils demonstrate early hominin diversity in the Central Narmada valley with at least two hominins or distinct archaic populations during Middle to later Middle Pleistocene (250 to 150 Kya). The Narmada fauna show ancestral affiliations with the Siwalik, Peninsular Indian, East/SE Asian and African fauna (Badam & Sankhyan, 2009). This is due to a mid-continental location and eco-climatically most ideal region in South Asia with plenty of food resources during a larger part of the Pleistocene. In addition, numerous rock-shelters in the Vindhyan and Satpura mountains provided natural abodes for the safe survival of the prehistoric man and also aided him in the "volcanic winter". This attracted many contemporary archaic hominins populations from all directions and continents. Thus, Central Narmada valley served as a "paradise" for the prehistoric man in South Asia (Sankhyan, 2010).

Thus, Central Narmada valley was an important haven and refuge of the Quaternary hominins and fauna, and a corridor of intra and inter-continental migrations. Possibly therefore, the 'large-bodied' *Homo heidelbergensis* populations received considerable admixture from the evolving *H. erectus, H. neanderthalensis* and other archaic *H. sapiens* populations. This explains its unique mixed/mosaic biological character, which gave a selective advantage enhancing adaptability to colonize diverse eco-zones of South Asia, reflected through variations in Acheulian industries.

The 'short-bodied' hominin with Mode 3 technology was most likely a later new entrant to Narmada Valley at the terminal stage of Acheulean, and the humerus documents its continuity beyond the Toba Eruption ~74 Kya. It probably served as an ancestral stock of small-bodied later Pleistocene descendants among the Andaman-Nicobar pygmies as well as among the Proto-Australoid /Austric Pauri Bhuya/Munda populations. The recent genomic signatures of 60 Kya (Bark et al., 2008; Chandrasekar et al., 2009) indicate *in situ* evolution of the earliest mtDNA haplogroup 'M' in Indian mainland and support the present hypothesis. Further, it is not unlikely that Indian heartland also inhabited more 'small-bodied' later Pleistocene *H. sapiens* population which could be the ancestor of Indonesian micro pygmies, the Hobbits – *Homo floresiensis* (Sankhyan & Rao, 2007). Another corollary is that the Upper Palaeolithic and Microlithic cultures have evolved or arrived in Indian heartland quite early than thought previously. About 30 Kya fossil evidences in Darri-I-Kur in northeastern Afghanistan (Angel, 1972) and in the Fa Hien cave of Sri Lanka adds to this support (Sankhyan, 2010b).

Acknowledgements

The author thanks to the Indian Council of Historical Research, New Delhi for sponsoring his visit to U.K. which enabled him to organise the BH 13 Panel of the 17th IUAES at Manchester and also made the present work possible conducted at the Palaeo Research Society. The fossil and archaeological material was collected under the Narmada Project of the Anthropological Survey of India and lies in its Palaeoanthropology repository. The views expressed are of the author who was the Principal Investigator of the project and his team of researcher and other scholars have been credited and acknowledged in various earlier publications also cited in this work.

References

ACHARYYA, S.K. & BASU, P.K. 1993. Toba Ash on the Indian subcontinent and its implications for thecorrelation of Late Pleistocene Alluvium. *Quat. Research* 40: 10-19.

AMBROSE, S.N. 1998. Late Pleistocene human population bottlenecks, volcanic winter, and differentiation of modern humans. *J. Hum. Evol.* 34: 623-651.

ANGEL, J.L. 1972. A middle Paleolithic temporal bone from Darra-I-kur, Afghanistan. In: L. Dupree (ed), Prehistoric research in Afghanistan (1959-1966), *Trans. Am. Philos. Soc., Special issue* 62: 54-56.

ATHREYA, S. 2007. Was *Homo heidelbergensis* in South Asia? A test using the Narmada fossil from central India, in M.D. Petraglia and B. Allchin (eds) *The evolution and history of human populations in South Asia,* pp. 137-170, Springer Press, New York.

BADAM, G.L. & SANKHYAN, A.R. 2009. Evolutionary Trends in Narmada Fossil Fauna, in A.R. Sankhyan (ed), *Asian Perspectives on Human Evolution*, pp. 92-102. Serials Publications, New Delhi.

BADAM, G.L., GANJOO, R.K., SALAHUDDIN, M., R.K.G. & RAJAGURU, S.N. 1986. Evaluation of fossil hominin – the maker of Late Acheulean tools at Hathnora, Madhya Pradesh, India. *Curr. Sci.* 55: 143-145.

BARIK, S.S., SAHANI, R., PRASAD, B.V.R., ENDICOTT, P. et al. 2008. Detailed mtDNA genotypes permit a reassessment of the settlement and population structure of the Andaman Islands. *Am. J. Phys. Anthrop.* 136: 19-27.

CAMERON, D., PATNAIK, R. & SAHNI, A. 2004. The phylogenetic significance of the Middle Pleistocene Narmada Hominin Cranium from Central India. *Internat. J. Osteoarchaeol.* 14: 419-447.

CHANDRASEKAR, A., KUMAR, S., SREENATH, J. et al. 2009. Updating phylogeny of Mitochondrial DNA Macrohaplogroup M in India: Dispersal of modern human in South Asian Corridor, *PLoS ONE*, e7447 4:1-13. www.plosone.org.

CHATTERJEE, B.K. 1955. A comparative study of the different body proportions of the Onges of Little Andamans. *The Anthropologist 2*(2): 12-21.

CHESNER, C.A., ROSE, W.I., DRAKE, A.D.R. & WESTGATE, J.A. 1991. Eruptive history of earth's largest Quaternary Caldera (Toba, Indonesia) clarified. *Geology* 19: 200-203.

CHURCHILL, S.E. 2007. Endocrine models of skeletal robusticity and the origins of gracility. In *Human Origins, Genome and People of India* (eds. Sankhyan, A.R. and Rao, V.R.), Allied Publishers, New Delhi, pp. 337-368.

DENNELL, R.W. & ROEBROEKS, W. 2005. An Asian perspective on early human dispersal from Africa. *Nature* 438, 1099-1104.

HETEREN, A.H. VAN & SANKHYAN, A.R. 2009. Hobbits and Pygmies: Trends in Evolution. In A.R. Sankhyan (ed), *Asian Perspectives on Human Evolution*, pp. 172-187. Serials Publications, New Delhi.

JAMES, H.V.A. & PETRAGLIA, M.D. 2005. Modern human origins and the evolution of behavior in the later Pleistocene record of South Asia. *Curr. Anthropol.* 46: 3-27.

KHAN, A.A. & SONAKIA, A. 1992. Quaternary deposits of Narmada with special reference to the hominid fossil. *J. Geol. Soc. India*, 39: 147-154.

KENNEDY, K.A.R. 2000. *God-Apes and Fossil men: The Paleoanthropology of South Asia.* The University of Michigan Press, Michigan.

KENNEDY, K.A.R. 2007. The Narmada fossil hominid. In *Human Origins, Genome and People of India.* A.R. Sankhyan and V.R. Rao (Eds), pp. 188-192. New Delhi: Allied Publishers.

KENNEDY, K.A.R. & DERANIYAGALA, S. 1989. Fossil remains of 28,000 year old hominids from Sri Lanka. *Curr. Anthrop.* 30: 394-99.

KENNEDY, K.A.R., SONAKIA, A., CHIMENT, J. & VERMA, K.K. 1991. Is the Narmada hominid an Indian *Homo erectus*? *Am. J. Phys. Anthropol.* 86: 475-496.

LUMLEY, H. DE & SONAKIA, A. 1985. Contexte stratigraphique et Archéologique de L'Homme de le Narmada, Hathnora, Madhya Pradesh, Inde. *L'Anthropologie* 89: 3-12.

LUMLEY, M.-A. DE & SONAKIA, A. 1985. Premiere de coaverte d'un *Homo erectus* sur le continent Indien A. Hathnora, dans la Moyenne Vallee de la Narmada. *L'Anthropologie 89*: 13-61.

MALLASSE, A.D. 2009. Cranial embryogeny and hominin phylogeny. In *Asian Perspectives on Human Evolution* (ed. Sankhyan, Anek R.), Serials Publications, New Delhi, pp. 103-121.

OPPENHEIMER, C. 2002. Limited global change due to largest known Quaternary eruption, Toba ~74 Kyr BP. *Quat. Sci. Rev* 21: 1593-1609.

OPPENHEIMER, C. 2003. Ice core and palaeoclimatic evidence for the great volcanic eruption of 1257. *Internat. J. Climat.* 23: 417-426.

PATNAIK, R., CHAUHAN, P.R, RAO, M.R., BLACKWELL, B.A.B. et al. 2009. New geochronological, palaeoclimatological and Palaeolithic data from the Narmada Valley hominin locality, central India. *J. Hum. Evol.* 56: 114-133.

PETRAGLIA, M.D. & ALSHAREKH, A. 2003. The Middle Paleolithic of Arabia: Implications for modern human origins, behaviour and dispersals. *Research* (Online).

PETRAGLIA, M.D. 2007. Middle Paleolithic assemblages from the Indian Subcontinent before and after the Toba Super-Eruption. *Science* 317: 114-16. www.sciencemag.org.

ROSE, W.I. & CHESNER, C.A. 1990. Worldwide dispersal of ash and gases from earth's largest known eruption: Toba, Sumatra, 75 Ka. *Palaeogeog., Palaeoclimat., Palaeoecol.* 89: 269-275.

SANKHYAN, A.R. 1997a. Fossil clavicle of a middle Pleistocene hominid from the central Narmada Valley, India; *J. Hum. Evol.* 32: 3-16.

SANKHYAN, A.R. 1999. The place of Narmada hominin in the Jigsaw puzzle of human origins. Gondvana Geol. Magz. Spl. Publ. 4: 335-345.

SANKHYAN, A.R. 1997b. A new human fossil from the Central Narmada Basin and its chronology; *Curr. Sci.* 73:1110-1111.

SANKHYAN, A.R. 2005. New fossils of Early Stone Age man from Central Narmada Valley; *Curr. Sci.* 88: 704-707.

SANKHYAN, A.R. 2007. Significance of Human Post-cranial Fossils from Narmada with Remarks on the Skullcap, In A.R. Sankhyan & V.R. Rao (eds) *Human Origins, Genome & People of India: Genomic, Palaeontological & Archaeological Perspectives*, pp. 193-217. Allied Publishers, Pvt. Ltd, New Delhi.

SANKHYAN, A.R. 2009a. On the status of Indian hominoid and hominid fossils, in R. Ray & V. Jayaswal (eds), *Status of Prehistoric Studies in the Twenty-first Century in India*, pp. 13-23. Proceedings of the 15th UISPP Congress, Lisbon, Portugal, 2006. BAR Int. Series 1924, BAR Publishing, England.

SANKHYAN, A.R. 2009b. Human Bio-Cultural Diversity in Prehistoric-to-Protohistoric India, in R. Ray & V. Jayaswal (eds) *Status of Prehistoric Studies in the Twenty-first Century in India*, pp. 67-74. Proceedings of the 15th UISPP Congress, Lisbon, Portugal, 2006. BAR Int. Series 1924, BAR Publishing, England.

SANKHYAN, A.R. 2010. *Pleistocene Hominins & Associated Findings from Central Narmada Valley bearing on the Evolution of Man in South Asia.* Unpublished Ph.D. Dissertation, Panjab University, Chandigarh.

SANKHYAN, A.R. & RAO, V.R. 2007. Did ancestors of the Pygmy of Hobbit ever live in Indian heartland? In Indriati, E. (ed.) *Recent advances on Southeast Asian Paleoanthropology and archaeology,* pp. 76-89. Laboratory of Bioanthropology and Paleoanthropology Faculty of Medicine Gadjah Mada University, Yogyakarta.

SANKHYAN, A.R. & TALWAR, I. 2009. Indian Palaeolithic Transitions, In A.R. Sankhyan (ed), *Asian Perspectives on Human Evolution*, pp. 163-171. Serials Publications, New Delhi.

SANKHYAN, A.R., SHEULI CHAKRABORTY, S., DEWANGAN, L.N. et al. 2009. Cultural, Faunal and Stratigraphic Contexts of Narmada Hominins. In A.R. Sankhyan (ed), *Asian Perspectives on Human Evolution*, pp. 73-91. Serials Publications, New Delhi.

SANKHYAN, A.R., BADAM, G.L., DEWANGAN, L.N., CHAKRABORTY, S. et al. 2012a. New Postcranial Hominin fossils from the Central Narmada Valley, India. *Advances in Anthropology* 2 (3), 125-131, Published Online August 2012 in SciRes DOI:10.4236/aa.2012.23015.

SANKHYAN, A.R., DEWANGAN, L.N., CHAKRABORTY, S. et al. 2012b. New human fossils and associated findings from the Central Narmada. *Current Science* 103 (12): 1461-1469.

SONAKIA, A. 1984. The skull cap of early man and associated mammalian fauna from Narmada Valley alluvium, Hoshangabad area, Madhya Pradesh, India; *Rec. Geol. Surv. India* 113 159-172.

TIWARI, M.P. & BHAI, H.Y. 1997. Quaternary stratigraphy of the Narmada Valley. *Geol. Surv. Ind. Spl. Pulb.* 46: 33-63.

TODD, R.Y. & CHURCHILL, S.E. 2006. Archaic and modern human distal humeral morphology. *J. Human Evol.* 51: 603-616. doi:10.1016/j.jhevol.2006.07.006.

TRINKAUS, E. 2007. Appendicular robusticity and the paleobiology of modern human emergence. *PNAS,* 94: 13367-13373. doi:10.1073/pnas.94.24.13367.

TRINKAUS, E. & RUFF, C.B. 1999. Diaphyseal Cross-sectional Geometry of Near Eastern Middle Palaeolithic Humans: The Femur. *J. Archaeol. Sci.*, 26:409-424. www.idealibrary.com.

WEAVER, T.D. 2009. The meaning of Neandertal skeletal morphology. *PNAS,* 106 (38): 6028–16033. www.pnas.org_cgi_doi_10.1073_pnas.0903864106.

WESTGATE, J.A., SHANE, P.A.R., PEARCE, N.J.C, PERKINS, W.T. et al. 1998. All Toba tephra occurrences across Peninsular India belong to 75000 years B.P. Eruptions. *Quat. Res.* 50: 107-112.

5.
GEOARCHAEOLOGICAL AND ENVIRONMENTAL ASPECTS OF THE CENTRAL NARMADA ALLUVIUM

Satya DEV
Aarhus University, Section for Prehistoric Archaeology, Aarhus University, Denmark
farksd@hum.au.dk

Anek R. SANKHYAN
Anthropological Survey of India, Kolkata, India
arsankhyan@gmail.com

Abstract: *Based on field data the present paper describes archaeological and geo-environmental processes at work in the alluvial deposits of Hoshangabad and Sehore area in a central Narmada valley where we recognized three distinct stratigraphic formations – the Dhansi (Lower Pleistocene), Surajkund (Middle Pleistocene) and Baneta (Upper Pleistocene). These alluvial strata have yielded evidence of vertebrate fossils and Palaeolithic industries with palaeoenvironmental signatures. Our observations indicate that the mottled palaeosols at Dhansi and Ganera represent carbonate precipitations in prolonged standing water conditions and yielded a few in situ flakes. On the other hand, the overlying Surajkund Fm had abundant calcretes which reflect seasonal moisture deficit, warm and humid conditions with seasonal pools. Thesed favoured the Lower Palaeolithic man and enhance his tool-making skills from the 'Middle Acheulean' to the 'Late Acheulean' types, who enjoyed mammalian faunal richness evident from numerous fossils of Equus, Elephas, Bos, Bubalus, Hexaprotodon, Stegodon, Rhinoceros, Cervus, etc., and indicate Middle Pleistocene age. The Dhansi/Surajkund palaeomag contact is correlated with the Brunhes-Matuyama boundary falling in the Lower Pleistocene, but the artefacts of such antiquity are indistinguishable from those of the overlying Surajkund Formation.*

Keywords: *Central Narmada Valley, Late Acheulian, Dhansi formation, Palaeoenvironment*

INTRODUCTION

The central Narmada valley is well studied from palaeontological, archaeological and geomophological points of view (detailed review in Sankhyan, 2010). Most of the earlier works in Narmada valley were confined to the areas in District Narsinghpur (Khatri 1961, 1962, 1963, 1966a, b, 1975; Salahuddin 1987, Supekar 1966, 1968; Misra, 1987, 1989, 1997). But, the greatest attraction of Narmada valley has been the discovery of hominin fossils from Hathnora and its vicinity by Sonakia (1984), Sankhyan (1997a, b, 2005, 2007, 2009a, b, 2010), and Sankhyan et al., 2012a, b). The striking feature is faunal and archaeological association of the hominin remains, perhaps one of the very rare examples which highlights the prehistoric significance of the region. This was the reason to make a brief exploratory trip to the area to make some added observations on lithstratigraphy, and collect some faunal and archaeological material.

The present study area (Fig. 5.1) falls in the district of Hoshangabad and Sehor between the coordinates 23°26'N: 77°24'E: 23°43'N: 77°21'E. The localities included for explorations are: Hathnora, Dhansi, Sardarnagar, Netankheri, Jet, and the Buddhi Narmada Palaeochannel within a distance of about 10 km. This paper is based on a short field work presents some added observations on the lithostratigraphical, archaeological and palaeoenvironmental aspects including an attempt to collect artefacts and faunal remains.

Tiwari & Bhai (1997) divided the sedimentary archive of the Narmada basin into seven formations based on geo-morphostratigraphy, litho-stratigraphy, tephrostratigraphy, erosional unconformities, palaeomagnetic signatures, mineralogy and granulometry. From the oldest to the youngest they are:

i. The Pilikarar formation – considered as the lowest Quaternary sedimentary unit overlying the Late Cretaceous Deccan Traps. Rao et al. (1997) dated it to Early Pleistocene.

ii. The Dhansi formation of dark brown and yellow-orange palaeosols showing significant oxidation dated to the Matuyama Normal Chron at the Brunhes-Matuyama boundary. Palaeomag studies by Agrawal et al. (1988) and Rao et al. (1997) suggest the lowest Narmada deposits fall within Brunhes Chron (<0.73 Mya).

iii. The Surajkund Formation, considered of the age between 128 to 500 Kya and based on magnetostratigraphy (Tiwari & Bhai 1997). It contained extensive records of Middle to Late Pleistocene fauna and Lower Palaeolithic artefacts, hominin fossils at Hathnora, a Struthio camelus eggshell fragment (Badam 2005; Patnaik et al. 2009). The basal part of the Surajkund formation known as Boulder Conglomerate is about 4 m in thickness (Khan & Sonakia 1992) and overlies on Deccan Trap. This and the cemented gravels layers are the main constituents of the Hathnora hominin site having

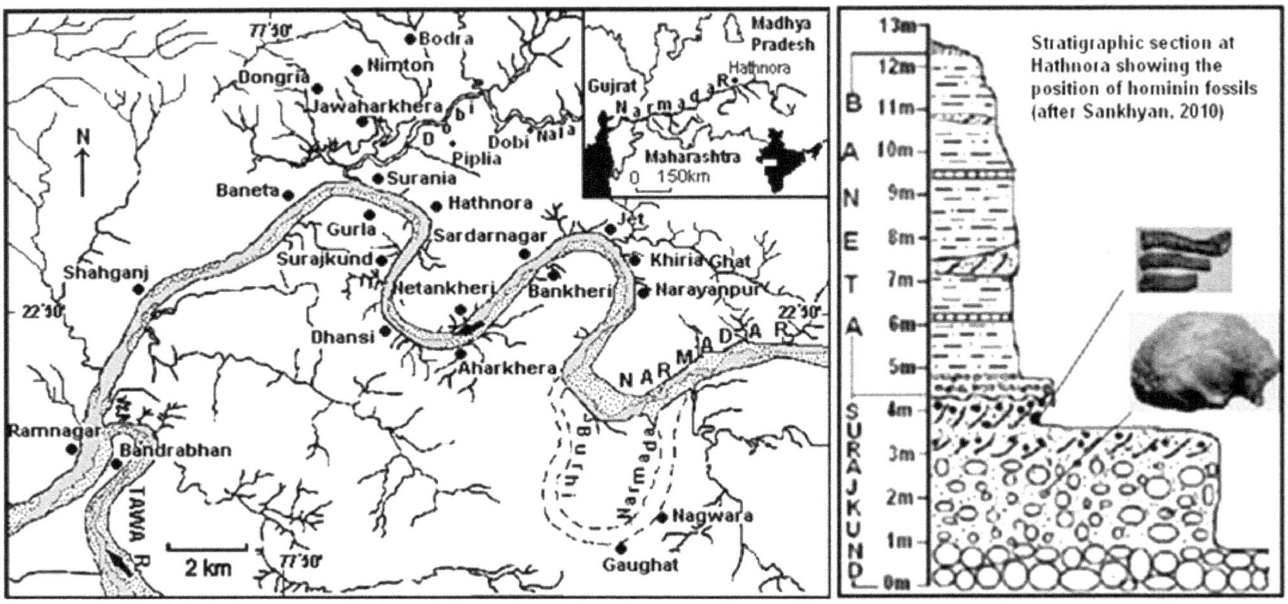

Fig. 5.1. The map of a portion of the study area in central Narmada basin and the stratigraphic section at Hathnora which has yielded the Palaeolithic artefacts, mammalian and hominin fossils (after Sankhyan, 2010)

yielded most of the Lower Palaeolthic and mammalian faunal material of the Middle Pleistocene) (Badam and Salahuddin, 1982; Badam, 1989; Khan & Sonakia, 1992; Biswas, 1997; Rao et al., 1997; Tiwari & Bhai, 1997; Sonakia & Biswas, 1998). Based on ESR dating, Patnaik et al. (2009) dated the Surajkund formation as early upper Pleistocene to late Middle Pleistocene, and the Baneta formation is dated to the Later Upper Pleistocene to Holocene.

iv. The Baneta Formation correlated with Late Pleistocene at ~128-70 Kya (Rao et al., 1997). At Hathnora the Baneta Formation is comprised of 9 m of clay and sand overlain on the Boulder conglomerate. Badam (1979a, b) assigned a time frame of 40 to 150 Kya. Khan and Sonakia (1992) assigned Holocene for the Younger Alluvium of the Upper Group.

v-vii. The Hirdepur, Bauras and Ramnagar Formations of Holocene period (less than 10 Kya). They contain little evidences of the mammalian fossils and archaeological contents.

MATERIAL AND METHOD

In our field survey, we explored the palaeochannel of the Narmada valley, including the villages of Budhni, Bikori, Gaughat, Ganera etc. Potential cliff sections along the banks of the central Narmada Valley (the region between Shahganj and Sardarnagar, including the villages: Sardarnagar, Jet, Netankheri, Dhansi, Surajkund and Hathnora on the bank of the river) were marked by the topographical sheets No. 55F/13 and then visited and studied. The lithostratigraphy of each prospective section on each type site was subject to detailed sketches as well as photographic documentation. The soil samples were also collected along with the artefacts and animal fossils. The point bars especially at the meanders and cliffs at confluences and palaeochannel exposures of older alluvium were explored for the better understanding of the present geomorphologic process. In our field methodology, we have recorded new artefact and fossil occurrences, made stratigraphic measurements.

The lab work comprised of cleaning of stone artefacts and fossils, labelled, and numbered site wise. The metamorphic attributes for stone were developed following Alam (1990), Bordes (1961), Corvinus (1983), Debenath and Dibben, 1991; Pappu (2001), and Abbas (2006). The length, breadth, thickness was measured by using the electronic clipper, whereas the weight was measured by using the electronic weighing machine. The stone artefacts were classified on the basis of artefact number, stratigraphic context, category of the artefact, the class of the artefacts, raw material, preservation (physical condition of artefacts), recent patina damage, cross section, plans form, the primary form of artefact, retouch (type, nature, character and location of retouch), the condition of cortex, type of striking platform and retouch type. The fossils were washed treated with dilute acetic acid, drying, labelling and lastly is preparing a catalogue as a reference point for the fossil assemblage. The discovered fossils were described in the following order: class, order, family, subfamily, genus and Species.

LITHO-STRATIGRAPHY OF THE SITES

Following the litho-stratigaphic sections delineated by previous works (Fig. 1a) provide excellent perspective: The quaternary exposures along the palaeochannel begin appearing from the Bikori-Budhni (22°24'N: 77°40'30"E). The horse-shoe shaped palaeochannel is well delineated by the satellite imagery. Over the period of time, the northern banks of the channel have given rise to expanded sheets of assorted gravel and modern soil that has come under regular agricultural practices. The region also accommodates a couple of villages (viz. Bikori and Tamcharu villages) today. This has been the ancient river bed which is cut off the main channel in the distant past of river history. The palaeochannel is abound by thick exposures of Quaternary alluvium. These exposures run laterally almost up to 12 km. The palaeochannel is commonly addressed as Buddhi Narmada (Old Narmada) by the local population. A brief account is given below for the individual sections sites.

1. Bikori-Budhni (22°24'N: 77°40'30"E)

The reference point of this section, i.e. the village Budhni is located on the right bank of Buddhi Narmada in Babai Tehsil of District Hoshangabad, Madhya Pradesh. The section represents two sedimentary units. The bottom most is represented by the rusty brown to dark red silt which has a general thickness of about 6 to 7 m. It is followed by yellow brown silt and clay with large amount of calcretes. Earlier works of Tiwari & Bhai (1997) classify the lower formation as Dhansi while upper as Suarajkund. The overlying unit (commonly referred to as Surajkund contains high concretion and has a striking resemblance with fossil yielding gravels of Vardha valley & V. Sathe: pers. comm. to Dev). The lower unit (Dhansi Fm. of Tiwari & Bhai, 1997) is a highly oxidized member and comprises of red sandy silt, yellowish brown silt sand, medium to coarse grained yellow sand and cross bedded fine to medium grey sand. The observation of the author about the so called Dhansi formation at the Bikori section is quite different. The author found only the dark red sandy silt in the lower unit of the Bikori section where several 'fresh artefacts' were collected such as varieties of flakes, scrapers, cleavers etc. on quartzite. Keeping in view their concentration on slopes suggest them deriving from the supposedly Surajkund formation. This formation does not represent thickness of more than 5 m. The colour matching was done with the help of Munsell Chart for the soil selection. The bottom unit colour = dark red (Munsell chart: Hue 7.5 R: Dark red, 316 = value 3 and chroma 6).

It is a badlands topography where the continued efforts to check its lateral extension reveal 1.6 km as the approximate lateral extension where the uniformity in the sedimentary constituents is well maintained. Sooner the 'Dhansi elements' disappear and yellowish brown silt and clay takes over as a single unit from the Bikori village to Gaughat. Cankers increase enormously with size ranging from 1 mm to as large as the pebble. The top soil is more blackish with higher representation of clay practically caps this formation. The Gaughat is located at the centre of the palaeochannel.

2. Ganera (22°18'30"N: 77°43'30"E)

The section stands about 7 meters high and for about a couple of hundred meters of the Ganera village on the right bank of the palaeochannel of the Narmada River. The village is also located at the palaeochannel of the Narmada valley, around 15 to 20 km from the Babai in the Hoshangabad district of M.P. The Ganera section is a distinct sedimentary member hitherto known from the rest of the palaeochannel. This is the very interesting section which predominantly lacks the grain size being more than 5 mm. This is divided into three startigraphic units The lower most soils has compact clay with silt that resembles its compaction to mottled clay with the height of 2.5 m. Followed by yellow silt, with the height of 3 m, the homogeneity of yellow silt is strikingly unique to pale channel. A cursory glance reveals it to be would be unimplemtiferus and unfossilferous. This unit is overlain by yellow brownish Kankary silt and clay with the height of 2.5 meters. The lateral extension of this section stretches for about hundred meters. This yellow silt has the similar appearance at Devakachar of Narsinghpur district which has not so far yielded archaeological or fossil material from the similar horizon.

3. Sardarnagar (22°25'N: 77°37'30"E)

The village Sardarnagar (also called Gharjala is located on the left bank of the Narmada, around 25 km from the Shahganj in the Hoshangabad district in M.P. The basal unit is represented by the loose pebbly gravel overlain by cementing sheets of sand that covered almost 5 meters of exposures. The entire cemented sheet underlies thick sand with rhizo concretion and calcareous nodules representing the earlier classification of Surajkund formation. Embedded right on the bank of Narmada in a sandy pebbly gravel, the candidate documented horizontally aligned cleaver on quartzite and isolated lower tooth of a bovid. This horizon is alone member carrying archaeological and fossil material. About hundred meters south of the spot of bovid tooth discovery, a long bone embedded in the sandy matrix was found aligned to about 40° of the channel direction. It is highly covered and hardened by the matrix.

4. Jet (22°52'N: 77°41'30"E)

The location of the village Jet is about 4 km from the Sardarnagar and 21 km from Shahganj in the Hoshangabad district of M.P. Downstream of Jet (next to Sardarnagar) massive quaternary exposures on the left bank can be seen in the lateral extent of about 500 meters. These deposits are extensively exposed and provide an ample data bank for its fluvial history. Tiwari & Bhai (1997) identified two major litho-units namely lithified pebbly sand overlain by yellowish brown silt (Baneta Formation) followed by soil no. six as the top soil. The author aimed to relocate the stratigraphy so as to ascertain the precise litho-stratigraphy with respect to possible occurrences of the lithic material. On the top of conglomerates, right jaw fragment of a large-sized Bos namadicus was discovered. Very close to this find isolated tooth of yet another large bovid has been found. It is very interesting to mention that tip of tusk of an elephant with a core was embedded in the lithified or the conglomerate gravel. This section is divided into three units. The top soil with the height of 30 cm caps this section. Tiwari & Bhai (1997) address it as the Baneta formation. It is succeeded by the second unit comprises of yellow brown silt (cross bedded) and calcareous nodules with the maximum height of 8.19 m. The third unit is made up of the lithified pebbly gravel with sand with 3.20 meter height. The author has collected cleavers and various kinds of flakes and mandibular ramus (ascend) isolated with scapula of bovids from this horizon.

5. Netankheri (22°23'N: 77°35'30"E)

Netankheri is situated on the right bank of Narmada, around 5 km from the village Jet and around 18 km from the Shahganj, in the Sehore district of M.P. Four litho units are observed at Netankheri. The top unit is followed by the 2nd unit which is comprised of rhizo concretion with the approximate height of 9.80 m. It is succeeded by the third unit having the constituents like silt with calcareous nodules. The basal unit, made up of the pebbly sandy gravel with the height of 3 m. A long abraded bone fragment was found in the pebbly gravelly sandy sheets.

6. Dhansi (22°50'N: 77°51'20"E)

Dhansi is located on the left bank of Narmada, around 3 km from Netankheri and around 15 km from the Shahganj in the Sehore district of M.P. Huge sections of the Dhansi formation (supposedly of Lower Pleistocene) characterize the site of Dhansi. The predominant red silt and underlying pebbly bed are devoid of any archaeological and fossil material at this level. The section of Dhansi is also divided into four units (Tiwari & Bhai 1997). The top most units are assigned as brown soil. It is followed by the silt unit that comprises yellow, red, brown silt and sand at the top of the unit and succeeded by the calcareous nodules. The 3rd unit consists of sand and pebbly bed. It is succeeded by the mottled palaeosoil followed by the dark mustered yellowish silt with cankers. It is followed by the pink red (4/6 of Munsell Color Chart) silt dark reddish brown (3/4 of Munsell Color Chart) with the complete absence of cankers. The bottom-most exposed unit is made up of the pebble bed, mottled palaeosol, reddish brown (4/4 of Munsell Color Chart), canker sandy gravel and cemented sandy sheets.

7. Surajkund (22°50'40"N: 77°51'20")

The village of Surajkund is inhabited on the left bank of the Narmada river, around 2 km from the Dhansi village and 13 km from the Shahganj in the Sehore district in M.P. The observations made by this author at Surajkund section do not differ from the previous workers like Tiwari and Bhai (1997). Interestingly, the exposed lithified cemented sheets of sands yielded enormous yet amorphous nodules that often mimic a fossil and even at times the tools! A couple of cleavers and half a dozen flakes and couple of fossils embedded in the cemented sheets of sands were collected from the site. The erosional surface of the left bank shows this very stratum underlain by the silt and sand within calcrete nodules. The former is covered by the yellow silt and fine sand. The general thickness of sediments does not go beyond 10 meters. The color of each unit is systematically identified with Munshell chart confirming the formations like Dhansi, Surajkund, and Baneta. About a hundred meters down streams, near the steps of the temple, at Surajkund, a couple of sections gave a continuous sequence of 14 meters of sediments. The author does not find any bedrock exposures here.

The author has collected three dozen of stone tools, including cleavers, flakes and cores (from both in situ as well as surface) and couples of fossils from the exposed conglomerate sheets. The fossils found so far this site include elephants, Hippos, and large bovids, often with rolled edges and moderate to severe breakage. The tools found here are not as fresh as those found at Sardarnagar. The assemblage of lithic and faunal evidently show reworking and transportation.

THE RECOVERED ARTEFACTS

The present study a modest brief work, however attests the great richness of the Lower Palaeolithic artefacts in central Narmada valley. The tool types recovered include: bifacial cleavers, bifacial chopping tools and unifacial choppers, various types of scrapers, core scrapers and different types of flakes, and a few possible Middle Palaeolithic artefacts in stratified contexts from Sardarnagar (Gharjala), Jet, Surajkund, Hathnora, and Bikori Budhani. However, it is a matter of chance that no true handaxes could be found during the trip, but, some Kombewa flakes could be found. Quartzite is used abundantly as the raw material for the manufacturing the tool for the present assemblage. The sources of the raw material are around eight km from the study area. Most of the tools have been recorded from the basal lithified pebbly gravely horizon (basal conglomerate) of the Surajkund formation. The tool kit presents typical characteristics of the Acheulian culture of peninsular India. The present cleaver assemblage comprises 14 specimens recovered from Surajkund, Jet, Hathnora, Bikori, and Sardarnagar. Most of the cleavers have been recorded from the basal lithified pebbly conglomerate sections except the two from Hathnora above the pebbly gravel horizon. The raw material of these cleavers is quartzite. The average weight, maximum length, width and thickness of cleavers are: 300 Gm-1.75 kg, 116-191 mm, 81.10-148 mm, 32-79.81 mm, respectively.

THE RECOVERED FAUNA

The collection comprises a total of six NISP which are in different state of preservation. Undoubtedly the assemblage is too small but the region being the palaeontologically promising makes this study worthwhile.

As regards the preservation, it is obvious that it has been exposed to fluvial transportation and reworking. The bones are highly fossilized and there seems to no new addition of any particular species in addition to the known assemblage of the Narmada Valley. Except one

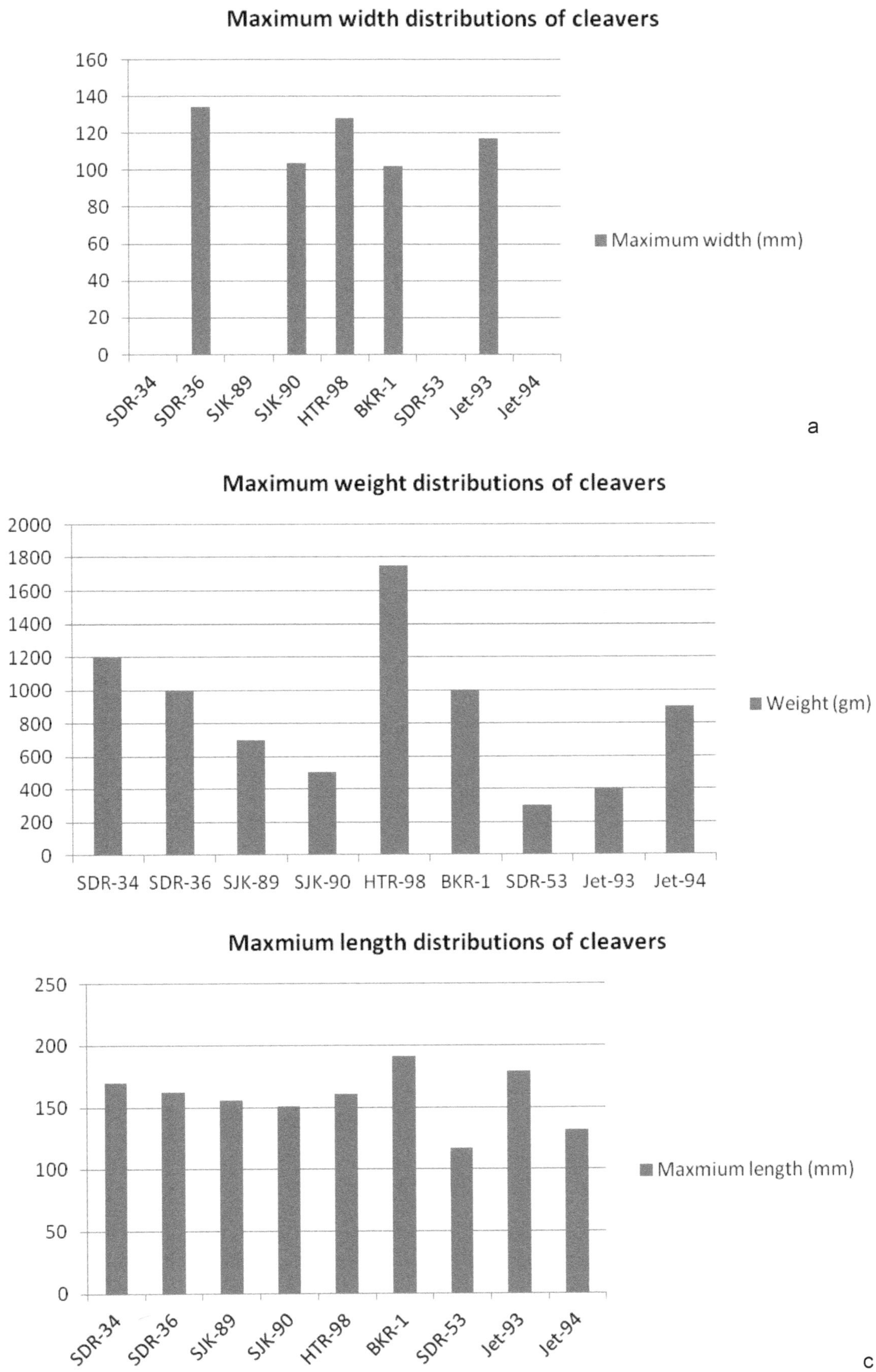

Fig. 5.2. Maximum weight, length and width of the cleavers in the sample

specimen rest all have no archaeological association. A tusk fragment of an elephant draws our attention to its association with a core on quartzite which is lithified with the matrix that embeds the specimen.

DISCUSSION

The present study though yielded little faunal material, the area is already well known for rich faunal contents. The riddles of palaeoecological reconstruction are severely marred by the taphonomic bias. However, researches in Narmada (Badam et al., 1986) help us to identify taphonomic co-ordinates with great precision. Eventually the picture that emerges about Palaeoecology can be employed on the larger scale with regard to microvertebrates and mega herbivores that lived and roamed in the vicinity of the Central Narmada valley. The review of faunal abundance in the area (Badam & Sankhyan) with respects to its interdisciplinary analysis from the Pleistocene of the Narmada reveals a diverse faunal spectrum that stands unique in the context of Indian prehistory. In spite of inherent flaws in assessing the time averaging and locating precise homes of animals that lived and perished in this area, recent discoveries have enabled reconstruction of the palaeoecocommunity structure in certain pocket of the Narmada valley with far greater precision (Patnaik et al., 1995). The situation is often met with when there are not microvertebrates (Patnaik et al., 1995), yet the reconstruction of faunal system may be complimented by very habitat specific mega herbivores like hippo, which is a potential climate indicator (Jablonski 2004, Sathe 2005a, b). The herbivores were perhaps were able to inhabit areas with diverse vegetation The vast distribution of these animals throughout the study area implies vast expanses of pools supporting perennial grass cover that defines the basic demands for survival for the amphibious but large herbivores. Geological observations often support the existence of stagnant water bodies.

The pollen evidence though confined to the level of Upper Pleistocene carbonaceous clay from the basal layer, compliment the faunal evidence. "That the ground vegetation was in proximity to the deposition site was dominated by grass and Cheno-Amarnpha type of plants. The abundance of fern spores along with fungal spores would suggest humid climate and existence of moist niches in the habitats of the habitats of the pteridophytic plants in depositional site was dominated by grass and Cheno-Amarnpha type of plants. The abundance of fern spores along with fungal spores would suggest humid climate and existence of moist niches in the habitats of the habitats of the pteridophytic plants in the source area" (Nandi 1997, p. 90).

There are mottled palaeosols underlying the Dhansi or even at Ganera represented by carbonate precipitation and prolonged standing water conditions. the calcretes are the most common occurrences at Surajkund formed from the cementation of calcium carbonate and are a prominent feature in climatic zones where a seasonal moisture deficit occurs allowing the $CaCO_3$ to accumulate. Palaeo-calcretes found here are in a distinctive alluvial setting with shallow water carbonate systems. Evapotranspiration is a major process in semi-arid calcretes and a major cause of rhizocretion formation. The rhizocretions are accumulations of mineral matter around the roots of the plants; these are accompanied by cementation, which may occur during the life or after the death of the plant roots. Aerial roots or former vegetation may have decomposed or been eroded, whereas the subterranean root system remains (Wright & Tucker 1991). These rhizocretions and sheet like structures formed below the surface therefore increasing their preservation potential. The rhizocretions within the Surajkund Formation are up to 1m in length and up to 0.15 m in diameter. These areas are interpreted as being episodes of vegetation cover in and around the close proximity to the standing aquatic sources. The depth of the roots is indicative of taproots, which penetrate deep into the sediment to reach the main water table; these plants are typical of semi-arid to arid channel-bar or floodplain deposits. The types of plant responsible were probably shrubs and possibly even trees. Groundwater cementation is common in arid and semi-arid alluvial settings. The sheet like structures is thought to have originated at or just above the water table in the capillary fringe zone. The occurrence of numerous sheets shows the episodic rises in the water table as the sediment surface was aggraded.

All stone specimens were mostly associated with fine-grained sediments indicating the use and discard of artefacts in an ancient floodplain environment. The majority of the collected material seems to belong to the late Acheulian. Most of the cleavers belong to the Late Acheulian, indicated by their overall level of trimming, retouch, number of flake scars, and related attributes. Some of the Kombewa-based specimens may be Early Acheulian but this is not definite. The choppers appear to be universal in all Palaeolithic assemblages and are made on the river-worn cobbles. In addition to cleavers, the other common type is flakes and flakebased scrapers. The virtual absence of handaxes in the study area denotes a probable techno-functional factor and probably a sampling artifact. If the absence of handaxes is confirmed, it may reflect specific types of activities which were being carried out here that did not require handaxes, discoids, points, picks, and other absent artifacts. Only meticulous excavations can reveal the past functions of the hominins in this part of the valley during this time frame. The diversity of the collected assemblage further signifies the variety of manufacturing techniques and desire for diverse morphological forms. Surajkund yielded the least number of specimens while the other four localities yielded higher numbers, Sardarnagar being the highest. The artefacts are very similar to other Acheulian and Middle Palaeolithic assemblages in other parts of the Old World. The majority of them has been initiated as a flake blank off a cobble or boulder via hard hammer percussion and then possibly reduced through both hard and soft hammer percussion. The cores are small as they originate from the river-worn cobbles and pebbles. Boulders appear to be absent in the vicinity of the study area, and such large clasts are generally found away from the river near the base of the Vindhyan hills. The occurrence of several typical Acheulian flakes in supposedly Dhansi formation (which is found overlain by the Surajkund formation at Bikhor Budhni) if confirmed, may push the antiquity of human activity down to the Lower Pleistocene period. However, the possibility of their being rolled down from the overlying Surajkund context may not be ruled out at this point of our observations. It may draw the palaeochannel into much significant position than that of Hathnora where eventually most of the antiquities appear allochthonous in nature.

The artefacts which have been recovered in situ from pebbly sandy gravel underlying the lithified sheets of sand are so fresh and in a 'mint condition' that their preservation defies their occurrence in such a reworked horizon. A systematic study of tool taphonomy in several geomorphic settings is indeed warranted at most of these tool bearing sites to understand their response to fluvial behaviour, especially after they have been discarded after utilization. However, none of the fossils show any convincing association with the parent litho-unit. Save in a case of an elephant's tusk fragment which has a core attached to it and yet the broken specimen entails an interesting history of its deposition that needs to be ascertained in the light of recent advances in fluvial taphonomy. There are a few interesting observations. Faunally, animals like *Equus*, *Stegodon*, *Bos*, *Hexapotodon*, have been some the most common members of Pleistocene in the Narmada point to warm and humid conditions with seasonal pools away from the main channel. Recent studies (Sathe 2005b; Jablonski 2004) have shown that even the large herbivore like hippopotamus can yield important information with regard to the palaeoenvironment if the taphonomic assessment of their discovery is met with better resolution. Discovery of microvertibrates from the Late Pleistocene deposits at Devakachar (Distt. Narsingpur) (Patnaik 1995; Patnaik et al., 1995) have enabled a high enabled the reconstruction of palaeocommuntiy structure. Its supports the presence of seasonal savannah types of grass land with subsoil moisture content, wooded grassland with patches of sandy plains. The floral diversity suggested for carbonaceous clays at the basal parts of the Upper Pleistocene deposits supports faunal interoperations (Nandi 1997).

INFERENCES

Some of the interesting observations of the present study are:

i. The mottled palaeosols found in the Dhansi or even at Ganera represented by carbonate precipitation and prolonged standing water conditions.

ii. The calcretes in greater frequency in the Surajkund are important climate indicators of a seasonal moisture deficit, distinctive alluvial settings of shallow water carbonate systems.

iii. The rhizocretions-mineral accumulations around the roots of the plants due to evapo-transpiration indicate semi-arid conditions, accompanied by cementation favoured bone preservation in Surajkund. The rhizocretions within the Surajkund Formation are up to 1 m in length and up to 0.15 m in diameter.

iv. The sheet structures formed due to groundwater cementation in arid and semi-arid alluvial settings just above the water table in the capillary fringe zone. The occurrence of numerous sheets shows the episodic rises in the water table due to aggradations of the sediment.

v. The collected artefacts (e.g. cleavers represented in the sample) are mostly of the late Acheulean typology indicated from their overall level of trimming, retouch, number of flake scars, and related attributes. Some Kombewa-based flakes thought of Early Acheulian do continue. The choppers are made on the river-worn cobbles. The scrapers are generally flakebased. The diversity of the collection signifies a variety of manufacturing techniques and desire for diverse morphological forms.

vi. The presence of typical Acheulian flakes at Bikhor Budhni, supposedly Dhansi formation, may push the antiquity of human activity down to the Lower Pleistocene, if confirmed; but they appear likely re-deposited rolled down material from the overlying Surajkund.

vii. Several artefacts have been recovered in situ from the pebbly sandy gravel underlying the lithified sheets of sand are so fresh and in a 'mint condition' that their preservation defies their occurrence in such a reworked horizon.

viii. A systematic study of tool taphonomy is indeed warranted to understand their response to fluvial behaviour, discarding after utilization.

References

ABBAS, R. 2006. Prehistoric Investigations in the Parvati Basin. Unpublished Ph.D. Dissertation. Pune: Deccan College Post-graduate and Research Institute.

AGRAWAL, D.P., KOTLIA, B.S., KUSUMGAR, S. 1998. Chronology and significance of Narmada Valley formations. Proceedings of the Indian National Science Academy 54, 418-424.

ALAM, M.S. 1990. A Morphometric Study of the Paleolithic Industries of Bhimbetka, Central India. Unpublished Ph.D. Dissertation, Poona: University of Pune.

BADAM, G.L. 1979a. Pleistocene Fauna of India. Deccan College Postgraduate and Research Institute, Pune.

BADAM, G.L. 1979b. Quaternary Paleontology of Central Narmada valley and its Implications in the Prehistoric Studies, Geological Survey of India Miscellaneous Publication No. 45, 311-320.

BADAM, G.L. 1989. Observations on the fossil hominid site at Hathnora, Madhya Pradesh, India. In: Sahni, A., Gaur, R. (Eds.), Perspectives in human evolution. Renaissance Publishing House, Delhi, 153-172.

BADAM, G.L. 2005. A note on the ostrich in India since the Miocene. Man and Environment 30 (2), 97-104.

BADAM, G.L. & SANKHYAN, A.R. 2009. Evolutionary Trends in Narmada Fossil Fauna, in A.R. Sankhyan (ed), Asian Perspectives on Human Evolution, pp. 92-102. Serials Publications, New Delhi.

BADAM, G.L. & SALAHUDDIN, R.K.G. 1982. Some New Fossils Sites on the Central Narmada Valley, Madhya Pradesh, India. Current Science 51 (18), 898-899.

BADAM, G.L., GANJOO, R.K., SALAHUDDIN, R.K.G. & RAJAGURU, S.N. 1986. Evolution of fossil hominid—the makerof Late Acheulian tools at Hathnora, Madhya Pradesh, India. Current Science. 55 (3), 143-145.

BISWAS, S. 1997. Fossil mammalia of the Quaternary sequence of Narmada Valley: their affinity, age and ecology. G.S.I. Special Publication No. 46, 91-104.

CORVINUS, G. 1983. A Survey of Parvara River System in Westeran Maharastra, India, Vol. II: The Excavation of the Acheulian Site of the Chirki-on-Parvara, India. Tubingen: Institute for Urgeschichte.

DEBENATH, A., & DIBBLE, H.L. 1991. Handbook of Paleolithic Typology, Vol. I, Lower and Middle Paleolithic of Europe.Cypher Press.

JABLONSKI, N.G. 2004. The hippos' tail: how the anatomy and physiology of Late Neogene Hexaprotodon shed light on Late Neogene environmental change. Jacobson, J. 1975. Early Stone Age Habitation Sites in Eastern Malwa. Procceding of the American Philosphical Society. 119.

KHAN, A.A. & SONAKIA, A. 1992. Quaternary deposits of Narmada with special reference to the hominid fossils. Journal of Geological Society India 39, 147-154.

KHATRI, A.P. 1961. Stone Age and Pleistocene chronology of the Narmada Valley (Central India). Anthropos 56, 519-530.

KHATRI, A.P. 1962. Mahadevian: an Oldowan pebble culture of India. Asian Perspectives 6(1), 186-197.

KHATRI, A.P. 1963. "Mahadevian: An Oldowan Pebble Culture of India." Asian Perspectives VI, 186-197.

KHATRI, A.P. 1966a. Origin and evolution of hand-axe culture in the Narmada Valley (Central India). In: Sen, D., Ghosh, A.K. (Eds.), Studies in Prehistory: Robert Bruce Foote Memorial Volume. Firma K.L. Mukhopadhayay, Calcutta.

KHATRI, A.P. 1966b. The Pleistocene mammalian fossils of the Narmada River Valley and their horizons. Asian Perspectives 9, 113-133.

KHATRI, A.P. 1975. "Handaxe and chopper-chopping tool radiation: the ideas of two culture theory of Palaeolithic in Asia." Presented in the Pre-Congress symposium on Early Palaeolithic in South and East Asia. August 28 – Sept. 8, 1973, 3-14.

MISRA, V.N. 1987. "Middle Pleistocene adaptations in India." In (O. Soffer, Ed.) The Pleistocene Old World – Regional Perspectives. New York: Plenum Press, 99-119.

MISRA, V.N. 1989. Stone Age India: An Ecological Perspective, Man and Environment 14, 1764.

MISRA, V.N. 1997. Early Man and his environment in Central India. Journal of Paleontological Society of India 42, 1-18.

NANDI, A. 1997. Palynological Studies on Quaternary Sediments of Central Narmada Valley, Madhya Pradesh. Quaternary Geology of the Narmada Valley GSI Special Publication No. 46, 87-90.

PAPPU, S. 2001. A Re-Examination of the Paleolithic Archeological Record of Northern Tamil Nadu, South India, Bar International Series 1003.

PATNAIK, R. 1995. Narmada Valley Microvertebrates: Systematic, Typhonomy, and Palaeoecology. Man and Environment 20 (2), 75-90.

PATNAIK, R., BADAM G.L. & SATHE, V. 1995. Discovery of Microvertebrates from the Pleistocene Deposites of the Central Narmda Valley, India. Current Science 68 (8), 828-830.

PATANAIK, R., CHAUHAN, P.R., RAO, M.R., BLACKWELL, B.A.B. et al. 2009. New geochronological, paleoclimatological, and archaeological data from the Narmada Valley hominin locality, central India. Journal of Human Evolution 59, 114-133.

RAO, V. K., CHAKAROBORTI, S., RAO, K.J., RAMANI, M.S.V. et al. 1997. Magnetostratigraphy of the Quaternary fluvial sediments and tephra of Narmada Valley, Central India. Special Publication of Geological Survey of India 46, 65-78.

SALAHUDDIN, R.K.G. 1986-87. On the archaeological association of the fossil hominid from Hathnora, Madhya Pradesh, India. Asian Perspectives 27(2), 193-203.

SALAHUDDIN, R.K.G. 1987. Late Quaternary Ecology, Fauna and Human Cultures of the central Narmada Valley, Madhya Pradesh. Ph. D. Dissertation. Puna: University of Puna.

SANKHYAN, A.R. 1997. A new human fossil from the central Narmada basin and its chronology. Current Science 73, 1110-1111.

SANKHYAN, A.R. 2005. New fossils of early stone age man from central Narmada valley. Current Science 88(5), 704-707.

SANKHYAN, A.R. 2007. Significance of human post-cranial Fossils from Narmada with remarks on the Skullcap, In A.R. Sankhyan & V.R. Rao (eds) Human Origins, Genome & People of India: Genomic, Palaeontological & Archaeological Perspectives, pp. 193-217. Allied Publishers, Pvt. Ltd, New Delhi.

SANKHYAN, A.R. 2009a. On the status of Indian hominoid and hominid fossils, in R. Ray & V. Jayaswal (eds), Status of Prehistoric Studies in the Twenty-first Century in India, pp. 13-23. Proc. 15th UISPP Cong. Lisbon, Portugal, 2006. BAR International Series 1924, BAR Publishing, England.

SANKHYAN, A.R. 2009b. Human Bio-Cultural Diversity in Prehistoric-to-Protohistoric India, in R. Ray & V. Jayaswal (eds) Status of Prehistoric Studies in the Twenty-first Century in India, pp. 67-74. Proc. 15th UISPP Cong. Lisbon, Portugal, 2006. BAR International Series 1924, BAR Publishing, England.

SANKHYAN, A.R. 2010. Pleistocene Hominins & Associated Findings from Central Narmada Valley bearing on the Evolution of Man in South Asia. Unpublished Ph.D. Dissertation, Panjab University, Chandigarh.

SANKHYAN, A.R., BADAM, G.L., DEWANGAN, L.N., CHAKRABORTY, S. et al. 2012a. New Postcranial Hominin fossils from the Central Narmada Valley, India. Advances in Anthropology 2 (3): 125-131, Published Online August 2012 in SciRes DOI:10.4236/aa.2012.23015.

SANKHYAN, A.R., DEWANGAN, L.N., CHAKRABORTY, S., PRABHA, S. et al. 2012b. New human fossils and associated findings from the Central Narmada. Current Science 103 (12): 1461-1469.

SATHE, V. 2005a. Neogene – Quarternary Fossil Vertebrates in the Indraprastha Museum of the Indian Archaeological Society, New Delhi: A preliminary Report. In Mukhanob, G.S. and Nazipoba (Eds.). National Museum Republic of Tatarstan, Kazan (Russia).

SATHE, V. 2005b. Contexts of faunal assemblages and changed perspectives. Paper presented in British Museum and Institute of Archaeology, University College London UK from 4-8 July 2005.

SONAKIA, A. 1984. The skull cap of early man and associated mammalian fauna from Narmada Valley alluvium, Hoshangabad area, Madhya Pradesh, India. Recordings of Geological Survey of India 113, 159-172.

SONAKIA, A. & BISWAS, S. 1998. Antiquity of the Narmada Homo erectus, the early man of India. Current Science 77(4), 391-393.

SUPEKAR, S.G. 1966. Pleistocene Stratigraphy and Prehistoric Archeology the Central Narmada Basin. Ph.D. Dissertation. Poona: University of Puna.

SUPEKAR, S.G. 1968. Pleistocene stratigraphy and prehistoric archaeology of the Central Narmada Basin. Ph.D. Dissertation. Deccan College Postgraduate and Research Institute, Pune.

TIWARI, M.P. & BHAI, H.Y. 1997. Quaternary stratigraphy of the Narmada Valley. Geological Survey of India Calcutta Special Publication No 46: 33-63.

WRIGHT, V.P. & TUCKER, M. 1991. Calcretes, Blackwell Scientific Publications, 352.

6.
THE ROLE OF BALKANS IN PEOPLING OF EUROPE: NEW EVIDENCE FROM SERBIA

Mirjana ROKSANDIC
University of Winnipeg, 515 Portage Avenue, Winnipeg. Manitoba, Canada R3B 2E9
m.roksandic@uwinnipeg.ca

Abstract: *Systematic excavations of the Mala Balanica cave, Sicevo, (Serbia) yielded a left semi mandible: BH-1, the only specimen from the Balkan Peninsula securely dated to the Middle Pleistocene. The primitive morphology of the mandibular body and the lack of derived Neanderthal traits place this specimen outside the variation of Middle Pleistocene European hominins. The specimen's primitive morphology is more consistent with the new radiometric age estimate that places it into the earlier part of the Middle Pleistocene. Here I examine the significance of this specimen for the role that the Balkan Peninsula – the only refugium that never experienced isolation – could have played in maintaining gene flow and allowing primitive traits to remain present in the population for a longer period of time.*

Keywords: *human evolution, mandible, morphology, Middle Pleistocene, Balkan, palaeoanthropology*

INTRODUCTION

The Middle Pleistocene fossil record plays a crucial role in later human evolution as it documents greater encephalization and the emergence of modern morphology and behavioural repertoire (Marean et al. 2007; Roebroeks 2001). In Europe, the Middle Pleistocene is generally associated with *Homo heidelbergensis* (Schoetensack 1908), commonly accepted as ancestral to Neanderthals, since all of the European specimens included in the *H. heidelbergensis* hypodigm present some Neanderthal traits. Given the ambiguity associated with the species definition for African, Asian and European Middle Pleistocene specimens, Cartmill and Smith (2009) opt for the generic term "Heidelbergs" when discussing all Middle Pleistocene archaic humans; here I use the term to indicate only European "Heidelbergs" because of their essential connection with Neanderthals. In order to acknowledge the extent of variation in the Middle Pleistocene, and compare evolutionary trajectories across the Old World, paleoanthropologists are increasingly turning towards the concept of "paleo-deme" or "p-deme" (Howell 1999) which allows us to distinguish between local populations and discuss their possible phyletic relationships without implying (or rejecting) speciation events. Against this background, any Pleistocene hominin fossils from the Balkans could represent important contributions to our understanding of hominin evolution in Europe.

WHAT IS SO SPECIAL ABOUT THE BALKANS?

At the gates to the continent, the Balkan Peninsula, already identified as the confirmed route of animal migrations in the Early Pleistocene (O'Regan et al. 2011) represents the most logical route of migration into Europe. This is supported by the 1.4 Mya old Kozarnika cavein Bulgaria (Sirakov et al. 2010), which is contemporaneous with Ubeidiya in Israel (Belmaker et al. 2002). Furthermore, paleontological evidence supports successive movement of animals from Africa / Southwest Asia (SWA) into Europe in the Early and Middle Pleistocene (see issue 295 (5) of Qi (2013) devoted to the question). This notion of successive movement has recently been explored by Dennell et al. (2011) and Bermudez del Castro and Martinon-Torres (2013) using a demographic "sinks and sources" model. The authors postulate a demographic source population in SWA, which would have repopulated Europe in successive migrations, intermixing to an extent with humans that were present in southwestern European refugia at the time.

For all their importance in allowing the movement of animals and people into Europe, the Balkans should not be conceptualized exclusively as a transit zone. More than just a migratory route, together with the Iberian and Apennine Peninsulas (Carrión et al. 2011) the Balkans played the role of a refugium (Hewitt 2011; Griffiths et al. 2004) for temperate deciduous forests and associated biota (Eastwood 2004; Tzedakis 2004). The Balkans acted as a southern refugium for plant and animal species during Pleistocene glaciations when ice cover made most of Northern Europe inhospitable (Ehlers and Gibbard eds. 2004). Unlike the more western peninsulas, this region maintained open contact with the rest of the inhabited world. The lack of geographic isolation of the Balkans and their position between the continents links this area to SWA to form what I would like to refer to as the "eastern Mediterranean" geographic entity, a region at the crossroads of the continents that should be conceptualized as the fertilization zone between different populations and their technological traditions.

We still do not have answers to some of the key questions about human evolution in Europe: who the first

Fig. 6.1. a. The mandible from the site of Mala Balanica in Serbia (BH-1); b. the excavations at the cave of Mala Balanica in 2013

inhabitants were where they came from, how they relate to earlier African/Asian and later European populations. We are only starting to recognize the possibility of more than one migration and more than one direction for it. Many of these questions will remain unanswered for as long as the Balkans represent this *terra incognita* of Pleistocene research. Fortunately there is new research coming out of Balkan countries that promises to address these questions (Havarti and Tourlakis, 2013).

FOSSIL EVIDENCE FROM THE BALKANS

Despite the area's incontestable importance and the strong tradition of archaeological research in the region, the Balkan Palaeolithic record is sparse. In addition to the securely dated BH-1 mandible from Balanica, Serbia, the only other Middle Pleistocene hominins come from Greece (Fig. 6.1 & 6.2). Both Apidima and Petralona demonstrate Neanderthal traits in varying combinations with archaic and more modern traits, in contrast to the Balanica mandible which does not exhibit any Neanderthal traits. The mandible was excavated at Mala Balanica cave, which, together with Velika Balanica, forms the Balanica cave complex in the Sićevo Gorge (43°20.211'N: 22°05.115'E). This complex has been the focus of systematic archaeological excavations since 2004 (Mihailovic 2009); the excavations are ongoing and the bedrock has not been reached in either of the caves. Middle Palaeolithic artefacts have been uncovered in the upper levels of both caves, while the mandible was recovered from the lower stratigraphic levels of Mala Balanica, 1.5 m below the artefact bearing levels. Inferences drawn from the morphology of the BH-1 mandible (Roksandic et al. 2011) places it outside currently observed variation of European *Homo heidelbergensis*.

Radiometric ages based on electron spin resonance (ESR)/uranium series dating of associated faunal dentition, and infrared stimulated luminescence (IRSL) dates obtained on superimposed sediments, date the BH-1 mandible to older than 395 to 525 Kya (Rink et al. 2013). This age makes BH-1 the earliest dated hominin specimen in this part of Europe and among the oldest dated individuals in the Eastern Mediterranean, close in age to Quessem cave (Barkai et al. 2003). In comparison with specimens from Western Europe, the Balanica hominin is somewhat younger than Sima de los Huesos (600 ± 60 Kya) and Mauer (609 ± 40 Kya), similar in age to Arago (435 ± 85 Kya), Sima de los Huesos (450) and Visogliano (350-500 Kya), and somewhat older than Ceprano (353 ± 4 Kya) (Bischoff et al. 2007; Wagner et al. 2010; Falguères et al. 2004; Falguères et al. 1999; Falguères et al. 2008). With the detailed publication of the Sima de los Huesos material (Pablos et al. 2012; Gomez-Robles et al. 2012; Martinon-Torres et al. 2012; Bonmati et al. 2010) we see a model of population interaction in Europe different from the inherited wisdom of a gradual and mosaic evolution of acquired Neanderthal traits (Dean et al. 1998).

The presence of Neanderthals in Western Europe has substantially biased our views on human evolution in Europe. With their recognizable derived morphology, long duration in the continent and remarkable fossil record, Neanderthals dominate both popular imagination and scientific research, and our quest to understand these remarkable fossil humans has produced innumerable treatises, not all of which are to the credit of the science we practice. The large number of Neanderthals, in comparison to other fossil humans, has resulted in two uncompromising approaches: perceiving the difference between us (modern humans) and those (Neanderthals) as either non-existent or essentialised by those who support either their inclusion or exclusion (respectively) into our common ancestry.

Recent advances in aDNA studies and the sequencing of the whole Neanderthal and Denisovan genomes (Green et al. 2010; Meyer et al. 2012) have demonstrated interbreeding between different geographic populations, further emphasizing the need to leave speciation aside when examining the last two million years of human evolution (Holliday 2006). Neanderthals, for all the fascination that they hold, should be regarded as a European group that developed n glacial isolation with some intermixing with more eastern populations during

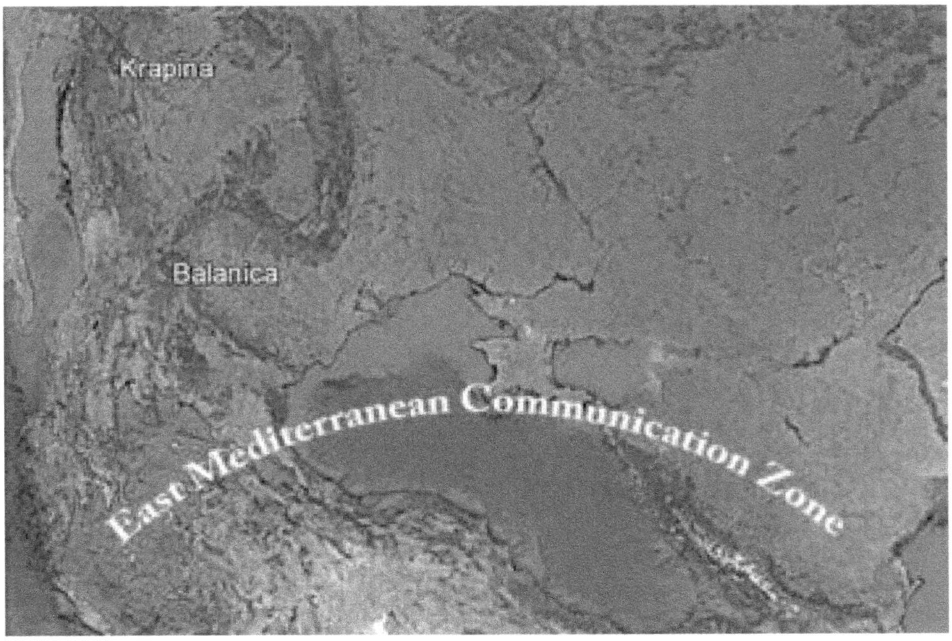

Fig. 6.2. A new look at an old map: a changed perspective on the Eastern Mediterranean zone connecting the Balkan Peninsula and the Southwest Asia

interglacial times, when migrations, contact and genetic exchange could have taken place in different degrees and modalities. The exchange of genes and tool kits need not always have taken the same direction and degree. It is enough to evoke the spread of Neanderthals in MIS 5 from Europe into the Near East (Condemi 1985, 1991) resulting in less prominent Neanderthal morphology in Krapina, Saccopastore, Amud and Tabun than in the "classical Neanderthals" of the MIS 4 and 3 in the Western Europe.

If the Middle Pleistocene variability in Europe is examined in the context of geographically and chronologically defined p-demes as suggested by Howell (1996), and we accept several successive migrations into Europe on the basis of lithic (Lycett 2009) and palaeontological evidence (Carrión et al. 2011), one could postulate a core demographic area in SWA (Dennell et al. 2011; Bermúdez de Castro and Martinón-Torres in press). With western source populations as bearers of derived Neanderthal morphology, attenuation of Neanderthal traits in the more easterly or later populations could be explained by admixture with a group from outside of the isolated glacial refugium, i.e. a population from SWA. Under this model, we would expect that Southeast Europe (SEE) – the Balkan Penninusla – which would have remained in contact with SWA during glacial times belonged to this core demographic area. While isolation represented the major mechanism of evolutionary change in the west of the continent (Rightmire 1998), causing a bottleneck and fixation of derived traits, the Balkan Peninsula did not experience the effects of isolation. As already suggested (Roksandic et al. 2011) the population that inhabited the Balkan Peninsula and maintained contact with SWA throughout glaciations could have retained a number of primitive (i.e., non-Neanderthal) traits, without precluding morphological changes associated with encephalization and tooth reduction observed in Middle Pleistocene populations on all three continents. The BH-1 mandible would fit well with this explanation. Considering the Balkans as part of the larger area open to communication throughout the Pleistocene is not only warranted, but necessary. It will, however, require a shift in our communal perception of the geography of the region. We might need to do away with the perception of the Aegean and the Black seas as barriers for movement of populations and view them as a geographic centre of the Eastern Mediterranean geographic entity that includes Southeast Europe and Southwest Asia (SEE/SWA), which could have maintained population contact and gene exchange throughout human evolution.

References

BARKAI, R., GOPHER, A., LAURITZEN, S.E., & FRUMKIN, A. 2003. Uranium series dates from Qesem Cave, Israel, and the end of the Lower Palaeolithic. *Nature* 423 (6943): 977-979.

BELMAKER, M., TCHERNOV, E., CONDEMI, S. & BAR-YOSEF, O. 2002. New evidence for hominid presence in the Lower Pleistocene of the Southern Levant. *J. Hum. Evol.* 43 (1):43-56.

BERMÚDEZ DE CASTRO, J.M., MARTINÓN-TORRES, M. (in press). A new model for the evolution of the human Pleistocene populations of Europe. *Quat. Int.* 0. doi: http://dx.doi.org/10.1016/j.quaint.2012.02.036.

BISCHOFF, J.L., WILLIAMS, R.W., ROSENBAUER, R.J., ARAMBURU, A. et al. 2007. High-resolution U-series dates from the Sima de los Huesos hominids yields 600-66 kyrs: implications for the evolution of the early Neanderthal lineage. *Journal of Archaeological Science* 34 (5):763-770.

BONMATI, A., GOMEZ-OLIVENCIA, A., ARSUAGA, J.L., CARRETERO, J.M. et al. 2010. Middle Pleistocene lower back and pelvis from an aged human individual from the Sima de los Huesos site, Spain. *PNAS* 107 (43):18386-18391. doi:10.1073/pnas.1012131107.

CARRIÓN, J.S., ROSE, J. & STRINGER, C. 2011. Early Human Evolution in the Western Palaearctic: Ecological Scenarios. *Quaternary Science Reviews* 30 (11-12):1281-1295.

CONDEMI, S. 1991. Some considerations concerning Neandertal features and the presence of Neandertals in the Near East. *Rev. Antropol* 69:27-38.

CONDEMNI, S. 1985. Les hommes fossiles de Saccopastore (Italie) et leurs relations phylogénétiques. Doctoral dissertation: Université de Bordeaux I, Bordeaux, France.

DEAN, D., HUBLIN, J.-J., HOLLOWAY, R. & ZIEGLER, R. 1998. On the phylogenetic position of the pre-Neandertal specimen from Reilingen, Germany. *J. Hum. Evol.* 34 (5):485-508.

DENNELL, R.W., MARTINÓN-TORRES M. & BERMÚDEZ DE CASTRO, J.M. 2011. Hominin variability, climatic instability and population demography in Middle Pleistocene Europe. *Quaternary Science Reviews* 30 (11-12):1511-1524. doi:http://dx.doi.org/10.1016/j.quascirev.2009.11.027.

EASTWOOD, W.J. 2004. East Mediterranean vegetation and climate change. In: Griffiths, H.I., Krystufek, B., Reed, J.M. (eds) Balkan Biodiversity: Pattern and Process in the European Hotspot. Kluwer Academic, Dordrecht, pp. 25-48.

EHLERS, J, & GIBBARD, P.L. (Eds.) 2004. Quaternary Glaciations – Extent and Chronology, Part I: Europe. Elsevier Science.

FALGUERES, C., BAHAIN, J.-J., TOZZI, C., BOSCHIAN, G., DOLO, J-M. et al. 2008. ESR/U-series chronology of the Lower Palaeolithic palaeoanthropological site of Visogliano, Trieste, Italy. *Quaternary Geochronology* 3(4): 390-398. doi:http://dx.doi.org/10.1016/j.quageo.2008.01.007.

FALGUÈRES, C., BAHAIN, J.-J., YOKOYAMA, Y., ARSUAGA, J.L. et al. 1999. Earliest humans in Europe: the age of TD6 Gran Dolina, Atapuerca, Spain. *J. Hum Evol* 37 (3/4):343-352.

FALGUÈRES, C., YOKOYAMA, Y., SHEN, G., BISCHOFF, J.L., KU, T-L. & DE LUMLEY, H. 2004. New U-series dates at the Caune de l'Arago, France. *Journal of Archaeological Science* 31 (7):941-952. doi:http://dx.doi.org/10.1016/j.jas.2003.12.008.

GREEN, R.E., KRAUSE, J., BRIGGS, A.W., MARICIC, T., STENZEL, U., KIRCHER, M., PATTERSON, N., LI, H. et al. 2010. A draft sequence of the Neandertal genome. *Science* 328 (5979):710-22. doi: 10.1126/science.1188021.

GOMEZ-ROBLES, A., BERMUDEZ DE CASTRO, J.M., MARTINON-TORRES, M., PRADO-SIMON, L. & ARSUAGA, J.L. 2012. A geometric morphometric analysis of hominin upper second and third molars, with particular emphasis on European Pleistocene populations. *J. Hum. Evol.* 63 (3):512-526. doi:10.1016/j.jhevol.2012.06.002.

GRIFFITHS, H.I. KRYSTUFEK, B. & REED, J.M. (Eds) 2004. Balkan Biodiversity: Pattern and Process in the European Hotspot. Kluwer, Dordrecht.

HEWITT, G.M. 2011. Mediterranean Peninsulas: The Evolution of Hotspots. In: Zachos, F.E., Habel, J.C. (eds) Biodiversity Hotspots: Distribution and Protection of Conservation Priority Areas. Pp. 123-147.

HOLLIDAY, T.W. 2006. Neanderthals and modern humans: an example of a mammalian syngameon? In: Hublin, J-J., Harvati, K, Harrison, T (eds) Neanderthals Revisited: New Approaches and Perspectives. Vertebrate Paleobiology and Paleoanthropology. Springer Netherlands, pp. 281-297. doi:10.1007/978-1-4020-5121-0_16.

HOWELL, F. 1996. Thoughts on the study and interpretation of the human fossil record. In: Meikle W, Howell F, Jablonski, N. (eds) Contemporary issues in human evolution, vol. mem 21. California Academy of Sciences, San Francisco, pp. 1-46.

LYCETT, S.J. 2009. Understanding ancient hominin dispersals using artefactual data: a phylogeographic analysis of Acheulean handaxes. *PloS One* 4 (10):e7404. doi:10.1371/journal.pone.0007404.

MAREAN, C.W., BAR-MATTHEWS, M., BERNATCHEZ, J., FISHER, E. et al. 2007. Early human use of marine resources and pigment in South Africa during the Middle Pleistocene. *Nature* 449 (7164):905-908. doi:10.1038/nature06204.

MARTINON-TORRES, M., BERMUDEZ DE CASTRO, J.M., GOMEZ-ROBLES, A., PRADO-SIMON, L. & ARSUAGA, J.L. 2012. Morphological description and comparison of the dental remains from Atapuerca-Sima de los Huesos site (Spain). *J. Hum Evol* 62 (1):7-58. doi:10.1016/j.jhevol.2011.08.007.

MEYER, M., KIRCHER, M., GANSAUGE, M.T., LI, H et al. 2012. A high-coverage genome sequence from an archaic Denisovan individual. *Science* 338: (6104):222-226. doi:10.1126/science.1224344.

O'REGAN, H.J., TURNER, A., BISHOP, L.C., ELTON, S. & LAMB, A.L. 2011. Hominins without fellow travellers? First appearances and inferred dispersals of Afro-Eurasian large-mammals in the Plio-Pleistocene. *Quaternary Science Reviews* 30 (11):1343-1352.

PABLOS, A., LORENZO, C., MARTINEZ, I., BERMUDEZ DE CASTRO, J.M. et al. 2012. New foot remains from the Gran Dolina-TD6 Early Pleistocene site (Sierra de Atapuerca, Burgos, Spain). *J. Hum. Evol.* 63 (4):610-623. doi:10.1016/j.jhevol.2012.06.008.

RIGHTMIRE, G.P. 1998. Human evolution in the Middle Pleistocene: the role of *Homo heidelbergensis*. *Evolutionary Anthropology* 6 (6):218-227.

RINK, W.J., MERCIER, N., MIHAILOVIC, D., MORLEY, M.W. et al. 2013. New Radiometric Ages for the BH-1 Hominin from Balanica (Serbia): Implications for Understanding the Role of the Balkans in Middle

Pleistocene Human Evolution. *PloS One* 8 (2): e54608. doi:10.1371/journal.pone.0054608.

ROEBROEKS, W. 2001. Hominid behaviour and the earliest occupation of Europe: an exploration. *J. Hum. Evol.* 41 (5):437-461. doi:10.1006/jhev.2001.0499.

ROKSANDIC, M., MIHAILOVIC, D., MERCIER, N., DIMITRIJEVIC, V. et al. 2011. A human mandible (BH-1) from the Pleistocene deposits of Mala Balanica cave (Sicevo Gorge, Nis, Serbia). *J. Hum Evol* 61 (2):186-196. doi:10.1016/j.jhevol.2011.03. 003.

SCHOETENSACK, O. 1908. Der Unterkiefer des Homo heidelbergensisaus den Sanden von Mauerbel Heidelberg: ein Beitragzur Palaöntologie des Menschen. Engelmann, Leipzig.

SIRAKOV, N., GUADELLI, J.L. IVANOVA, S., SIRAKOVA, S. et al. 2010. An ancient continuous human presence in the Balkans and the beginnings of human settlement in western Eurasia: A Lower Pleistocene example of the Lower Palaeolithic levels in Kozarnika cave (North-western Bulgaria). Quaternary International 223: 94-106.

TZEDAKIS, P.C. 2004. The Balkan as prime glacial refugial territory of European temperate trees In: Griffiths, H.I., Krystufek, B., Reed, J.M. (eds) Balkan Biodiversity: Pattern and Process in the European Hotspot. Kluwer Academic, Dordrecht, pp. 49-69.

WAGNER, G.A., KRBETSCHEK, M., DEGERING, D., BAHAIN, J.J. et al. 2010. Radiometric dating of the type-site for Homo heidelbergensis at Mauer, Germany. *Proc. Natl. Acad. Sci.* USA 107.

7.
THE ROLE OF LANDSCAPES IN SHAPING HOMININ HABITATS IN AFRICA

Sally C. REYNOLDS
Institute for Human Evolution, University of the Witwatersrand (South Africa) and
School of Natural Sciences and Psychology, Liverpool John Moores University (United Kingdom)
S.Reynolds@ljmu.ac.uk; reynolds@ipgp.fr

Abstract: *Here I briefly review palaeoenvironmental evidence from sites repeatedly used by hominins in eastern and southern Africa, such as Sterkfontein (Cradle of Humankind, Gauteng, South Africa). Common 'mosaic' habitat reconstructions involve the presence of a lake or river setting, with a combination of forest or woodland with savannah grasslands in close spatial proximity. The Tectonic Landscape Model [Bailey et al. pp. 257-280 & Reynolds, et al., pp. 281-298 (JHE 60, 2011)], is a new hominin habitat model which explains why certain sites appeared to have been repeatedly used by our ancestors over millions of years. Specific geomorphological processes, such as tectonic faulting, would have created complex topography and thereby encouraged heterogeneous habitats to form, and sustained such features through time. The Plio-Pleistocene is characterised by several key climatic transitions that would have presented unique challenges for hominins and other fauna. Therefore, a more complete appreciation of how geomorphological processes affect landscapes, surface water and vegetation is critical to the characterisation of the hominin niche and also of the strategies employed by our ancestors to adapt to past climatic changes. Furthermore, the use of complex topography by hominins may explain aspects of their postcranial anatomy, diets and possible routes that they may have used to disperse to other regions.*

Keywords: *geomorphology, endemism, speciation, human evolution, Africa, Plio-Pleistocene*

INTRODUCTION

Faunas are affected by landscapes in various ways and on a number of spatial and temporal scales. This is true also of human beings and their ancestors. Studies examining the role of landscape processes in shaping floral and faunal habitats and influencing evolutionary patterns over time are becoming more commonplace (Sepulchre et al., 2006; Trauth et al, 2010 Ashley et al., 2010; Bailey et al., 2011; Reynolds et al., 2011). Examining the role played by landscapes enables us to consider the following questions: How did habitats favoured by hominins promote their evolution of specific traits such as bipedality, tool use and larger brains? And to what extent did large-scale geomorphological processes influence the creation of the habitats sought out by hominins? And finally, do landscape processes create landscapes that modern man finds attractive?

This contribution focuses on the role played by landscape processes at different scales in shaping hominin habitats, and also creating opportunities for faunal populations to become separated and for speciation to occur over time. The first aspect is the role of geomorphological processes in creating 'mosaic' habitats, which are loosely defined as a lake, pan or river setting with gallery forest and patches of open savannah in close proximity. The Tectonic Landscape Model (Reynolds et al., 2011; Bailey et al., 2011) explains how specific geomorphological processes in the presence of ground water will create and maintain mosaic habitats for as long as the faults remain active (Reynolds et al., 2011). Here I discuss tectonic and faunal features from the site of Makapansgat (South Africa) to illustrate how the geomopholology and re-assessments of the faunal lists can be explained by the Tectonic Landscape Model, and how these types of data can be applied to other fossil sites.

The second aspect, which acts on far longer timescales (i.e. thousands and millions of years), is the role of landscape features in promoting speciation by acting as a vicariant agent, separating populations and allowing genetic divergence to occur. For this aspect, I use the example of the Eastern Arc Mountains in Kenya and Tanzania as a case study (Burgess, et al., 2007). A third aspect must be noted, that is the deliberate utilisation of landscape features for hunting purposes, specifically during the Later Stone Age. Evidence from Greece and Kenya show how humans successfully incorporated landscape features into their hunting strategies and patterns of land-use (Bailey et al., 1993; Marean, 1997). While the discussion of deliberate and tactical use of the landscape is well-known in military and hunting contexts, this use falls outside of the scope of the present paper.

Developments in remote sensing and satellite imagery during the 1970's and 1980's facilitated the recording and interpretation of landscape features. This, together with several landmark studies of how earthquakes deform the Earth's crust (e.g. King and Vita-Finzi, 1981) allowed scientists to understand how geomorphological changes affect landscapes at various scales. From there, this understanding of the dynamic interplay between faulting and human habitats could be extrapolated back into the past, as was done for Palaeolithic sites in Greece, such as Klithi (Bailey et al., 1993). Later, this essential

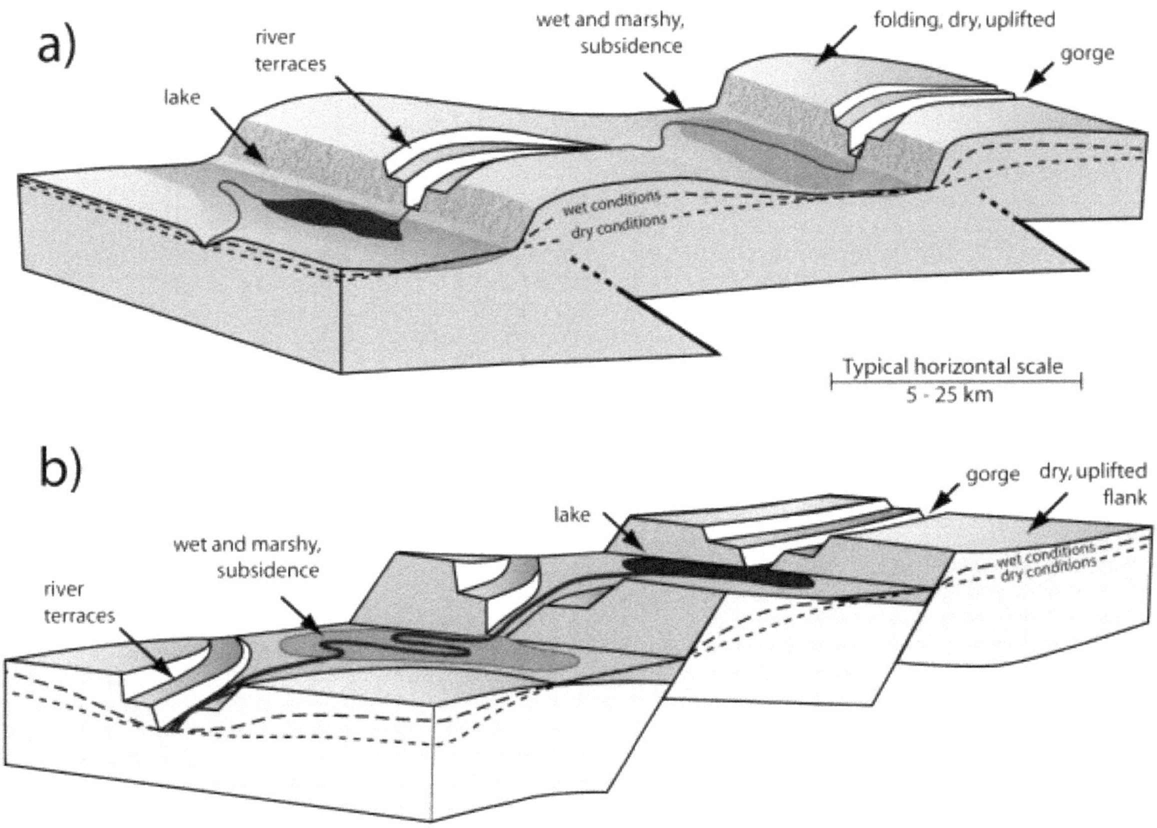

Fig. 7.1a & b. Tectonic Landscape Model (after Bailey et al., 2011; Reynolds et al., 2011) illustrating the range of landscape features and vegetation types that occur within a relatively restricted spatial area as a result of reverse (shown in 1a) and normal faulting (shown in 1b) in the presence of surface water

understanding was reinterpreted for the australopithecine sites of eastern and southern Africa (Reynolds, et al., 2011). As well as creating attractive habitats for hominins, the use of tectonically altered landscapes may also explain the routes followed by hominins as they dispersed from Africa (Bailey et al., 2011).

METHODS

The theory underlying the landscape reconstruction in eastern and southern Africa follows different methodological rationales, and is discussed in depth in Bailey et al. (2011). The landscape features of South African sites are preserved by hard rocks and relatively low levels of tectonic activity and thus can be reconstructed within the present site locations, but the eastern African sites are in regions were erosion, faulting and volcanism destroy evidence of older landscapes. Landscapes in eastern Africa (such as sites in Ethiopia) are therefore reconstructed by the use of analogy with areas presently closer to the active Rift margin, which is where the ancient sites were located at the time of hominin habitation (Bailey, et al., 2011; Reynolds et al., 2011).

From a practical perspective, a combination of satellite images and digital elevation models are used to examine the landscape and interpret the salient features (e.g. Bailey et al., 2011). For example, images deriving from Landsat ETM (Enhanced Thematic Mapper) in different spectral colour bands which correspond to different features (such as different rocks or vegetation types) can be used. These images may be overlain with digital elevation models (DEMs), thereby creating composite images of the landscapes, the salient features of which are then interpreted (as in Figure 2a). Field work and visual inspection are used to validate these interpretations (Fig. 7.2b & c).

Examining the landscape uses combinations of satellite imaging methods and old-fashioned ground truthing to examine and understand the landscape on several spatial scales. Examining changes in the landscape over temporal scales partly involves the reinterpretation of faunal data recovered from previous excavations as a means to identify the types of landscape features (such as rocky cliffs and marshy areas) present at the sites during deposition (e.g. Reynolds et al., 2011).

Tectonically modified landscapes & creation of hominin habitats

Early publications on human evolution inferred that our ancestors left the forest and became bipedal in the African savannah. But from the 1970's onwards, palaeoenvironmental reconstructions pointed to the mosaic environment as an important habitat associated with hominins (Kingston, et al., 1994), and indicative of

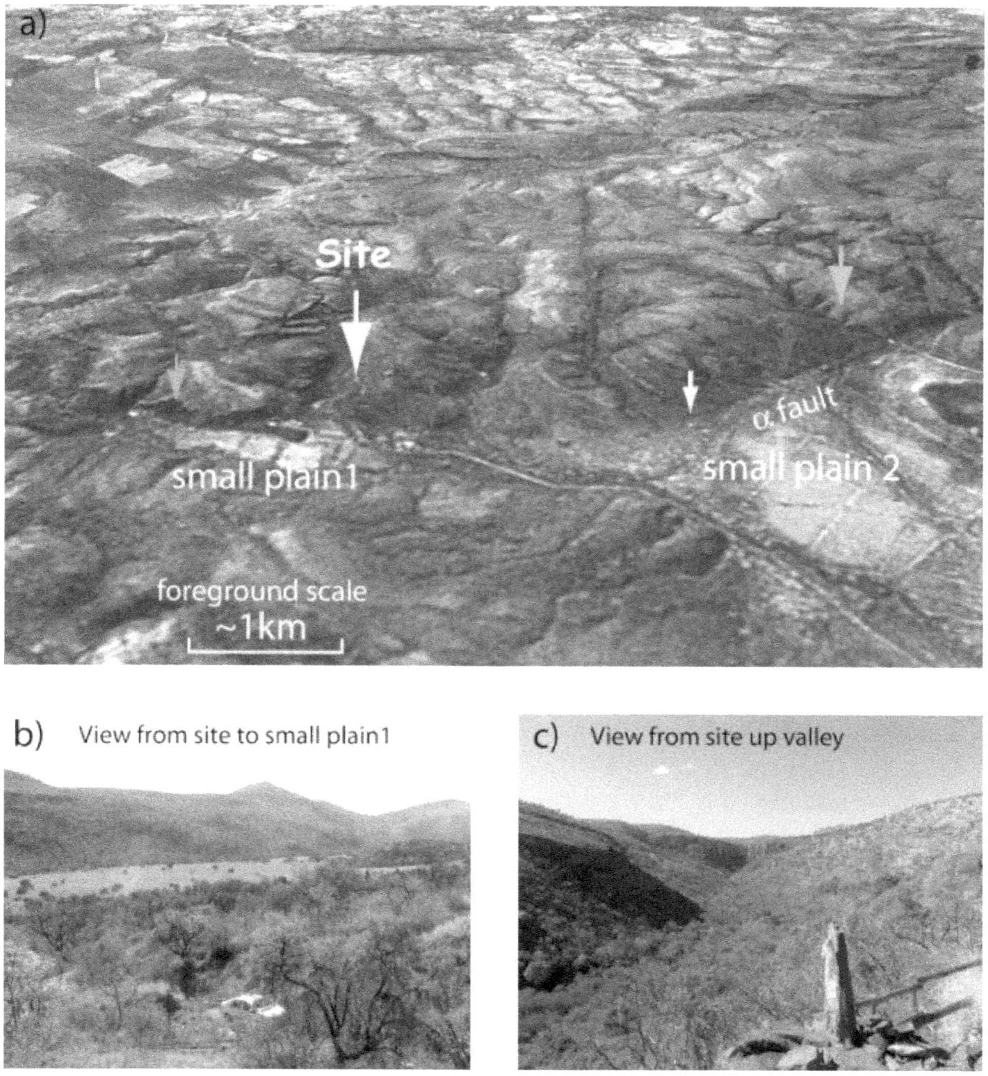

Fig. 7.2a-c. An interpreted landscape around the Makapansgat fossil site in South Africa illustrates how key landscape features attributed to tectonic activity can be identified within the site catchment area, specifically the sedimentary plains and rugged cliffs that are typically present in tectonically-altered landscapes. Observations of the key landscape features are made using Landsat enhanced thematic mapper (ETM) images overlain with digital elevation model (DEM) data (a) and visual inspection of key features on the ground (b, c) (after Bailey et al., 2011; Reynolds et al., 2011)

specific adaptations to mixed habitats and variable environments (e.g. Potts, 1998). 'Mosaic habitats' are loosely defined as a lake, pan or river setting with gallery forest and patches of open savannah in close proximity. The faunal assemblages of the majority of hominin sites in central, eastern and southern Africa show a combination of arboreal species, browsing and grazing species, along with certain aquatic animals, such as hippopotamus, crocodiles and otters, which are suggestive of these mixed habitats (e.g. Reynolds et al., 2011). A possible explanation could be time-averaging in the fossil record means that either fossils from wet, forested, climate phases are being mixed with species from separate, temporally distinct drier, more open grassland climatic episodes, making the habitats appear more mixed (heterogeneous) then they really were in the past. This explanation would imply that hominins were not adapted to live in variable habitats (sensu Potts, 1998).

An alternative explanation is the Tectonic Landscape Model, which explains how specific geomorphological processes in the presence of ground water will create and maintain mosaic habitats (Reynolds et al., 2011). Faulting can be of several different types, but in the cases of reverse and normal faulting, they create similar landscape features (Fig. 7.1a & b). Even in environments where there is no lake or river to begin with, tectonic activity can alter the drainage of the area to create surface water features where there were none before. Seeps and springs can form when faults intersect with the water-table (Reynolds, et al., 2011; Fig. 7.1a & b). A secondary effect of faulting is also the formation of grabens (a valley created by two parallel faults, such as the East African Rift Valley) and fault-bounded basins, which can then collect water that drains down from higher elevations, examples of which are the Olduvai Basin (Tanzania) and the Okavango Delta (Botswana). Fault-bounded basins can often host large, but shallow, seasonal lakes, such as the present-day lakes of Eyasi and Manyara (Tanzania). Shallow lakes with large surface areas are subject to high levels of evaporation, and are therefore often highly alkaline and saline. They do, however, create important and attractive habitats for bird and mammal faunas, such as flamingos.

In this way, faulting has the potential to transform a flat, relatively featureless landscape into one with cliffs and river gorges, water features, fertile sedimentary plains and uplifted, drier flank areas (Figure 1). This will increase the types of vegetation and fauna in a given region, and this effect can be seen in the high levels of biodiversity observed in the tectonically-controlled Okavango Delta (Botswana), or the Nylsvlei wetland (South Africa) for instance. This suite of landscape features will persist for as long as the fault motions continue. This effect will occur whether the tectonics are small, frequent motions, or large magnitude events which occur only very rarely (e.g. Bailey, et al., 2011).

Of particular interest is the creation of swampy, or marsh areas, a sign of disturbed drainage that signals tectonic alteration of the landscape. Swampy areas host high levels of biodiversity (for example, the Okavango Delta in Botswana) and are indicated in the fossil record by types of fauna unique to marshy habitats (discussed in Reynolds, et al., 2011). In sum, the tectonic alteration to a landscape creates zones of drier, uplifted savannah areas, with downdropped sedimented plains, steep cliffs and swampy areas; in short – habitat heterogeneity, or mosaic habitats. This type of habitat would meet the three essential habitat requirements of hominins, which are: 1) drinking water, 2) a range of vegetation zones for foraging and 3) places of refuge from predation, such as steep cliffs (Reynolds et al., 2011). Here I will summarise features from the South African site of Makapansgat to illustrate the geomopholology and re-assessments of the fauna and palaeoenvironmental reconstructions from these sites. Figure 2 shows the geomorphology of the Makapan Valley region and the cave sites, with the tectonically controlled Nylsvlei wetland indicated in Figure 2b). The river Nyl runs close to the fault location, which may indicate that tectonic activity is ongoing or that sediment from the east (left), is pushing the river course westwards (Bailey et al., 2011). There is a close agreement between present habitat heterogeneous conditions and those inferred from the fauna of Member 3 and 4 (Reynolds et al., 2011, and references therein).

The high biodiversity of present and past species suggests that the wetland and rocky cliff areas were present during the time of hominin occupation, while studies of stable isotope of australopithecines from Limeworks Cave reveal a significant C4 (grass and sedge) signal, suggesting these vegetations were spatially accessible (Sponheimer and Lee-Thorp, 1999). The landscape evidence from Makapansgat shows the combination of features, such as the river and the sedimentary plains, which would have offered foraging areas close to the relative safety of the river gorges, caves, and the characteristic cliffs of the Makapan Valley (Fig. 7.2c).

Within the site region, there are also plains of varying sizes and wetlands within ranging distance of the valley itself. Presence of species such as the klipspringer (Oreotragus oreotragus) as well as several primate species indicates the presence of rocky outcrops and trees for these animals (Reynolds et al., 2011 and references therein). Similar landscape geomorphology and tectonic activity is also identified at the fossil sites of Taung and the Sterkfontein Valley (South Africa), suggesting that tectonic motions created attractive, heterogeneous habitats which Australopithecus most likely exploited preferentially (Bailey et al., 2001; Reynolds et al., 2011). It is very likely that sites repeatedly occupied over longer timescales (such as Sterkfontein) were tectonically-altered, and that this maintained the suite of attractive features. Over longer timescales, other types of landscape changes have the potential to influence the genetic structure of populations, and this is dealt with in the next section.

Influence of landscape changes on populations

Eastern Africa experiences high levels of geomorphological processes such as rifting, faulting and volcanism due to the splitting of the Nubian and Somalian plates. The distribution of species is affected by altering topography, altitude, and surface drainage.

Tectonic activity can inhibit the movements of animals, by creating surface water features (such as rift and craterlakes) and landscape features such as steep fault scarps and lava flows. Large-scale climatic patterns can be mediated by these tectonically created lakes, as has been suggested for the so-called 'amplifier lakes' in eastern Africa (Trauth et al., 2010).

The overall effect of the geomorphological changes is that the physical location of habitats of plant and animal communities are altered, leading to habitat fragmentation, leading to population fragmentation. Populations subdivision is an important precursor to speciation, but when smaller, vulnerable species are further subdivided, the risks of inbreeding, disease and predation increase dramatically. Thus, areas where fragmentation and rates of endemicity are high are also at higher risks of extinction (e.g. Reynolds, 2007).

While changes in landscape geomorphology over time can have significant impacts on the populations, with long-term consequences for speciation and extinction, these effects have only recently become quantifiable by measuring levels of divergence across populations (e.g. chimpanzees; Eriksson et al., 2004; wildebeest; Arctander et al., 1999 and gelada baboons; Belay and Mori, 2006 among others).

Such studies clearly indicate that genetic structuring can be significantly altered by landscape factors, acting in concert with climatic changes. The most common effect of landscape changes on species is to separate populations, either completely or partially. This process, called vicariace, in turn can lead to speciation (e.g.

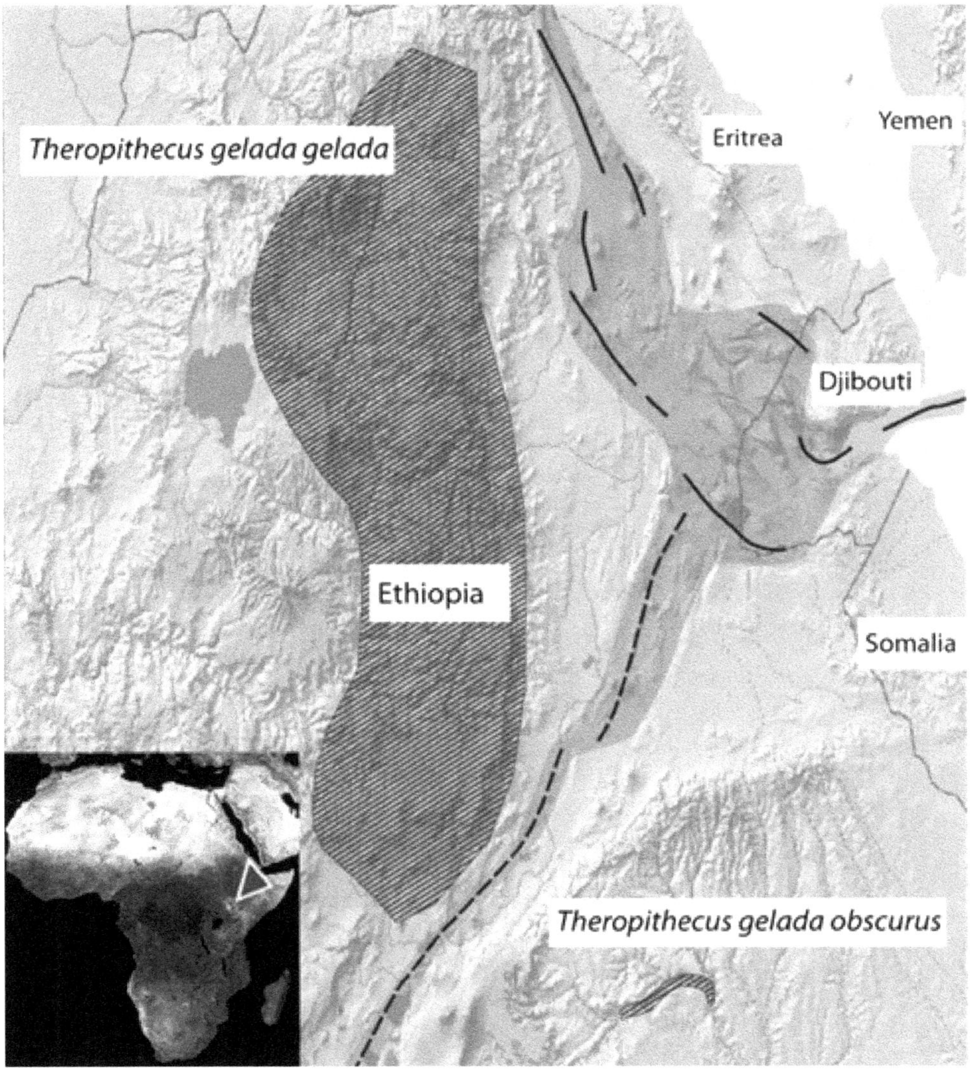

Fig. 7.3. Influence of Rifting in Ethiopia. It has influenced the vicariance and genetic divergence within several species of vertebrates, including geladas (from Reynolds et al., 2011). Over time, the landscape geomorphological processes have significantly altered the genetic population structure of species living on the uplifted flanks of the Rift

Measey and Tolley, 2011). Obvious barriers to gene flow include deserts, mountain ranges and river courses. For example, the population genetic structure of the bonobo chimpanzees (Pan paniscus) in the Congo is affected by river flow, which separates populations from each other (Eriksson et al., 2004). In general, landscape vicariance can take several forms and act on widely differing spatial and temporal scales. Certain large-scale landscape changes dramatically alter almost every species living within the region. An excellent example of this is the uplift and spreading of the Rift, which has split populations of montane primates, bovids, and frogs with clearly identifiable genetic divergence in all cases (Fig. 7.3).

Rift-splitting of populations have been documented in the gelada baboon subspecies (*Theropithecus gelada gelada* and *T.g. obscurus*), in the wildebeest (*Connochaetes taurinus*), and also in the African clawed frog (genus Xenopus, also known as the platanna), suggesting that this large-scale ecological pattern is likely to be identified from other species in future studies (Arctander et al., 1999; Belay and Mori, 2006; Evans et al., 2011; Reynolds et al., 2011). African clawed frog populations are estimated to have diverged between approximately 1-3.5 million years ago (Evans et al., 2011). Migration between flanks of the Rift appears to be limited because the floor of the Rift Valley is much drier and hotter than the prevalent conditions on the Ethiopian highlands, thus acting as a barrier between the two flanks (Evans et al., 2011).

In the contribution, I have outlined the Tectonic Landscape Model (Bailey, et al., 2011; Reynolds et al., 2011) and how this may explain why so many hominin sites appear to have been located in heterogeneous (i.e. mosaic) habitats. The Model explains how such habitats are created and sustained, and real-life examples of tectonically-controlled regions, such as the Okavango Delta illustrate that faulting can create areas of extremely high biodiversity (Ramberg, et al., 2006). These areas also offer a certain level of buffering against climatic aridification (Reynolds et al., 2011). The presence of sustained heterogeneous environments in specific areas

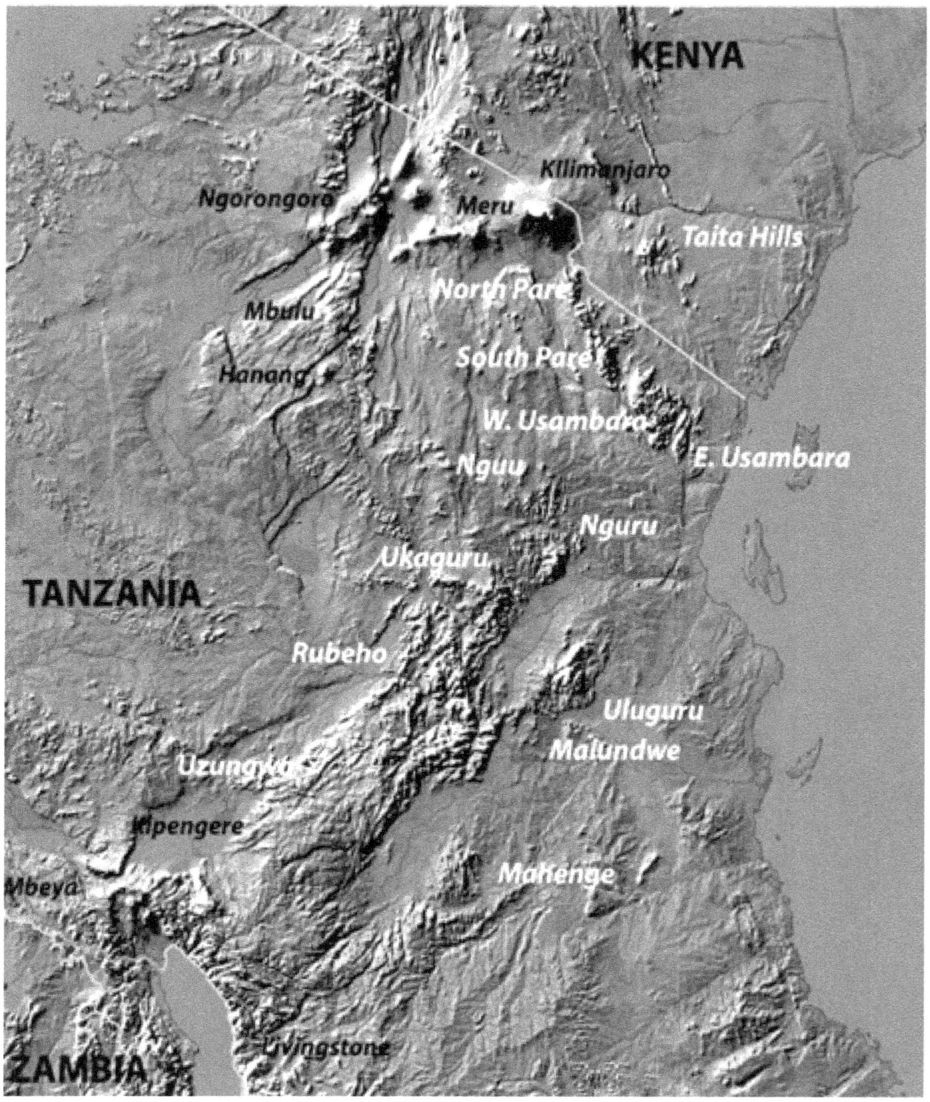

Fig. 7.4. The Eastern Arc Mountains of Kenya and Tanzania illustrate the relationship between rough topography, climate change and endemism. The 13 mountain blocs which comprise the Eastern Arc range are labelled in white, while other important forested mountains are indicated in black

may explain why certain archaeological sites are used only once, while others, such as Sterkfontein (South Africa) was repeatedly used by several species of hominins over millions of years (Reynolds, et al., 2011).

Finally, large-scale climatic changes act in concert with topography to fragment forest populations, such as the forest chameleons of the Eastern Arc Mountains of Kenya and Tanzania (genus Kinyongia; Measey and Tolley, 2011). Studies of the forest chameleon indicate that forested areas on the montane regions of the eastern Arc Mountains was stable over long timescales, but that Pleistocene drying fragmented these forest refugia. Forested mountaintops were isolated from each other by unforested, deep valleys which isolated the chameleon populations very effectively and leading to genetic divergence (Measey and Tolley, 2011).

The Eastern Arc Mountains are an excellent case study of an area with rough topography, high levels of endemism and also with high levels of endangered species. These are a chain of 13 mountain blocks in southern Kenya running into eastern Tanzania (Fig. 7.4). This region has been highlighted as an important area of extremely high levels of endemism, i.e. containing species unique to this area (Burgess, et al., 2007, Measey and Tolley, 2011). So far, 96 endemic vertebrate species have been recorded, including four primate species. However, over 70 of these species are endangered, which underlines the close links between centres of endemism, fragmented populations and high risks of extinction (Burgess, et al., 2007; Reynolds, 2007).

DISCUSSION AND CONCLUSIONS

Military historians and hunters have always appreciated the tactical advantages of landscape features as lookout points or ambush locations. In human evolution studies, however, the role played by the landscape was largely overlooked in favour of the role played by vegetation and large-scale climate shifts as drivers of evolution and adaptation (e.g. de Menocal, 1995).

Therefore, a more complete appreciation of how geomorphological processes affect landscapes, surface water and vegetation is critical to the characterisation of the hominin niche and also of the strategies employed by our ancestors to adapt to past climatic changes. Over longer timescales, landscapes influence the fauna that inhibit them, causing vicariance and becoming centres of endemism, such as in the Eastern Arc Mountains of Tanzania and Kenya. However, the possibility of fragmenting populations also increases the risk of extinction. Thus, the study of the landscape can aid our understanding of how certain African regions have influenced the creation and preservation of biodiversity, both now, and in the past (Reynolds, 2007).

Acknowledgements

Many thanks to Geoffrey King for providing the images and landscape figures.

References

ARCTANDER, P., JOHANSEN, C. & COUTELLEC-VRETO, M.-A. 1999. Phylogeography of three closely related African bovids (Tribe Alcelaphini*). Mol. Biol. Evol.* 16: 1724-1739.

BAILEY, G.N., KING, G.C.P. & STURDY, D. 1993. Active tectonics and land-use strategies: a Palaeolithic example from northwest Greece. *Antiquity* 67:292-312.

BAILEY, G.N., REYNOLDS, S.C. & KING, G.C.P. 2011. Landscapes of human evolution: models and methods of tectonic geomorphology and the reconstruction of hominin landscapes. *Journal of Human Evolution* 60:257-280.

BELAY, G. & MORI, A. 2006. Intraspecific phylogeographic mitochondrial DNA (D-loop) variation of Gelada baboon (Theropithecus gelada) in Ethiopia. *Biochemical Systematics and Ecology* 34:554-561.

BURGESS, N.D., BUTYNSKI, T.M, CORDEIRO, N.J. DOGGART, N.H. et al. 2007. The biological importance of the Eastern Arc Mountains of Tanzania and Kenya. *Biological Conservation* 134:209-231.

NASHANDA, E., PERKIN, A., ROVERO, F., STANLEY, W.T. & STUART, S.N. 2007. The biological importance of the Eastern Arc Mountains of Tanzania and Kenya. *Biological Cconservation* 134: 209-231.

DEMENOCAL, P.B. 1995. Plio-Pleistocene African climate. *Science* 270: 53-59.

ERIKSSON, J., HOHMANN, G., BOESCH, C. & VIGILANT, L. 2004. Rivers influence the population genetic structure of bonobos (Pan paniscus). *Molecular Ecology* 13: 3425-3435.

EVANS, B.J., BLISS, S.M., MENDEL, S.A. & TINSLEY, R.C. 2011. The Rift Valley is a major barrier to dispersal of African clawed frogs (Xenopus) in Ethiopia. *Molecular Ecology* 20:4216-4230.

KING, G.C.P. & VITA-FINZI, C. 1981. Active folding in the Algerian earthquake of 10 October 1980. *Nature* 292:22-26.

KINGSTON, J.D., MARINO, B.D. & HILL, A. 1994. Isotopic evidence for Neogene hominid paleoenvironments in the Kenya Rift Valley. *Science* 264: 955-959.

MAREAN, C.W. 1997. Hunter-gatherer foraging strategies in tropical grasslands: model building and testing in the East African Middle and Later Stone Age. *Journal of Anthropological Archaeology* 16:189-225.

MEASEY, G.J. & TOLLEY, K.A. 2011. Sequential fragmentation of Pleistocene forests in an East Africa biodiversity hotspot: chameleons as a model to track forest history. *PLoSONE* 6 (10):e26606 doi:10.1371/journal.pone.0026606.

POTTS, R. 1998. Variability selection in hominid evolution. *Evolutionary Anthropology* 7:81-96.

RAMBERG, L., HANCOCK, P., LINDHOLM, M., MEYER, T. et al. 2006. Species diversity of the Okavango delta, Botswana. *Aquat. Sci.* 68: 310-337.

REYNOLDS, S.C. 2007. Mammalian body size changes and Plio-Pleistocene environmental shifts: implications for understanding hominin evolution in eastern and southern Africa. *Journal of Human Evolution* 53:528-548.

REYNOLDS, S.C., BAILEY, G. & KING G.C.P. 2011. Landscapes and their relation to hominin habitats: case studies from Australopithecus sites in eastern and southern Africa. *Journal of Human Evolution* 60 (3):281-298.

SEPULCHRE, P., RAMSTEIN, G., FLUTEAU, F., SCHUSTER, M. et al. 2006. Tectonic uplift and eastern African aridification. *Science* 313: 1419-1423.

SPONHEIMER, M. & LEE-THORP, J.A. 1999. Isotopic evidence for the diet of an early hominid, Australopithecus africanus. *Science* 283, 368e370.

TRAUTH, M.H., MASLIN, M.A., DEINO, A.L., JUNGINGER, A. et al. 2010. Human evolution in a variable environment: The amplifier lakes of Eastern Africa. *Quaternary Science Reviews* 29(23-24): 2981-2988.

8.
THE DENISOVA GENOME: AN UNEXPECTED WINDOW INTO THE PAST

John HAWKS
Department of Anthropology University of Wisconsin-Madison 5240 Social Science Building
1180 Observatory Drive Madison, WI 53706, USA
jhawks@wisc.edu

Abstract: *The recovery of whole genome evidence has begun to provide new evidence about the biology and dynamics of ancient populations. Denisova Cave, in the Altai Mountains of Russia, has produced several skeletal fragments with ancient DNA evidence, including one complete genome at high coverage. This individual represents a population previously unknown to science. Dating to roughly 50,000 years ago, the genome has approximately the same genetic difference from contemporary Africans as did Neandertals, but was quite distinct from them also. Here, I discuss the functional insights that have come from the Denisova genome, including details about the immune system, pigmentation, and genes involved in brain development. We can demonstrate that the Denisovans contributed some genes of important phenotypic effect to later populations, including genes related to immunity, while other genes exhibit distinctive changes in Denisovans that are not shared in living human samples. Today, people in island Southeast Asia including New Guinea, the Philippines, and Aboriginal Australians trace up to five percent of their ancestry to the ancient Denisovan population. I discuss the dynamics of this population, including its relationship with Neandertals and the subsequent population movements and interactions in East and Southeast Asia that may explain the present distribution of Denisovan ancestry.*

Keywords: *Denisovan, genome, ancestry, demography, Neandertals*

INTRODUCTION

In 2010, scientists from the Max Planck Institute for Evolutionary Anthropology reported a draft genome sequence from ancient skeletal remains from Denisova Cave, in the Altai Mountains of Russia (Reich et al., 2010). Following the description of the genomes of three Neandertals from Vindija, Croatia, earlier in 2010 (Green, et al., 2010), the Denisova find represented the second Pleistocene human population from which genome-wide evidence had been recovered. In 2012, a Denisova cavehigh-coverage genome from one of the individuals was made available to researchers (Meyer, et al., 2012). This has provided the best evidence yet discovered about the genetics of any Pleistocene-era individual.

Still, this exceptional specimen has provoked mystery.While the Denisova genomes closest similarity is with the Neandertal genomes, it is nearly as genetically different from Neandertals as both genomes are from those of living people. In addition to the nuclear genome, three mitochondrial genomes were sequenced from Denisova skeletal remains, all of which are also substantially different from the mitochondrial genomes of known Neandertal specimens, including Neandertal specimens from nearby sites in the Altai and across Central Asia (Krause et al., 2007). The Denisova genetic evidence appears to represent an ancient population substantially different from the Neandertals, adjacent to the easternmost end of the Neandertal range of occupation. Reich and colleagues (op.cit.) dubbed this ancient population the "Denisovans", and turned to consider its relationship with living and ancient populations.

ARCHAEOLOGICAL ASSOCIATIONS

The Denisova remains preserve very little anatomical detail, and it is impossible at present to attribute them to any previously-known group of fossil hominins. Most of the genetic evidence comes from a terminal fifth manual phalanx from an adolescent girl. This has no informative anatomy to base anatomical comparisons to other early human specimens. The first reported third molar has length and breadth dimensions within the size range occupied by australopithecines and early *Homo*, both *H. habilis* and *H. erectus*. There are no distinctive morphological characters that would allow it to be assigned to any taxon.

Adding to the mystery, the Denisova finger bone has anomalously good preservation of DNA. The cold temperature within the cave year-round likely contributed to the preservation, but nonetheless it is exceptional compared to other cold-temperature caves. There are no direct dates on the skeletal remains from Denisova. By inference (based on their genetic sequences), the skeletal individuals are likely to be older than the Upper Paleolithic (Meyer et al., 2012), but the stratigraphy does not require this.

The human remains come from layer 11 within the rear gallery of the cave. The stone artifacts in this layer come from a broad range of industries from Upper Paleolithic to a local Mousterian. The late Mousterian at the site includes a good number of blades, although it is technically different from initial Upper Paleolithic industries in the Altai, including the assemblages from Kara Bom, which exhibit strongly blade-based flaking techniques. It is not clear which of the stone tool industries present in Denisova may have been made by the Denisovan population.

How were the Denisovans related to Neandertals?

The Denisova, Neandertal and human genomes are close to a trichotomy in terms of their average relationship, with the Denisova genome being slightly closer to the Neandertals than to living people (Reich et al., op.cit.); Meyer et al., 2012). For any particular gene, of course, there may be sister pairings between any two of those three, and in many cases, between Denisovans and some living humans to the exclusion of other living humans. This gives rise to several tricky statistical issues as we consider particular gene loci.

The similarity of Denisova and Neandertal genomes suggests that they emerged from a single population. This possibly was the early Middle Pleistocene population of Eurasia. Green and colleagues (Green et al., 2010) derived Neandertals from a common ancestor with living Africans only 250-400 Kya years ago. A model including the Denisova data, reported by Reich and colleagues (op.cit.) puts the emergence of a Neandertal-Denisova clade at between 190 Kya and 520 Kya, and the divergence of the Neandertal and Denisova branches around 50 Kya-100 Kya later. This is later than the time of the last unequivocal *H. erectus* fossils, and more than a half million years after the Trinil individual — type specimen of *Homo erectus* — lived. Based on this chronology, the Denisova genome does not represent *Homo erectus* or any hominin population derived from the initial diversification of *Homo*.

A slower mutation rate might bring the most recent specimens of *Homo erectus* into the temporal range of the Denisovan population, by elevating the time of divergence of Denisovans, Neandertals and contemporary African peoples. Whole-genome sequencing of human parent-offspring pairs, as well as re-sequencing of more limited regions of the genome, has suggested a slower rate of substitutions (Hawks, 2012; Scally & Durbin, 2012). This would be consistent with a much earlier diversification of Denisovan and Neandertal genomes, making it possible that they came from an early Middle Pleistocene adaptive radiation.

The Denisovans were too closely related to living humans to represent *Homo erectus*, as it is currently understood. *Homo erectus* occurred widely in Asia, including China and Java, and Africa during the span from 1.95 Mya to 750 Kya. In China and Java, fossils attributed to *Homo erectus* persisted until 200 Kya. There is no unequivocal fossil of *Homo erectus* after 200 Kya.

The mtDNA sequence of the Denisova genome is an outgroup to a clade including both humans and Neandertals, and appeared to branch from our ancestors roughly a million years ago. In addition to the finger bone, Reich and colleagues recovered the mtDNA of a second individual from Denisova Cave, represented by an isolated third molar (Reich et al., op.cit.). Reich and colleagues (op.cit.) showed that the mtDNA divergence between Denisova and the modern-Neandertal clade is deeper than expected given the nuclear genome genealogical divergence. They also showed that the nuclear genomes of Neandertals and Denisovans are somewhat closer than either is to the majority ancestors of living people. They discuss two possible explanations.

One scenario a mixture of the Denisovans with a more ancient Pleistocene population, followed by introgression of a more ancient mtDNA clade into the Denisovans. This would assert an ancient structured population preceding the origin of Denisovans, presumably from one of the Middle Pleistocene populations of Africa or Eurasia. A second scenario is incomplete lineage sorting, in which an earlier mtDNA divergence was captured by the Denisova and Neandertal populations at the time of their divergence and differentially lost from them.

Descendants of Denisovans

Some living humans trace a significant fraction of their ancestry to the population represented by the Denisova genome. As in the case of Neanderthals, different human populations show significantly different levels of similarity to the Denisova sequence. For Neanderthals, the similarities indicated between one and four percent Neanderthal ancestry for living people outside of Africa, including the peoples of Europe, East and South Asia, Australasia and the Americas (Green et al., 2010). In the case of the Denisova sequence, the greatest similarities are with living people in Melanesia, as represented by genome samples from Papua New Guinea and Bougainville. The similarities are consistent with approximately 4 percent contribution of a Denisova-like population to the ancestry of these living Melanesians (Reich et al., op.cit.). The paper estimates that together, the Denisova and Neanderthal-derived genes account for 8% of the ancestry of these living people.

The most plausible hypothesis to explain Denisovan ancestry of Australians and other Oceanian peoples is that the Denisova Cave individuals represent a much larger and more widespread population. A population of modern humans originating in Africa who were dispersing in the direction of island Southeast Asia may have encountered and mixed with a South or Southeast Asian population of Denisovans. The dispersing population would have absorbed some adaptive genes, which would have increased in frequency thereby increasing the apparent genetic contribution of the indigenous Pleistocene population.

Denisovan demography

It has become possible to make some good estimates of demographic history using only a single diploid genome, using a technique developed by Li & Durbin (2011). Meyer and colleagues (2012) applied this technique to the Denisova genome, finding that its genetic history contrasts with that of living human populations. Whereas African populations and living humans outside Africa share signs of population growth in the last 100,000 years, the Denisova genome seems to have come from a population that had long been small, and may have been shrinking.

We know that Neandertals of Europe and Central Asia lived in an environment that was relatively marginal for their technology and subsistence pattern. In contrast, the Denisovan population might well have lived in areas far from the Altai, such as South or Southeast Asia. These areas might have been expected to give rise to a larger and more successful human population. Several different hypotheses are possible explanations for the appearance of a Denisovan population restriction. Maybe Denisovans didn't live in South or Southeast Asia at all, or the Denisova genome itself may be a poor representative of a broader population. The demographic model of Li and Durban does not encompass admixture, just the probability of gene coalescence across time. We have recently begun to appreciate the reality of ancient African population structure. If those initial African populations were more divergent from each other than Neandertals and Denisovans, their later mixture would give rise to a picture of early population expansion, even if each of them had relatively low (Denisovan-like) diversity.

Human similarities and differences from Denisova

The Denisova genome lacks some genetic variations that are otherwise found universally in living human populations (Meyer et al., 2012). The number of such human-specific changes is relatively small, and they may give rise to a greater understanding of recent evolutionary change in human ancestors.

One interesting genetic detail about humans is our chromosome 2, which is an amalgam of two different chromosomes in the other great apes. Sometime in our evolution, two separate chromosomes fused into one, giving humans a karyotype of 46 chromosomes where chimpanzees, bonobos and gorillas have 48 chromosomes. The high-coverage genome was sufficient to show that Denisova shared the human fusion (Meyer et al., 2012). It remains unclear whether this fusion made any difference to any phenotype in ancient humans. However, considering that the fusion occurred prior to the divergence of Neandertals and Denisovans, we can conclude that this particular event did not contribute to any postzygotic isolation or chromosomal incompatibility between Neandertals, Denisovans, and modern humans.

CONCLUSIONS

The Denisova genome is the first high-coverage genetic evidence from a Pleistocene human population. It has already yielded substantial new evidence about the biology, demography and relationships of these ancient people. Of course, one exciting aspect of the discovery is as a proof of concept for future high-coverage archaic genomes. Recent work by Meyer and colleagues Meyer et al., 2012) make explicit that the Denisova phalanx simply has better endogenous DNA preservation than any other specimen known. That being said, new sequencing methods have greatly increased the sequence yield from the Denisova sample. Such techniques may bring higher coverage into reach for several specimens that are currently only sequenced at very low coverage, including the Vindija, Mezmaiskaya, and El Sidron Neandertals.

It has often been that paleoanthropology uses technological advances in other fields. But with ancient DNA, we really see an organic growth of technology along with research questions about our evolution. In our work on the ancient genomes, we're making some progress pushing forward knowledge about human biology by understanding human evolution. Evolution really is the fundamental principle of biology, but using evolution to learn about biology sometimes requires traveling through time. Ancient DNA gives paleoanthropologists a time machine to bring new insights into reach.

References

REICH, D., GREEN, R.E., KIRCHER, M., KRAUSE, J. et al. 2010. Genetic history of an archaic hominin group from Denisova Cave in Siberia. *Nature* 468:1053–1060. http://dx.doi.org/10.1038/nature09710.

GREEN, R.E., KRAUSE, J., BRIGGS, A.W, MARICIC, T. et al. 2010. A Draft Sequence of the Neandertal Genome. *Science* 328:710–722. http://dx.doi.org/10.1126/science.1188021.

MEYER, M., KIRCHER, M., GANSAUGE, M.-T., LI, H. et al. 2012. A high-coverage genome sequence from an archaic Denisovan individual. *Science* 338 (6104): 222-226.

KRAUSE, J., ORLANDO, L., SERRE, D., VIOLA, B., PRÜFER, K. et al. 2007. {Neanderthals} in Central {Asia} and {Siberia}. *Nature* 449:902-904. http://dx.doi.org/10.1038/nature06193.

HAWKS, J. 2012. Longer time scale for human evolution. *Proc Natl Acad Sci* U.S.A. 109: (39):15531-15532.

SCALLY, A. & DURBIN, R. 2012. Revising the human mutation rate: implications for understanding human evolution. *Nat. Rev. Genet.* 13 (10): 745-53.

LI, H. & DURBIN, R. 2011. Inference of human population history from individual whole-genome sequences. *Nature* 475:493-496. http://dx.doi.org/10.1038/nature10231.

9.
PRELIMINARY RESULTS ON THE FIRST PALEONTOLOGICAL, ANTHROPOLOGICAL AND ARCHAEOLOGICAL PLEISTOCENE LOCALITY IN ADRAR, MAURITANIA

Chérif Ousmane TOURE
Centre Africain de Recherches Archéologique et du Patrimoine, 91940 Les Ulis, France
ousmanecherif@gmail.com.

Anne DAMBRICOURT MALASSÉ
UMR 7194 CNRS, Department of Prehistory, National Museum of Natural History, Institute of Human Paleontology, Foundation Albert I Prince of Monaco, 1 rue René Panhard, 75013 Paris, France
iphadm@mnhn.fr

Abstract: *French naturalist Theodore Monod (1902-2000), founder of the Mauritanian Institute for Scientific Research (I.M.R.S.) inventoried many prehistoric sites in the Chinguetti area, at Guelb er Richat, one of the most concentrations of pebble tools in Western Sub-Saharan Africa, sometimes with pre-Acheulian facies. Few kilometres to the north, along the large desert namely Makteir (Adrar), the explorer noted high concentrations of handaxes and cleavers made on quartzite at El Beyyed, attributed to the Acheulian facies. All the sites were identified on the surface of ancient terraces. Since 2000 field works discovered lithic assemblages at El Beyyed with two stratified sites, especially Yeslem III which contained fossilised faunal and human bones. The Yeslem III locality is a set of small reliefs bordering a paleolake with two layers of human occupation. The first one in surface is a calcareous crust recently exposed by erosion, with stone artefacts assemblage and extinct fauna. The lithic industry is composed mainly of flakes, rare cores and handaxes. The second layer 80 cm below the surface corresponds to the Theodore Monod's level SB (white sand), observed in the nearby Acheulian sites Yeslem I, II and IV. In 2006 a first excavation of four square meters recorded in the second layer one hundred flakes, very few cores and handaxes without fauna. The fossil bones from the calcareous lay is highly fragmented and eroded, the assemblage is composed by aquatic species, ungulates, proboscides, large bovids and warthogs. The first fossils collected in 2002 correspond to four fragments of a human fibula. In 2004 a fifth fragment completed the bone. The preliminary analyses allow concluding a Pleistocene age without confusion with Neolithic occupations. The fossil fauna attests of a Sahelian environment with large permanent lakes, already observed in Zouerat – Hammami La défense in 1967 by Yves Coppens. The taxonomic identity of the human fibula is still unknown.*

Keywords: *Mauritania, Pleistocene, Acheulian, Faunal Assemblage, Hominid*

INTRODUCTION

Theodore Monod (1902-2000) inventoried in 1934 numerous prehistoric sites in the north of the Senegal River, particularly in the area of Guelb er Richat at oued Akerdil. Today this region of Mauritania is one of the greatest pebble tools concentrations of the Western Sahara. The huge desert of the Makteir (Adrar) begins 50 km to the north of the Guelb, the limit is marked by oasis, particularly at El Beyyed where the explorer raised important concentrations of handaxes and cleavers in quartzite, attribute to Acheulian "*From El Beyyed I went to Richât by the Dhar, covered here by a large reg, strewn by so numerous handaxes, that the plateau can be seen as a single layer of Acheulian all along the El Beyyed-Tililit axis, broad about twenty kilometers*".

In 1997, Theodore Monod initiated new field investigation to elaborate the chronology of the terraces, in association with the Laboratory of Prehistory, National Museum of Natural History, Paris, under the scientific direction of Henry de Lumley, and according to a protocol agreement defined by the Mauritanian government. Between 2000 and 2002, several localities were surveyed in the Guelb er Richat (pebbles of the Akerdil River) and El Beyyed borderlands of the Makteir (handaxes and cleavers) with Yeslem I and III. The first results were presented in 2004 at the International Conference on Human Evolution organized by the Academy of Sciences and Academy "des Inscriptions et Belles Lettres", Institute of France. But the most important locality, Yeslem III, remains unedited. The Guelb is a circular geological structure with a succession of annular reliefs centered on a solid mass of stromatolithes marked by old hydrothermal activities, its diameter exceeds 45 kilometers. The Makteir consists of paleolake and detrital quaternary deposits in discordance on an Ordovician schistous platform, covered by dunes.

All identified sites were in surface, on old colluvial or alluvial terraces, depriving the technological facies of

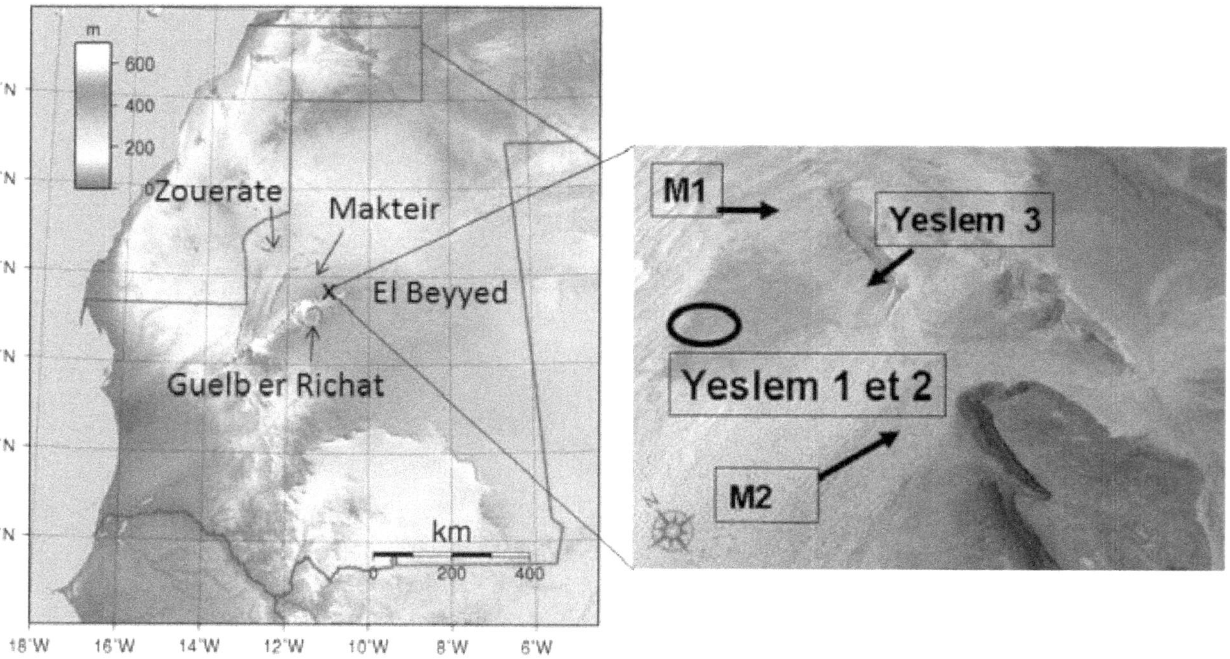

Fig. 9.1. Archaeological and geological localities (A: source Wikipedia and Google Earth)

chronological data. In 1998 Theodore Monod in agreement with the Director of the National Museum of Natural history, Henry de Lumley, lead again research in Mauritania intend to discover new sites in stratigraphy.

HISTORY

Theodore Monod establishes in Dakar, continued his research in Western Africa as director of the I.F.A.N. (Institut Français d'Afrique Noire), in close collaboration with the National Museum of Natural History. He surveyed Mauritania and could not pass beside the kilometers of sands strewn with lithics industries. Richard-Molard and Mauny (1953) contributed then to the research in Northern Mauritanian, both Adrar and Makteir areas. From the sixties, it was necessary to engage surveys on the inventoried sites, Theodore Monod entrusted this program to Pierre Biberson, titular of the Quaternary geology and paleontology chair of the Institute of Human Paleontology in Paris and specialist of the Moroccan Paleolithic. In 1967 Biberson located two sites at the mouth of the oued El Beyyed in the Makteir, M1 and M2, covered by sand (figure 1), with the recurrent problem to establish a chronology in endoreic context, since the level of the lakes and rivers does not depend on the oceanic fluctuations more easily identifiable (Miskovsky et al., 1991).

One second major difficulty was due to the important fossiliferous gap "the *absence of fossil faunas in the quaternary formations of Adrar makes until now difficult the tests of dating the sedimentary layers*" (Biberson 1969). Only one paleontological and archaeological site allowed knowing the ecological and climatic conditions of the Sahel during the Middle Pleistocene. This one was discovered, definitely more in north, in Zouerate (Douerat) by Le Floch, in the new mining city (Fig. 9.1). Yves Coppens studied the bones and the teeth and presented the faunal inventory at the sixth Panafrican congress of Prehistory in Dakar (Coppens et al. 1967). The extinct fauna of Zouerate consists of one species close to *Hippopotamus amphibius*, one warthog, one crocodile, different bovines and one Probocidian *Palaeoloxodon iolensis*, which made it possible to fix fauna at the end of Middle Pleistocene (Biberson 1969, Aumassip and Chaid-Saoudi 2004) "*the presence of hippopotamus imposes the conviction of the existence of great permanent water levels. The occurrence of Elephants in the same site confirms well the radical transformation of vegetable cover, more especially as Elephas iolensis was regarded until now as an Elephant of forest*" (Biberson 1969). The lithic industry associated with fauna corresponds to evolved Acheulian similar to the lithic assemblage collected by Biberson at El Beyyed in the two sites M1 and M2.

Theodore Monod published a synthesis in 1975 "Three layers of Pebbles tools of Adrar of Mauritania" underlining the real issue: no site could have been studied in its stratigraphical and paleoecological context, whereas a pre-Acheulian facies noted in Guelb er Richat would attest human presence in Western Africa since at least one million years (Monod 1975). Twenty years were passed when in 1997 on the initiative of its director Henry de Lumley, the National Museum of Natural History (NMNH) pays homage to Theodore Monod at the time of his 95 years. In 1998, the naturalist turns over in Guelb, exactly in the Oued Akerdil where the first assemblages of choppers have been located on the surface, in order to identify sites in stratigraphic position. In 1999 a first protocol is signed with the Mauritanian Institute of Scientific research (I.M.R.S.). The naturalist dies on November 22nd, 2000 after seventy six years devoted to the large Western Sahara.

Fig. 9.2. Yeslem III: flakes and core (arrows) in the carbonated encrusting (photo A. Dambricourt Malassé 2004)

The Franco-Mauritanian team continued the investigations; the Mixed Unit of Research 6569 CNRS – Laboratory of Prehistory, NMNH, engaged new fieldwork in the Northern Adrar under the scientific supervision of Henry de Lumley. Coordination was entrusted to Ousmane Chérif Touré, the intendance taken of load by Somasert and transport by "Point Afrique". In 2002 Anne Dambricourt was invited as speaker for conferences and proposed to enlarge the field work to paleontological surveys. Between 2002 and 2004, several localities were inventoried in stratigraphy with pebble tools in the Guelb er Richat (oued Akerdil) and handaxes and cleavers at El Beyyed in borderland of the Makteir (Yeslem I, II, III, and IV).

THE SITE OF YESLEM III

The discovery and the first collections: 2002 & 2004

The site was discovered by Dambricourt 2 kilometers north from Yeslem II (Figure 1) during a short halt on the pist. The first bones were collected several meters from the vehicle and correspond to four fragment of a thin fibula and few small fragmented bones. The day before leaving the base camp, Dambricourt collected new fossils on 4 square meters with 16 small flakes in quartzite. The flakes, quite distinct from Neolithic industry, and the important mineralization of bones, supported a Paleolithic facies. In conformity with the French-Mauritanian agreement, the fossilized bones were stored at the Department of Prehistory, Institute of Human Paleontology, Paris.

In 2004 the fieldwork was partly devoted to a new site on surface located one hundred meters far from Yeslem III, with handaxes, cleavers and flakes. Yeslem III was studied more attentively by Dambricourt, eight kilograms of fossils were collected, including a fifth fragment of the fibula, the majority in the immediate periphery of the small relief, while several artefacts (cores, bifaces and bigger flakes) were collected in the encrusting (Fig. 9.2). The locality visited by Henry de Lumley is the base of an old dune eroded by the wind, reduced to small basins covered by carbonate cement crust. No Neolithic polished axes and grinding stones were collected in the carbonate crust.

The Mauritanian collection of 2002 was stored at the Tautavel European Prehistoric Research Centre (C.E.R.P.) and the fibula restored by Anne Marie Moigne, classified as primate then as Homo genus (Fig. 6.3). The collection of 2004 was stored at the laboratory of Prehistory in Paris for a taxonomic determination jointly with Lucien Jourdan, then at the C.E.R.P. The paleo-biodiversity is typical of a lacustrine environment surrounded by savanna and grassland, with aquatic species such as softshell *turtles*, crocodile, *Hippopo-*

Fig. 9.3. Cast of the restored human fibula on both sides (UMR 7194 CNRS – C.E.R.P.) and the original fifth fragment (photo A. Dambricourt Malassé, 2004, all rights reserved)

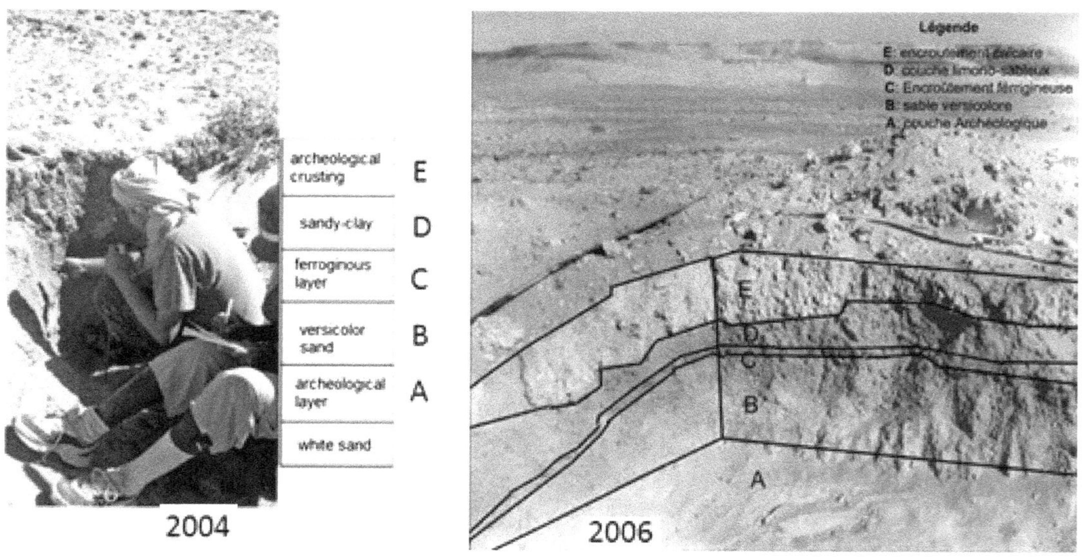

Fig. 9.4. Yeslem 3: first digging (2004) and excavation (2006)

tamus, and Sahelian fauna: *Phacochoerus aethiopicus*, *Alcelaphus buselaphus, Hippotragus*, rhinoceros, Proboscidians and large Bovids. By comparison, the site of Zouérate corresponds to 1 Hippo, 1 *Phacochoerus*, 1 crocodile, Bovids, Elephant with *Palaeoloxodon iolensis* typical of Middle Pleistocene (Coppens et al., 1967).

The first excavation (2006) and the stratigraphical context

In 2004, a first digging in limit of the carbonate crust revealed two levels of human occupation (Fig. 9.4). In January 2006, a first excavation on four square meters was carried out by a team of the C.E.R.P. under the supervision of Ousmane Chérif Touré at the limit of the relief in the locality named "mound 1" which confirms two archaeological levels: i) the carbonate cement crust with lithic and paleontological materials, and 80 cm below the surface ii) an archaeological layer with more than 317 artefacts including shaping flakes, cores and a few handaxes but without fossil bones.

Theodore Monod identified a succession of layers illustrating the Quaternary deposits in the Majabat al Kubraune, northeast of the Makteir and at Tazamout, few kilometers western of El Beyyed in border of the Makteir

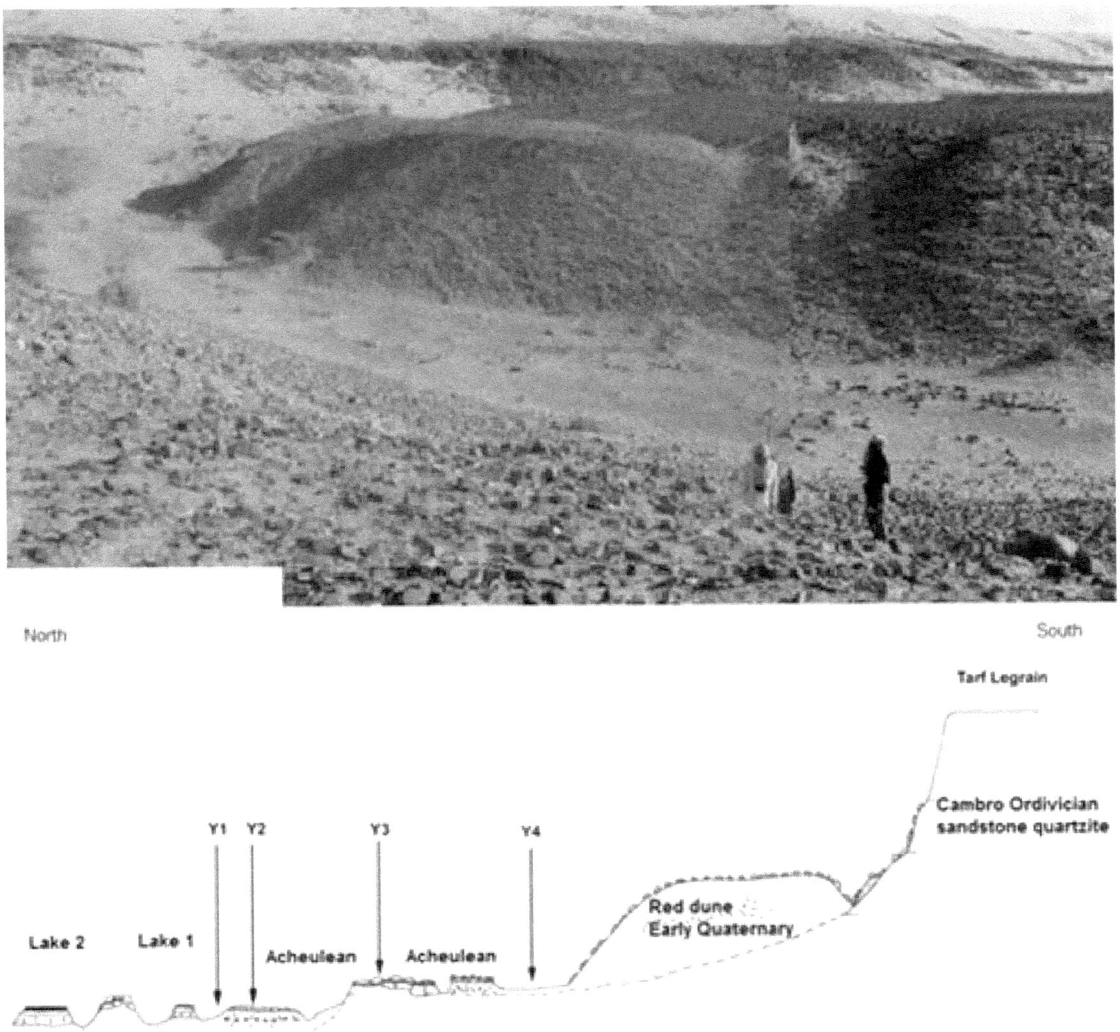

Fig. 9.5. The large red dune bordering the Tarf Legrain and a geological section of the limit between the Cambro-Ordovician plateau and the desert Makteir with the four archeological sites

Tab. 9.1. Lithostratigraphy of Majabat al Kubraune, Tazazmout and El Beyyed and archaeological layers

Majabatal Koubra *Th. Monod*	**Tazazmout** *Th. Monod*	**El Beyyed** *Th. Monod*	**Yeslem I**	**Yeslem II**	**Yeslem III**
GB 2 Lacustrine Sandy Clay		Missing			
CL Milk Chocolate Sand	Milk Chocolate and				
GB 1 Sand Clay Lacustrine Carbonate	Carbonate	Red gravels?	Archaeo layer	Archaeo layer	Archaeo layer
SB White sand	Polychromatic sand with one calcareous bank	Missing	SB White sand	SB White sand	SB White sand

(Monod 1958). He reported cultural facies to distinct sedimentary levels: the Acheulian in white sand formations (SB) and sands, clays and water limestones (lacustrine) (GB 1); the Neolithic in "milk chocolate" sands (CL) and sandy clays (GB2).

At El Beyyed a large fossil dune bording the Cambro-Ordovician plateau namely Tarf Legrain located 1000 meters upstream the four Yeslem sites, contains Acheulian tools (Fig. 9.5). The correlations have been established between the dune and the sites, then with Monod's stratigraphy (Tab. 9.1).

The stratigraphic section of Yeslem III from the inferior level is: A) white sand (SB) on which the flakes, handaxes and cores were collected, B) a vermiculated greensand been, C) a ferruginous level and D) the sandy level with the carbonate cement on several centimetres. The two archaeological layers can be correlated with the GB1 Monod's level including lacustrine carbonate, and the SB level can be correlated with the basal sequence of Yeslem II. The conservation of the bones, even particularly eroded, testifies to a fast covering of the ground shortly after the human occupation. The carbonate cement indicates water circulation per infiltration. The patine of artefacts collected on the surface and of fossil bones around the eroded zone of the mound 1, is limited to the external surface. This indicates that the archaeological ground was recently covered by sand dunes. Indeed, the winds are very powerful in El Beyyed.

CONCLUSIONS: LITHIC ASSEMBLAGES AND FAUNA OF YESLEM III

The preliminary study of the industry in both layers shows two Acheulian assemblages mainly composed of shaping flakes. The excavated surface is too limited to conclude the lack of fauna.

The first analyses support the Pleistocene age with a Sahelian fauna composed of aquatic species (tortoises, crocodiles, Hippo) and small and large terrestrial herbivores. The first ground would have been covered first by the lake, and again occupied after the lowering of its level. The exceptional conservation of the fauna indicates a new and fast elevation of the lake level. The diversity of lacustrine and surface among artefacts, suggests the presence of a durable habitat with culinary activities, changing with seasonal fluctuations of the lake.

References

AUMASSIP, G. & CHAID-SAOUDI, Y. 2004. Préhistoire du Sahara et de ses abords. Au *temps* des *chasseurs*: *Le Paléolithique.* Tome 1, Éd. Maisonneuve et Larose.

BIBERSON, P. 1969. Etat des recherches pléistocènes de l'Adrar mauritanien. Bulletin de l'Association française pour l'étude du Quaternaire, 6-1: 13-34.

COPPENS, Y., GOUZES, R., LE FLOCH, R. & PAQUET, M. 1967. Découverte d'un gisement de vertébrés fossiles avec industrie acheuléenne près de Zouérat en Mauritanie. Actes du IVe congrès panafricain de Préhistoire, Dakar, Sénégal: 467-451.

LUMLEY, DE H., TOURE, C.O., OULD RACHID, M., MOIGNE, A.M., DAMBRICOURT MALASSE, A, et al. 2006. Découverte d'un assemblage lithique sous un encroûtement calcaire à El Beyyed Yeslem II, Mauritanie. Comptes Rendus Académie des Sciences, Palevol, 5, 1-2: 263-271.

MISKOVSKY, J.C., SAOS, J.-L. & BONKEL, A. 1991. Paléohydrologie et paléoclimatologie des bassins endoréiques du sud de la Mauritanie depuis 18 000 ans. Bull. Soc. Geol. France, 162-6: 1141-1150.

MONOD, T. 1958. Majâbat-al-koubra. Contribution à l'étude de l' « Empty Quater » ouest-africain. Mémoires de l'I.F.A.N., Dakar, 52, 407 p.

MONOD, T. 1975. Trois gisements de galets aménagés dans l'*Adrar de Mauritanie. Provence historique,* 99: 87-97.

RICHARD-MOLARD, J. & MAURY, R. 1953. Contribution à la Préhistoire de l'Adrar mauritanien septentrional et du Makteir. Bulletin de l'IFAN, 15, 3: 1229-1241.

SAO, O., GIRESSE, P., LUMLEY, (DE) H., CHAPON, C., FAURE, O. et al. 2008. Les environnements sédimentaires des gisements pré-acheuléens et acheuléens des wadis Akerdil et Bamouéré (Guelb er-Richât, Adrar, Mauritanie), une première approche. L'Anthropologie, 112 (1): 1-14.

TOURE, C.O. 2006. Le Paléolithique inférieur en Mauritanie: étude technologique et typologique des industries lithiques la région de l'Adrar. Ph D dissertation, University of Provence.

TOURE, C.O. 2006. Contribution à l'étude du Paléolithique inférieur de l'Adrar mauritanien: oued Akerdil et El Beyyed. In Descamps, C. & Camara, A. (Ed.), Etudes sur le patrimoine Ouest Africain. Hommage à Guy Thilmans. Senegalia, Sépia.

TOURE, C.O. & DAMBRICOURT MALASSE, A. 2009. Un nouveau site paléontologique et archéologique du Pléistocène, Afrique saharienne occidentale, Mauritanie. Extended Abstract. International Symposium "Africa, Cradle of Humanity: Recent discoveries", Sétif Octobre 26-28.

10.
THE ORSANG MAN: A ROBUST *HOMO SAPIENS* IN CENTRAL INDIA WITH ASIAN *HOMO ERECTUS* FEATURES

Anne DAMBRICOURT MALASSÉ
UMR 7194 CNRS, Department of Prehistory, National Museum of Natural History,
Institut de Paléontologie Humaine, Paris, France
iphadm@mnhn.fr

Rachna RAJ
Department of Geology, The Maharaja Sayajirao University of Baroda, Vadodara, India

S. SHAH
Aster Diagnostic & Therapeutic Radiology Services, Vadodara, India

Abstract: *A Homo sapiens calvarium recovered in a fluvial deposit of the Orsang River give evidence of genetic continuity between late Asian Homo erectus suggesting an Asian "like-Cromagnoid" stadium in the evolutionary process. IRSL dating of the host sediments provided an age ranging from 50 to 30 ka and 4.6 ka 14C date for the cranium. The interesting features observed are a well-developed asteriac process or gracile torus angularis, a well-developed supramastoid crest, a very broad frontal sinus in the glabellar area, the greatest width located in the lower part of the cranium. CT-scan shows no pneumatization of the mastoid process as in Homo erectus, but two temporal bosses. The study establishes that in the absence of new contradictory data, the Orsang man can be seen as the oldest Homo sapiens of the subcontinent*

Keywords: *Orsang,* Homo erectus, Homo sapiens, *Cromagnoid*

INTRODUCTION

Hominin occupation in the Indian subcontinent by the middle Pleistocene was widespread as confirmed by substantial archaeological evidences from south Asia (Mishra et al., 1995; Petraglia, 2001; Paddayya et al., 2002) and origin of Anatomically Modern Human (AMH) in the mainland Asia (Wolpoff et al., 1984) is again in debate (Coppens et al., 2008; Dambricourt Malassé 2009, 2013; Wu et al., 2010; Chamyal et al., 2011; Sankhyan et al., 2012; Demeter et al., 2012). The details of the human calvarium discovered in Gujarat from the Quaternary alluvial sediments of Orsang River, at Ratanpur in the lower Narmada basin, have been reported by Chamyal et al. (2011). Infra-red stimulated luminescence (IRSL) dating was carried out on coarse grained feldspar extracted from the host sediments and those within the skull and the results were considered as the minimum ages. The host sediments at the same level as that of the skull gave an age of 38±5 Kya and 45±7 Kya. The sediments inside the skull yielded an age of 30±7 Kya and 27±4 Kya, suggesting, that the exposed sediment sequence at Ratanpur was deposited during the upper part of late Pleistocene. 14C dating of the skull using AMS method provided an age of 4600±200 Kya. The skull was identified as *Homo sapiens* with features which threw new light on the biological relationships between Asian *Homo erectus* and Early Asian *Homo sapiens*. The Orsang calvarium was compared with 130 skulls from lower to late Pleistocene (collection of the Institute of Human Paleontology, Paris) and its age debated. In continuation with the previous study by Chamyal et al. (2011), Beck et al. (2012) have undertaken micro CT reconstruction (X ray microtomography) and Ion Beam Analysis (IBA) to check the preservation of bone structure and to understand the diagenetic alteration of the bone tissue. This article analyses the computed tomography (CT scan) of the skull which supports the hypothesis of a quick burying of the body in the late Pleistocene sediments and establishes that in the absence of new contradictory data, the Orsang man can be seen as the oldest *Homo sapiens* of the subcontinent.

Diagenetic alteration of bone tissues

Initially the first sample of bone studied for dating was taken on the part damaged by exposure to the air. This exposure has probably accelerated the process of degradation of collagen and biased radiocarbon dates. A second sample was taken from a skull part protected by the sediments in order to check the structural preservation of the bone. A piece from the left orbital cavity was made available for investigations on fine-scale structure. A high resolution micro-CT (X-ray microtomography) reconstruction of the bone sample was obtained at the Natural History Museum of London using a Nikon Metrology HMXST CT System. The fossil shows increased normal porosity with some internal deposits of sedimentary origin compared to the corresponding recent human bone. In an another study, a new experimental

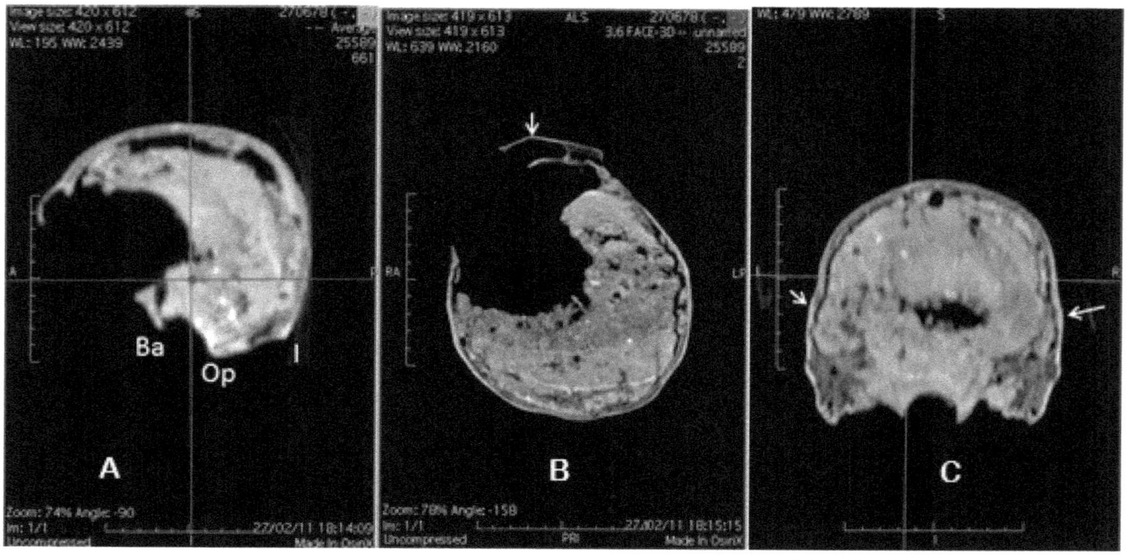

Fig. 10.1. A: CT scan of Orsang calvarium. A: midsagittal section with Ba: basion, Op: opisthion, I: inion. The clivus is visible; B: longitudinal section showing the broad frontal sinus (arrow) and the stratification of the sediments; C: coronal section in the posterior part of the calvarium with the two temporal bosses (white arrows). CT scan, Dr. S. Shah

protocol have been carried out for chemical composition measurements of Orsang and of the skull from the rockshelter Abri Pataud (Late Upper Paleolithic, Dordogne, France) by using the external proton micro-beam of AGLAE (The Centre for Research and Restoration of the Museums of France, C2RMF, Paris) applied simultaneously with PIXE (particle induced x-ray emission) and RBS (Rutherford backscattering) in order to assess the collagen content. By using 3 MeV protons, the PIXE spectra give access to the determination of element Na to U, and backscattering spectrometry is well adapted for the quantification of carbon, nitrogen and oxygen. To validate the procedure, nuclear reactions have been undertaken at the JANNUS laboratory (CEA-Saclay, France) with a 2 MeV Van de Graff accelerator. Carbon, nitrogen and oxygen concentrations have been measured. The results showed a rate of preservation sufficient for a new 14C dating of the skull with this sample (Beck et al. 2012).

The skull, morphology

The morphological characters were studied at the M.S. University of Baroda, while most part of comparative anatomy was done at the Department of Prehistory, National Museum of Natural History, with the anthropological collection of the Institute of Human Paleontology (IPH), Paris. The CT scan was realized in 2011 by Dr. S. Shah, Aster Diagnostic & Therapeutic Radiology Services, Vadodara. The analyses in Paris allow comparative anatomy, and measurements were verified directly on the scan.

The skull was partially emptied of its fill in the laboratory. A first observation of the sediment shows a stratification attesting to a quiet hydraulic regime and its horizontal orientation indicates that the body was on the back with the head inclined on the right side (Fig. 10.1A, B, C). The preservation of fragile parts i.e. the root of the styloid apophysis visible on the right side, the clivus in connexion with the petrous pyramid, the occipital condyles, the auditory canal and the temporo-mandibular joints; indicate that no displacement of the skull occurred since its quick burial.

The two mastoid processes are quite well developed and support the calvarium when it is placed on a horizontal surface, this argues in favour of a male individual. The degree of synostosis of the cranial sutures is high, the lambdoid and sagittal sutures attesting a mature adult stage. Besides these, other features worth noting are the remarkable anatomical characteristics. A very broad frontal sinus of 16 mm is visible between the inner and outer compact tables in the glabellar area without supra-glabellar depression; the pneumatization of the frontal sinus extended laterally on 174 mm, the external orbital apophysis not concerned is gracile (Fig. 10.1B and Fig. 10.2A, B, C). On the superior part of the vault there are a post-bregmatic depression and an important ossification of the coronal synostosis visible only with the CT scan indicating a coronal craniostenosis. The trace of a metopic suture is visible. The nuchal plane and the exocranial surface of the cerebellar fossa is short, and above them the supra-iniac zone is flat and vertical; the profile presents an important planoccipitaly; there is no *torus occipitalis*, nevertheless the external occipital protuberance (inion) and the superior nuchal line are developed. Muscular aponeuroses are well distinct on the left side of the cranium, especially those of the temporal muscle elongated vertically and rough, and those of the posterior belly of the digastric muscle. The right side does not show these features. This asymmetry could be compared to the one seen in inferior and superior views.

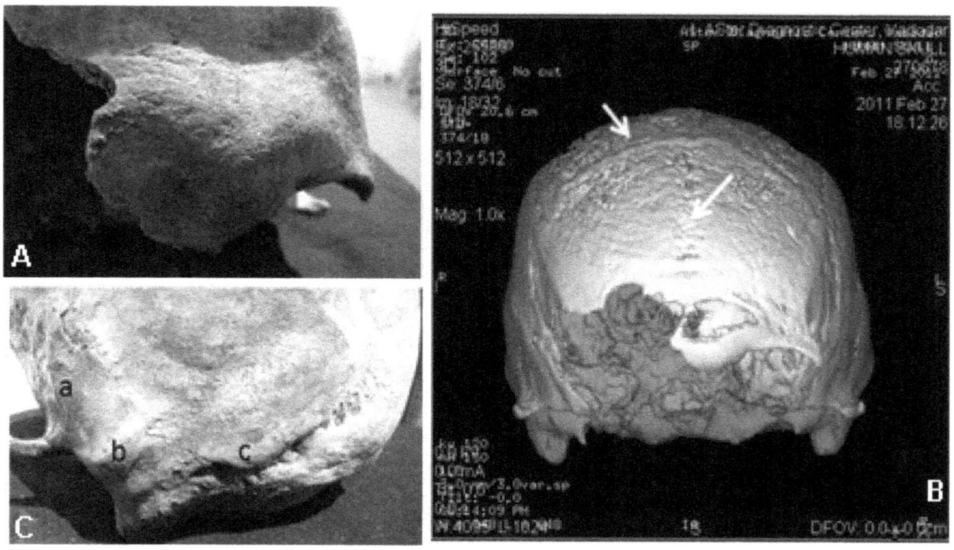

Fig. 10.2 A: the sinus frontal in the glabellar area; B: CT scan showing the strong coronal synostosis and the trace of a metopic suture (arrows); C: the left parietal and temporal bone with a: temporal bosse, b: supramastoidian crest, c: torus angularis. (CT scan, Dr. S. Shah)

A very well-developed asteriac process forms a thickening of the outer compact bone in the pars asterica of the parietal squama (Fig. 10.2B). Two vascular orifices delimit the relief which tends to overhang the occipital bone and can be defined as a gracile *torus angularis*. In *Homo erectus*, this protuberance is more developed in continuity with a marked superior temporal line which is not visible on the Orsang skull. In the approximate Frankfurt plane, its position is at the level of the zygomatic process and not above as in *Homo erectus narmadensis*, because of the downward and forward growth of the cerebellar fossa, which is very specific of the evolutionary grade of Anatomically Modern Human (AMH). The maximal width is in the lower part of the skull as in *Homo erectus* but two characters must be distinguished: i) a bulging of the temporal squama never seen in the normal growth of AMH (Fig. 1C) and ii) a well-developed supramastoidian temporal crest protuberance on a pneumatised petrous bone (Fig. 10.2C). In *posterior* view, the mastoid processes are not in a vertical alignment with the parietal eminences, a *Homo sapiens* conformation; they adopt an inclination resulting from this enlargement caused by both exocranial (crest) and endocranial (bulging) developments of the temporal squama. The glenoïd fossa of the temporal is better protected than the right one, which was exposed to air; they are asymmetrical, the left being appreciably wider and deeper. They are large-sized.

Biometry

Since several parts are missing to apply Martin and Seller measures, remarkable distances were taken with homologous anatomical points to be reproducible on other skull series. One measurement based on the CT scan is changing with the basi-bregmatic height; 126 mm and not 132 mm that modifies the cranial capacity estimation and the height/length ratio (Tab. 10.1). The maximum length (glabella-opisthocranion) is 170 mm. According to Martin and Seller, when symmetrical lateral expansion is excessive, euryon does not be located on the entire adjacent region above the external auditory meatus. Nevertheless with Orsang skull, the case is particular because the planoccipitaly can be correlated to a lateral expansion of the neurocranial cavity as seen below. Both cases have been taken into account, euryon on parietal and temporal squama. The greatest breadth is bitemporal, between the crests (146 mm), but also between the temporal bosses (145 mm), the horizontal cephalic index classifies the skull in the category of hyperbrachycranium. The ratio between the maximum height (basion-bregma, 126 mm) and the bitemporal squama (145 mm) accentuates the broad aspect of the skull (86, 03). Finally the height / length ratio shows a sharply longer skull (170 mm) than high (126 mm). The cranial capacity is estimated based on the exocranial measures according to the Martin's equation: 0.00037 (length x width (euryon-euryon) x basion-bregma height) + 328 ± 73. If the distance between the two euryons is located on the parietal bones, the capacity is 1342 ml. But the CT scan shows clearly that the distance between the two temporal bosses are greater than between the parietals. The estimation of the cranial capacity is 1485 ml. The foramen *magnum* is asymmetrical and wide (index = 85.7).

It is clear that the forward and downward development of the cerebellum is typically that of AMH. Whatever the cranial capacity, the posterior cranial fossa morphogenesis can never be confused with that of the oldest species defining the Homo grade (*Homo georgicus*, *Homo habilis*, *Homo erectus*) visible with their late evolved relatives such as *Homo neanderthalensis* in Western Europe and *Homo floresiensis* in Indonesia (Dambricourt Malassé 2011).

The study of the skull base growth takes into consideration the flexion of the spheno-occipital

Tab. 10.1. Orsang skull measurements in millimeters (number in the left column correspond to Martin and Saller 1956)

Orsang Cranium	Notes
1 maximum length glabella-opisthocranion	170 inion and opisthocranion coinciding
8 maximum breadth	146 location: supramastoid crest
bi-temporal breadth	145 between the two temporal bosses
bi-parietal breadth	128 between the two parietal bosses
17 basi-bregmatic height	126 CT scan measure
20 auriculo-bregmatic height	105
horizontal cranial index: temporal 100 x l/L parietal 100 x l/L	 85.88 hyperbrachycranium 75.29 mesocranium
index of height/length: auriculo-bregmatic height/max. cranial length basi-bregmatic height/max. cranial length ***index of height/breadth:*** auriculo-bregmatic height/max. cranial breadth basi-bregmatic height/max. cranial breadth	 61.76 orthocranium (low vault) 74.11 low skull 71.91 wide vault 86.30 wide vault
cranial capacity: maximum breadth bi-parietal breadth	 1485 ml ±73
frontal sagittal curvature: 26 nasion-bregma arc 29 nasion-bregma chord index	 116 104 89.65
divergence of the frontal: 9 minimum frontal breadth 10 maximum frontal breadth index of divergence	 110 by symmetry 112,5 97.77
nasion-inion arc	297
Parietal: length of sagittal arc/temporal arc: 27 Sagittal arc of parietal arc of parietal with temporal edge curvature index	 122 105 116.19 (high index probably caused by the planoccipitaly)
Upper temporal line to temporal edge	37 (between the highest point of the superior temporal curve and the highest point of the superior temporal line)
stephanion-bregma chord	superior stephanion: 60 mm inferior stephanion: 67 mm
upper stephanion arc	112
upper stephanion chord	103
nasion-frontomalar suture	47 (in horizontal plane)
45 bizygomatic breadth	128 anterior zygomatic tubercle
bi-asterion breadth	109
11 bi-porion breadth	132
bi-pterion breadth	100
bi-mastoid breadth	106
porion-asterion length	left: 45; right: 44
Length between the summit of the mastoid process and the centre of the carotid canal	28
Foramen magnum: length width index	 35 30 85.71 wide foramen magnum

synchondrosis, a very important factor in cranio-facial orthopaedic such as orthodontics (Maestripieri et al., 2002), in cranio-sacral therapy and osteopathy (Cook 2005, Downey et al., 2010). When the flexion is too active, it brings the clivus and the *tentorium cerebelli* forward and downward and hence limits the backward growth of the occipital squama. The maximal length tends to reduce so that the opisthocranion can merge to the inion resulting in planoccipitaly. The mid-sagittal section of the CT scan confirms a straightening of the sphenobasilar slope (clivus) and a short posterior fossa (Fig. 10.1A) of the sphenobasilar slope (clivus); CT scan has revealed the distance between the two temporal endocranial tables just above the petrous pyramid, is enlarged and a possible consequence of the impossibility for the brain to growth posteriorly. The hyperbrachycephaly conformation can be seen as the result of a neural growth and specific suture activities with a great flexion of the sphenooccipital synchondrosis, a shortening of the cerebellar fossa, early ossification of the coronal and quick synostosis of the lambdoïd sutures. This skull is not a strong case of craniostenoses, but the convergence with such abnormalities reveals congenital factors in favour of endemism.

The Asian regionalisation of the *torus angularis* (Dambricourt Malassé 2008, 2011, Chamyal et al., 2011) suggests that some populations were isolated from genetic flows coming from Africa, Near East and Europe, which mean that endemism played an important role for the differentiation of geographical diversities during at least 1.8 million years as shown by late *Homo florensiensis* in Indonesia (c.a. 18 ka). Only one African Pleistocene skull developed a *torus angularis*, Kabwe (*Homo rhodensiensis*, Middle Pleistocene) which supports evidences of a geographical arch of Asian-Arabo-African gene flows at that time.

For living populations the *torus angularis* (or process asteriacus in modern man), is still visible in certain Mongoloïd and Australoïd populations with high incidence of other bone thickenings like retromastoid process and supramastoid tubercle.

DISCUSSION AND CONCLUSION

The time gap between late evolved *Homo erectus* and Anatomically Modern Human, or *Homo sapiens Linnaeus,* has been filled recently with the discovery of the Chongzuo mandible (Zhiren cave) in South China dated of 110 ka (Liu et al., 2010). Contrary to critics questioning the possibility of a parallel emergence of modern anatomy in Asia, the release of the *trigonum mentale* under the incisor dental arch that shapes the first appearance of the *mentum osseum* is never visible on *H. habilis*, *H. erectus*, *H. georgicus, H. antecessor*, *H. heidelbergensis,* and on the late *Homo neanderthalensis* and *Homo floresiensis.* The oldest *mentum osseum* known before the Chinese discovery is Omo Kibish 1 (Ethiopia 200 ka) then Qafzeh (Near East 95 ka, for instance Qafzeh 7 original mandible at IPH).

The Pleistocene Anatomically Modern Human developed cranial capacity around 1400-1600 ml, for instance as in Iberomaurusian people greater than in Holocene people, but the brachycephaly is not visible before Mesolithic i.e. 9000 ka BP. Orsang cranial capacity (1485 ml) corresponds to the Paleolithic peoples, while its hyperbrachycranial conformation results of neural growth abnormalities. With its conspicuously developed glabellar pneumatization, its well-marked post-cranial superstructures, the skull from Orsang valley is closer to the robust *Homo sapiens* of the late Pleistocene than to the gracile Neolithic populations. Besides, with its well-developed *torus angularis*, this skull attests to the presence of Asian *Homo erectus* genes in its hereditary material and allows concluding that among ancestors of Orsang valleys populations were late Asian *Homo erectus* living in the Narmada basin, such as the large brained *Homo erectus narmadensis*. The paradigm of replacement by African anatomically modern humans arriving in Asia around 90-70 ka and extinction of autochthonous populations can no more be supported by recent paleontological data, the mandible of Chongzuo in South China dated of 110 ka and the Late Pleistocene skull of Orsang in Central India, with Early Pleistocene Eurasian feature, the *torus angularis*.

References

BECK, L., CUIF, J.P., PICHON, L., VAUBAILLON, S., DAMBRICOURT, MALASSÉ, A. & ABEL, R.L. 2012. Checking collagen preservation in archaeological bone by non-destructive studies (Micro-CT and IBA). Original Research Article. Nuclear Instruments and Methods in Physics Research Section B: Beam Interactions with Materials and Atoms, 273: 203-207.

CHAMYAL, L, DAMBRICOURT, MALASSE, A., MAURYA, D.M. et al. 2011. Discovery of a robust Homo sapiens in India (Orsang, Lower Narmada Basin, Gujrat). Possible continuity with Indian Homo erectus. *Acta Anthropologia Sinica* 2: 167-191.

COOK, A. 2005. The Sphenobasilar Synchnodrosis (SBS) Revisited – The Mechanics of Cranial Motion. Journal of Bodywork and Movement Therapies, 9, 3: 177-188.

COPPENS, Y., TSEVEENDORJ, D., DEMETER, F., TURBAT, T. & GISCARD, P.H. 2008. Discovery of an archaic Homo sapiens skullcap in Northeast Mongolia. *Comptes Rendus Palevol,* 7(1): 51-60.

DAMBRICOURT MALASSÉ, A. 2009. Embryogeny and Human phylogeny. Chapitre 8. In: *Asian Perspectives on Human Evolution*, New Delhi. A.R. Sankhyan (Ed.). New Delhi: Serial Publications, pp. 103-121.

DAMBRICOURT MALASSE, A. 2011. Permanent bipedal equilibrium, embryonic origin, morphogenesis, occlusal-postural balance, impact on psychomotor and behavioral hominid evolution (in French). Accreditation to Direct Research, Compiegne University, pp. 706.

DAMBRICOURT MALASSÉ, A. 2013. The emergence of gracile human anatomy during the Holocene in Asia: convergent acclimation and\or phylogenetic trend? The example of the occlusion. In: B.H. Kotlia (Ed), Holocene: Perspectives, Environmental Dynamics and Impact Events, Nova Sciences Publishers, Inc. New-York.

DEMETER, F., SHACKELFORD, L., BACON, A.M., DURINGER, P. et al. 2012. Anatomically modern human in Southeast Asia (Laos) by 46 ka, *Proc. Natl. Acad. Sci.* U.S.A., 109 (36):14375-14380.

OWNEY, P.A, BARBANO, T., KAPUR-WADHWA, R., SCIOTE, J.J. et al. 2010. Craniosacral therapy: the effects of cranial manipulation on intracranial pressure and cranial bone movement. Journal of Orthopaedic and Sport *Physical Therapy*: 845-853.

LIU, W., JIN, C., ZHANG, Y.Q., CAI, Y.J. et al. 2010. Human remains from Zhirendong, South China, and modern human emergence in East Asia. Proceedings of the National Academy of Sciences, *Proc. Natl. Acad. Sci.* U.S.A. 107 (45): 19201-19206. http://www.pnas.org/content/early/2010/10/15/1014386107.

MISHRA, S., VENKATESAN, T.R., RAJAGURU, S.N., SOMAYAJULU, B.L.K. 1995. Earliest Acheulian industry from peninsular India. *Current Anthropology* 36, 847-851.

PADDAYYA, K., BLACKWELL, B.A.B., JHALDIYAL, R., PETRAGLIA, M. et al. 2002. Recent findings on the Acheulian of the Hunsgi and Baichbal Valleys, Karnataka, with special reference to the Isampur excavation and its dating. *Current Science* 83: 641-647.

PETRAGLIA, M. 2001. The Lower Palaeolithic of India and its behavioural significance. In: Barham, L., Robson-Brown, K. (Eds.), *Human Roots: Africa and Asia in the Middle Pleistocene*. Western Academic and Specialist Press, Bristol, pp. 217-233.

SANKHYAN, A.R., BADAM, G.L., DEWANGAN, L.N. CHAKRABORTY, S. et al. 2012. New Postcranial Hominin Fossils from the Central Narmada Valley, India, *Advances in Anthropology* 2 (3): 125-131.

WOLPOFF, M.H., WU, X.Z., THORNE, A.G. 1984. Modern Homo sapiens Origins: A general theory of hominid evolution involving the fossil evidence from East Asia. In: *The Origins of Modern Humans: A World Survey of the Fossil Evidence*, F.H. Smith and F. Spencer (Eds.). Liss, New York: 411-483.

11.
GEOARCHAEOLOGY OF THE FLUVIAL TERRACES OF MIDDLE TAGUS RIVER, CENTRAL PORTUGAL

Satya DEV
Laboratory for Disaster Science (Archaeology), Department of Culture and Society, Aarhus University, Denmark
satyadevk@gmail.com

Abstract: *Alto Ribatejo is a region of central Portugal that extends along the Middle Portuguese Tejo River. Six river terraces have been recognised for the Tejo River in the Portuguese sector. The aim of the present research is to record the fluvial potential of the river at the lower terraces (T6/alluvial plain) and the middle terrace (T4), as well as to discover the archaeological potential in these sites. The present study brings very interesting results. Five main stratigraphical units were identified in the lower terraces; the second unit, which comprises boulders, cobbles, pebbles and some lithic artefacts, is identified as terrace T6. This terrace may belong to a late phase of the Upper Pleistocene or marine isotope stage 3 (MIS). Three main stratigraphical units have been identified in T4. The T4 is extremely significant for prehistoric human occupation in Portugal because it contains the earliest records of human occupation, which are dated to marine oxygen isotope stage 9-6 (MIS). The sedimentological analysis evidenced that T4 was deposited during temperate climatic conditions. On the basis of spatial analysis and geological context of the archaeological artefacts, they are divided into three categories. The first category groups the Lower Palaeolithic artefacts from the T4 terrace. The second category of artefacts recorded from the T6 at Outeiro du Pedro, thought to belong to the Upper Pleistocene. The third category is constituted of artefacts which were recorded as surface finds at various localities. Typologically and morphologically, they are very similar to artefacts from the Middle/Lower Palaeolithic and the Bronze Age.*

Keywords: *Fluvial terraces, Geoarchaeology, Central Tagus River, Lower Palaeolithic human occupation, Pleistocene, Central Portugal*

INTRODUCTION

The question of transformations (e.g. cultural, environmental, and geomorphological) and our observations for the evolution of past landscapes are the key questions in the geoarchaeological research area (Lourandos 1996) because they play an important role in understanding the differences in the environment and the formation of different types of landscapes which affected human behaviour throughout the Quaternary period. Prehistoric Hunter-gatherers exploited a landscape that was active and persistently transforming and where the activities of prehistoric human behaviours is preserved and recorded; therefore it is essential to understand the geological process that transformed the landscape for proper understanding of archaeological record (Waters and Khunen 1996). The present study focuses on the landscape changes in the form of Tagus River's fluvial terraces associated with prehistoric human occupation in central Portuguese Tagus River basin in Mação and Villa Nova Da Barquinha area in the Alto Ribatejo province (for geographical division of Tagus river see Cunha et al. 2005, 2008), especially the selected out crops of middle terraces and lower terraces of Tagus river (Fig. 11.1 & Fig. 11.2).

The Tagus River is longest river (1100 km) in Iberia Peninsula, with a catchment size of approximately 86,000 sq. km (Cunha et al. 2005, Cunha et al. 2008). It originates from Fuente de García in the Sierra de Albarracín in the Teruel, Aragón, Spain (40°22'N, 1°44"W). It flows westward across Spain and Portugal, drains a over Paleozoic basement in the Spanish area and over Tertiary sediments in the Portuguese sector (Cunha et al. 2005) and it debouches into the Atlantic Ocean near Lisbon (Cunha et al. 2005; Cunha et al. 2008; Martins et al. 2010; Cunha et al. 2012). Around 70% of the catchment area of Tagus River lies in Spain, whereas 30% area is located in Portugal (Cunha et al. 2008). Tagus River is characterized by deep V shaped valley, cut approximately 200 meter into the planiation surface (Cunha et al. 2012).

The sedimentary archives preserved in the fluvial terraces and strongly incised valley of lower Tagus river provides records of landscapes changes, variations in environment and evolution in the human behaviors during the Quaternary period (Breuil & Zbyszewski 1942, 1945, 1946; Mozzi et al. 2000; Rosina 2001; Rosina 2002; Rosina et al. 2004; Cunha et al. 2005; Santisteban and Schulte 2007; Grimaldi and Cura 2006; Rosina and Cura 2010; Martins et al. 2010; Osterbeek et al. 2010; Rosina et al. 2006; Cunha et al. 2012). Six river terraces are identified and mapped in the Portuguese catchment of the Tagus River (Martins et al. 2009; Martins et al. 2010; Cunha et al. 2012). These terraces are classified as (from oldest to youngest) Q1 (T1), Q2a (T2), Q2b (T3) (high terraces), Q3 (T4) (middle terrace, Q4 (T5) and Q4b (T6) (lower terraces) (Corral Fernandes 1998 a, b; Mozzi et al. 1999; Rosina 2002, 2004; Cunha et al. 2005; Cunha et al. 2008; Martins et al. 2010; Cunha et al. 2012). Many Middle Pleistocene archaeological site (Lower and Middle Palaeolithic sites) have discovered and excavated from these fluvial sediments and as well as in the

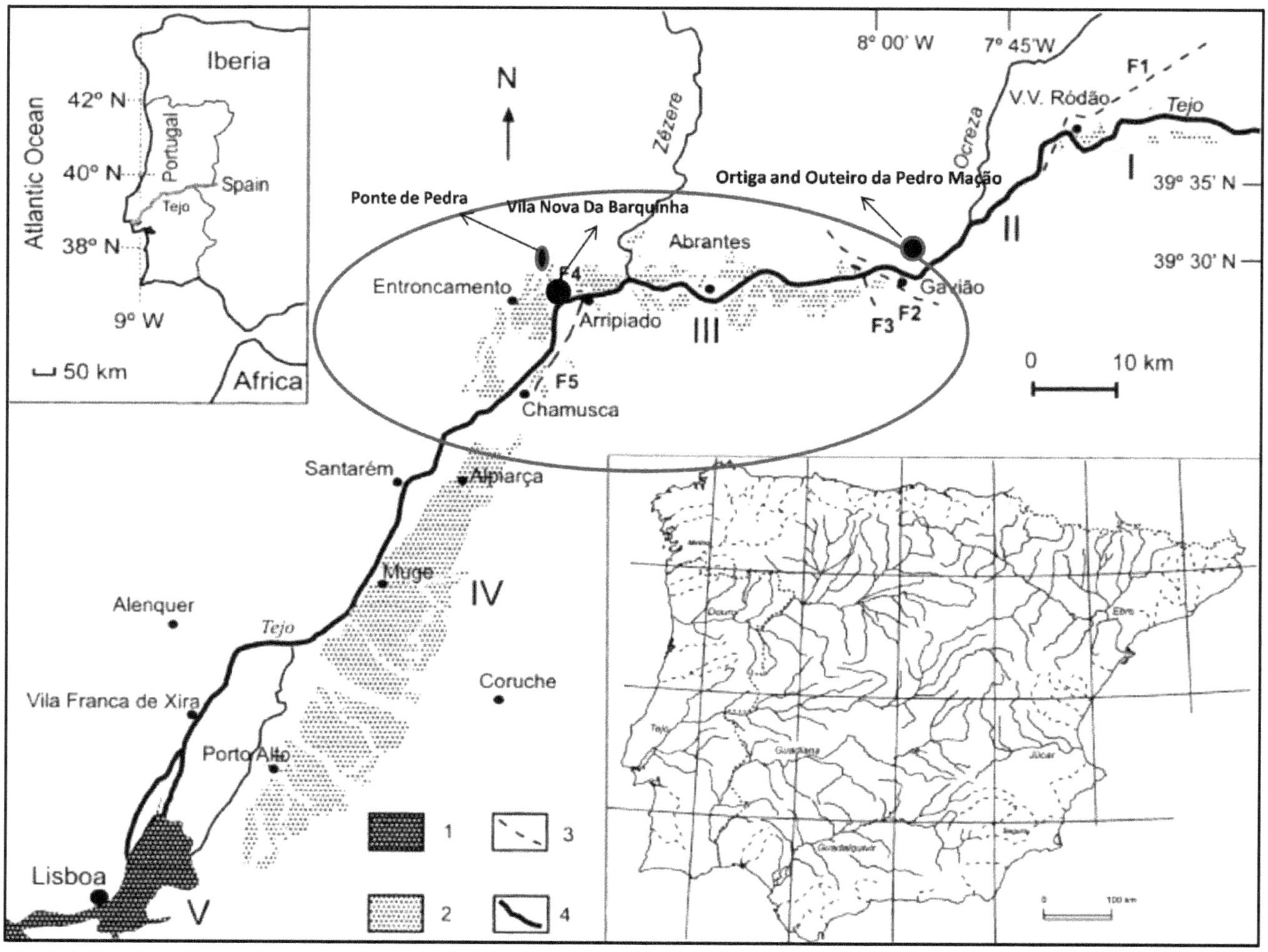

Fig. 11.1. Tagus river's geographical division in Portuguese section (Lower Tejo Basin): reach I – from the Spanish border to Ponsul fault (Ródão town); reach II – bound by the Ponsul and Gavião faults (from Ródão to Gavião); reach III – between the Gavião and the Vila Nova da Barquinha faults (from Gavião to Arripiado); reach IV – downstream of the Vila Nova da Barquinha fault, until the estuary; reach V – comprising the estuary. The study areas are part of reach III (Vila Nova da Barquinha and Mação) and reach IV (Chamusca area). The red circle is showing the study area in Portuguese central Tagus River (Modified Cunha et al. 2005)

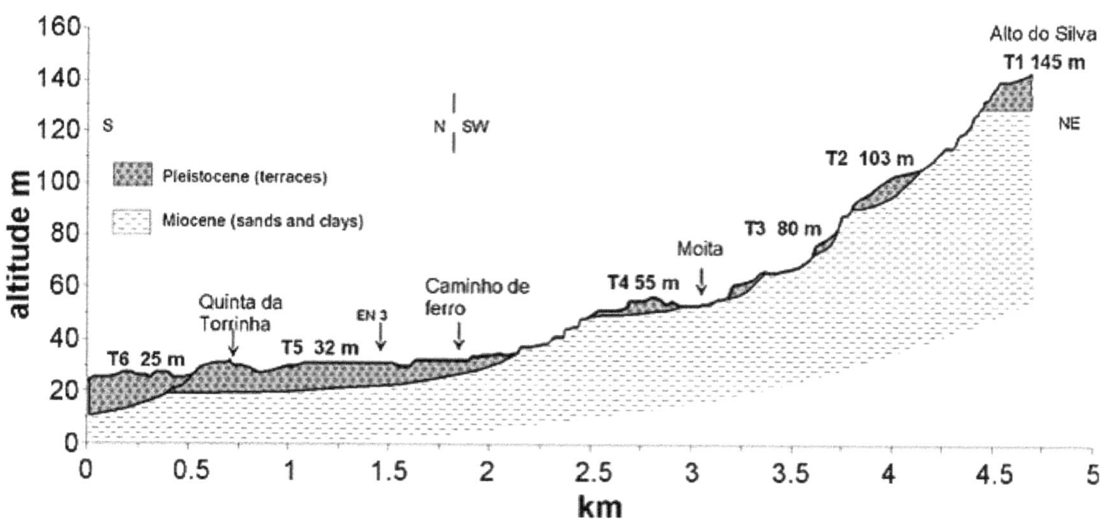

Fig. 11.2. Staircase of six fluvial terraces in the Vila Nova da Barquinha area (After Martins et al. 2010)

colluviums deposits (see, Rapaso et al. 1985; Rapaso 1986, 1987, 1995a, b, 2000, 2002; Mozzi et al. 2000; Grimaldi and Rosina 2001; Raposo 2002; Oosterbeek et al. 2004; Cura and Grimaldi 2006; Cura and Grimaldi 2009; Rosina and Cura 2010; Cura et al. 2010; Rosina et al. 2004; Rosina et al. 2010). Cunha et al. (2008, 2012)

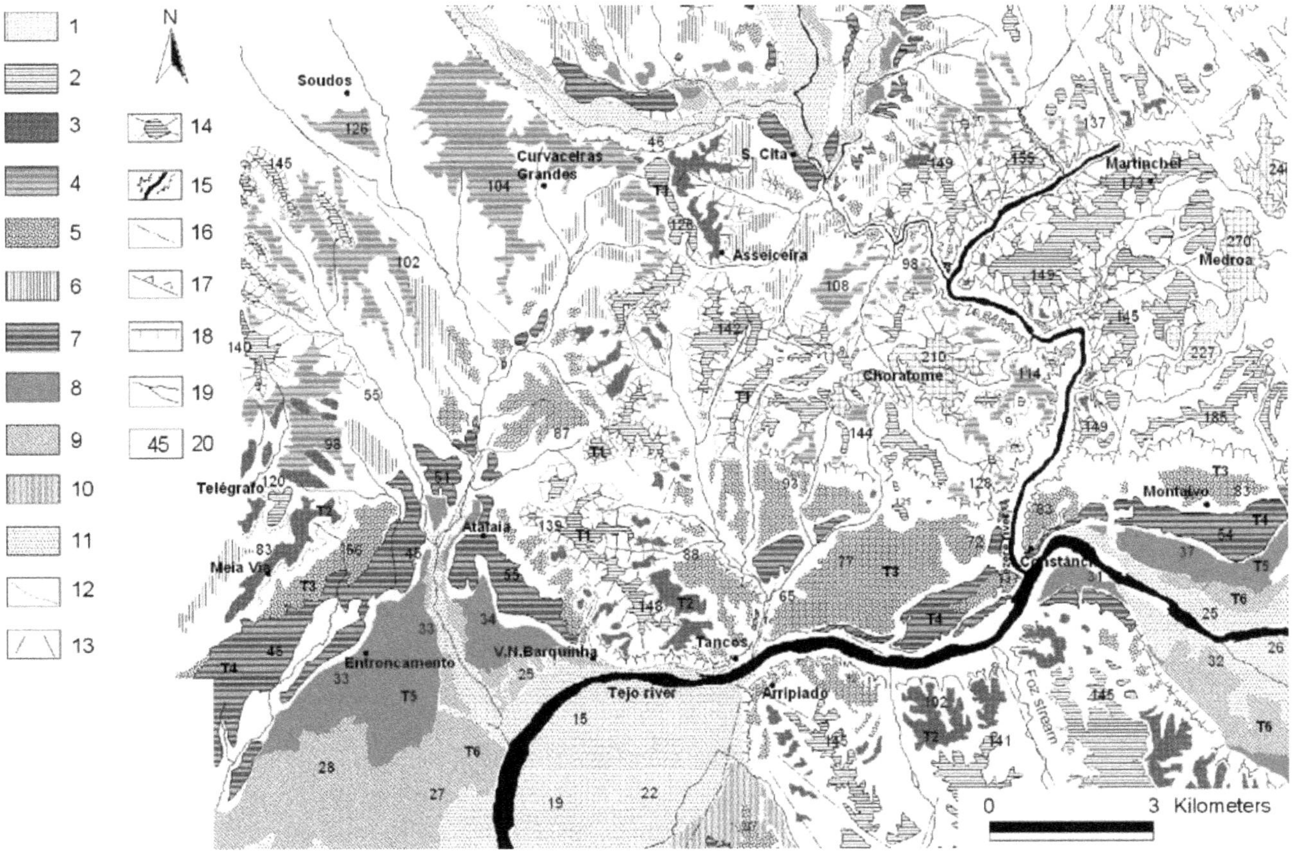

Fig. 11.3. Geomorphological map of the study area. Legend: 1: culminant surface of the sedimentary basin; 2: N1 (erosion surface) and coeval T1 terrace; 3: T2 terrace; 4: N2 (erosion surface inset in the N1 and T1); 5: T3 terrace; 6: glacis linked to the T3 terrace; 7: T4 terrace; 8: T5 terrace; 9: T6 terrace; 10: aeolian sands; 11: modern alluvial plain; 12: base slope of mesa relief; 13: steep slope; 14: residual relief; 15: entrenched valley; 16: probable fault; 17: scarp of asymmetric valley; 18: fault scarp; 19: water course; 20: altitude (meters) (After Martins et al. *2010)*

and Martins et al. (2009, 2010) provides the chronological sequences and elevation above the sea level for the Tagus river terraces in Ródão-Monte do Pinhal area and Vila Nova Da Barquinha in a following manner: no dates for T1 and T2, T1 +117 m.a.r.b. (183 m.a.s.l.); T2 +89 m.a.r.b. (155 m.a.s.l.); T3 +67 m.a.r.b. (133 m.a.s.l.) or 340 Ka; T4 +50 m.a.r.b. (116 m.a.s.l.) or 150 to 340 ka; T5 +24 m.a.r.b. (90 m.a.s.l.) or 75 to 136 ka; T6 +16 m.a.r.b. (82 m.a.s.l., 62 to 30 ka. Martins et al. (2010) provides synthesis of geomorphological and sedimentary details for Villa Nova da Barquina area.

It is because of many grounds, specially the raw material of quartzite pebble, and the attribution lithic industry is not clearly associated to particular Palaeolithic culture or chronological cultural framework or to the river terraces of Tagus River in Alto Ribatejo region. Therefore, the present research is aimed specifically at establishing the stratigraphical, sedimentological characterisation and palaeoenvironmental interpretations of the selected outcroups of T4 and T6 terraces of Tagus River in the study area. The main aims of the research are:

i. To characterize the fluvial terraces of archaeological importance. The objective s are to describe their stratigraphical and the geomorphological sequence (Fig. 11.3).

ii. To study the elements of the stratigraphy and associated human occupation in the lower terraces and the alluvial plain of the Tejo River in the study area at Outeiro du Pedro.

GEOLOGICAL SETTINGS

The geographical region of Alto Ribatejo is spread along the middle Tagus river valley in Central Portugal. The sedimentary records of the Alto Ribatejo is characterized by the "Estremenho" Limestone Massif, essentially Mesozoic, to the North West, and by the Schist-metamorphic Complex, Pre-Cambrian and Palaeozoic, to the East, while to the South it's contiguous with the Tejo-Sado Basin (Carvalho 1968; Cunha 1992; Cunha et al. 1993; Barbosa1995; Martins 1999; Rosina et al. 2004).

Lithologically, the detritic Tagus and Sado Tertiary basin are characterized by clay, silts, sands, and pebbles; the main constitutes of "Estremenho" Massif are limestones and marls (with weak occurrence of flint); and the main constituents of Ancient Massif are schist, greywacke, quartzite and granite (Rosina et al. 2010). The Quaternary sedimentlogical records in this region are mostly marked by the recent fluvial sediments, by wide Pleistocene fluvial terraces (covered by colluviums), and by karstic caves fillings (Rosina et al. 2004; Martins et al. 2010; Rosina and Cura 2010). The Zězere and the Nabão rivers is the main tributary of the Tagus River, which across the region from north to south (Martins et al. 2010). It is noteworthy to include the Ponte da Pedra stream that directly mingles into the Tagus River, Bezelga and the Pias streams in the north, and the Pias streams in the north (Oosterbeek et al. 2004). The Tagus River and its tributaries provides geological context to the Lower Palaeolithic sites of the Fonte da Moita, Ponte da Pedra and Santa Cita. Gonçalves (1969, 1970) described the geology of the zone. The studies of Breuil and Zbyszewski (1942, 1945) and Zbyszewski (1946) distinguish four terrace levels:

Q4 Terrace having deposits of 8 to 15 m, Q3 deposits of 25 to 40 m, Q2 of the 50 to 60 m and Q1 75-95 m. The stratigraphy of each terrace consists of a basal gravel unit, followed by sand and mud, covered by a colluvium. The formation of these four terrace levels was regarded as being due to alternating climate and eustatic sea-level fluctuations related to Pleistocene glacial/ interglacial cycles.

This model correlates the terrace sequence directly to the four-stage glacial-interglacial model of Penck and Bruckner (1909). Evidence from the ocean research has revealed the far greater complexity of Quaternary climate variation (e.g. Shackleton et al. 1990) and Cunha et al. 2005 discussed the consequences of these climate changes on terrace formation and deposition in the Lower Tagus Basin. In some areas, the correlation of the terraces may be badly addressed because they could not identify the vertical displacement affecting the terraces. Zbyszewski and his followers also could not identify the first incision (the older terrace and surface) that partially cut the Pliocene culminant surface of the basin.

Zbyszewski (1946) explained that the Pleistocene fluvial sediments of the Vila Nova da Barquinha region were deposited by the Tagus, Zêzere and Nabão rivers. The Tejo middle terrace level (Q3, T4) overlies on Miocene sands, in the following way from (bottom to top): a) conglomerate; b) medium sized sands with gravelly /or micro-conglomeratic layers; medium sized sands; loam layers; c) Colluvial deposit, approximately 1m in thickness. According to the theoretical model of Breuil and Zbyszewski (1945), stage A was formed during a cataglacial, "optimum" is recognised for the B stage and stage C formed in the new climate deterioration. The Q3 (T4) was ascribed to the penultimate Interglacial Mindel-Riss, and now Riss, now the Holsteinian (420-380 Kya) (Martins et al. 2010), and the last terrace level Q4 was deposited during the Inter-Glacial Riss-Wurm, now the Eemian (130-115 Kya) (Martins et al. 2010). Other hypothesis has been produced by Raposo et al. (1985) to the upper level of Middle terrace (Q3) at Alpiarça, a final Middle Pleistocene (Inter-Glacial Riss-Wurm). However, the Breuil and Zbyszewsky's hypothesis cannot be yet discarded. Carvalho (1968) defined the sedimentological units of Lower Tejo basin in the following order Atapulitic complex (Oligocene); Montmorillonitic complex (Miocene) Ilitic-Caolinitic Complex (Pliocene). Barbosa et al. (1989) defined lithostratigraphical units that correspond to an alluvial system with a dominant drainage towards the south-southwest, where the Serra de Almeirim Conglomerates passed downstream to sandy channels (Arenitos de Ulme). The middle terraces and lower terraces (T4, T5, T6) of Tagus River are linked to cyclic glacio-eustatic fluctuations overlie onto a long term regional uplift style (Martins et al. 2010). According to Fernandez (1998) and Rosina (2001, 2004) the main research question is to devise a chrono-stratigraphy of the fluvial terraces of the Tejo River and also the mechanism of formation and the interpretation of fluvial deposits. There are recent studies for the terraces of this region (Fernandez 1998; Mozzi 1998) that show a larger complexity in the terraces orders. The results of these investigations indicates that it is possible to subdivide the lower terraces, particularly the low terrace can be distinguished in two parts (named Q4-1 and Q4-2, Corral Fernandez 1998), as it was made in others areas of Portugal (for example, geological maps, Q4-a and Q4-b). Mossi et al. (1999) mapped charterised six terraces at the Nabão River. Cunha et al. (2005) recorded six river terraces on Radao area (Fig. 11.2).

Applying multidisciplinary research methods, recently, many efforts have been made to determine the fluvial terrace sequences and chronology archaeological material associated with these terraces in the Portuguese catchment area of the river by using geomorphological, sedimentological, paleontological and geochronological (Daveau 1993; Martins 1999; Santonja and Pérez-González 2000-2001; Gutiérrez-Elorza et al. 2002; Silva 2003; Cunha et al. 1993, 2005, 2008; Santisteban and Schulte 2007; Martins et al. 2009a), however the main focus of these research was in the Ribatejo region because of its unique geological and geomophrological landscape (Corral-Fernandez 1998, Mozzi 1998, Martins 1999, Rosina 2002; Rosina 2004, Cunha et al. 2005; Cunha et al. 2008, Rosina et al. 2010; Martins et al. 2010; Cunha et al. 2012). Coarse-grained siliciclastic gravels and sand are the main component of aggradational terraces of the Lower Tejo Basin (Cunha et al. 2008; Martins et al. 2009), unfortunately the organic material is not very well preserved in these sediments which restrict the use of radiocarbon dating method to date the cultural sequence associated with these terraces (Martins et al. 2010, Oosterbeek et al. 2010). In addition, unfortunately, the sedimentary and environmental records are facing dearth of application absolute or relative dating technique, where mostly the methods of morpho-stratigraphy and biostratigraphical for determining the cultural, environmental and terrace sequences, which are correlated with U-series dating, TL dating and palaeomegentic dating methods (Martins et al. 2010). Nonetheless, these above mentioned methods has methodological limitation that posed scrubby chances to date them spatially (Santisteban and Schulte, 2007).

The earliest human occupation is recorded from middle ter races (Q3 or T4) of Tagus river, and dated as as ca. 300 to 150 ka (Martins et al. 2010a). Moreover, cultural transformation between Lower Palaeolithic is recorded within (T4 or Q3) terraces (Martins et al. 2010). The main sites for T4 (Q3) are Vale do Forno and Vale da Atela in near the Alpiarça village (Mozzi et al. 2000), Monte Famaco site in Vila Velha de Rodaǒ (Raposo 1987), Portomaior site (Mendez Quintas et al. 2006), Pousia/Quinta do Cǒnego site (Cunha-Ribeiro 1999), Casal do Azemel site (Cunha-Ribeiro 1995, 2000), Galeria Pesada Cave (Marks et al. 1999, 2002a, 2002b), In order to determine earliest human occupation and its chronology in Alto Ribatejo, many field survey started in 1990's under the supervision of Prof. Osterbeek, and their systematic survey resulted into discovery and excavation at three main Palaeolithic sites (Fonte da Moita 1998, Santa Cita 1999-2000) and the third (Ribeira da Atalaia/Ribeira da Pedra) started in 1999 and is still ongoing (Grimaldi et al. 1999a, b; Lemorini et al. 2001; Grimaldi & Rosina 2000; Lussu et al. 2001; Rosina et al. 2004; Cura and Grimaldi 2009; Martins et al. 2010). The archaeological data recorded from these sites offered cultural sequences, and that is ranges from Clactonian-Abbevilian to Upper Acheulian assemblages (Breuil and Zbyszewski, 1942; Zbyszewski, 1943, 1946, 1958) or Lower Acheulian to Micoquian assemblages (Raposo et al. 1985; Ribeiro 1990; Raposo 1995a, b). The archaeological record from these sites exhibits continuity from late Lower Palaeolithic to earliest middle Palaeolithic (Raposo 2000). The middle Palaeolithic (Mousterian tools) and upper Palaeolithic artifacts of Iberian Palaeolithic sequences are recoded within T5 and T6 terraces (Raposo 1995a, b).

MATERIAL AND METHOD

The research material presented in this paper is derived from stratigraphical, sedimentological and archaeological data collected during systematic field observation in the T4 and T6 terraces in the study area, applying geoarchaeological field approach: a geomorphological study of every site, field description and stratigraphic correlation of the fluvial terraces, and collecting systematic soil sampling for grain size analysis. The field work component of this paper includes exploration using geological maps made by Portugal Geological Services of Portugal, digging of test trenches for evaluating the fluvial potential of lower terraces of Tagus river at Outeiro do Pedro in Mação municipality, these selected outcrops and trenches are described for characterizing the stratigraphic sequences, their vertical and lateral changes in stratigraphy and archaeological material. Informal geoarchaeological field units (GFU) were identified using a combination of lithostratigraphic, pedological or archaeological criteria. The geoarchaeological approach is adopted following methodology describes by Cunha et al. 2008; Martins et al. 2010; Rosina and Cura 2010; Cunha et al. 2012). The geological context of archaeological artifacts is recorded.

GRAIN SIZE ANALYSIS

In order to interpret the energy of deposition and the environment of deposition, grain size analysis of 50 sediments samples (Tab. 11.1) were carried out at the sedimentological laboratory in the University of Coimbra. The sedimentological samples have been collected from the Sondage at alluvial plain / T6 and T4 terraces of the river.

We distinguished the coarse elements (blocks, stones, gravels) of size higher than 2 mm of the fine elements (sands, silts, clays) lower than 2 mm. We used granulometry for separating the sediments less than 63 mm. Then, the fine elements are divided in three particle-size ranges according to following diameters: a) A sandy fraction whose diameter lies between 2 mm and 0,63 mm; b) A silty fraction whose diameter lies between 0,063 mm and 0,002 mm; c) A clay fraction whose

Tab. 11.1. Identification of the composition of the fraction <2µm. Legends: K-kaolinite; I-ilite; V-vermilite; S: smectite; Goet: gorthitite; – vestigial; abundant clay minerals represented by a capital letters

Sample Code	Site	Stratig.	Facies	Mean (um)	Mean Phi	Mean	Median (um)	<2 um Clay Minerals	Other
BARQ1	VNR Railway Stn	T4 t	Fm	56	4.16	coarse silt	53	K i (s)	Goet.
SOND 5A	Outeiro Pedro	Al. plain/T6	Sm	84	3.58	Fine dans	94	V i K	
SOND 5B	Outeiro Pedro	Al. plain/T6	Sm	233	2.10	Fine sand	675	I k v	
OP2A	Outeiro Pedro	Al. plain/T6	Sm	79	3.65	Very fine sand	108	S i k	
OP2D	Outeiro Pedro	Al. plain/T6	Sm	248	2.01	Very fine sand	332	S k I	
OP2H	Outeiro Pedro	Al. plain/T6	Sm	533	0.91	Coarse sand	691	S k i	
RPP1	A, R, Atalaia	T4	Sm	725	0.46	Coarse sand	800	V k I	
RPP2	A, R, Atalaia	T4	Sm	417	1.26	Medium coarse sand	500	V i k	
RPP3	A, R, Atalaia	T4	Sm	272	1.88	Medium coarse sand	243	K i s	Goet.
RPP8	A, R, Atalaia	T4	Sm	1109	-0.15	Very coarse sand	1580	I k s	Goet.
RPP10	A, R, Atalaia	T4	Sm	191	2.39	Fine sand	231	I k s	Goet.

	Median	Mean	Dispersion	Skewness	Kurtosis
Metric	$Md = P_{50}$	$M = \frac{P_{75} + P_{25}}{2}$	$QDa = \frac{P_{75} - P_{25}}{2}$	$Ska = \frac{P_{75} + P_{25} - 2Md}{2}$	$Kqa = \frac{P_{75} - P_{25}}{2(P_{90} - P_{10})}$
			$S_o = (P_{75}/P_{25})^{\frac{1}{2}}$	$Sk = \frac{P_{75} \cdot P_{25}}{Md^2}$	
Phi	$Md = \phi_{50}$	$M\phi = \frac{\phi_{16} + \phi_{84}}{2}$	$\sigma\phi = \frac{\phi_{84} - \phi_{16}}{2}$	$\alpha\phi = \frac{M\phi - Md\phi}{\sigma\phi}$	$\beta\phi = \frac{\frac{1}{2}(\phi_{95} - \phi_5) - \sigma\phi}{\sigma\phi}$
		$Mz = \frac{\phi_{16} + \phi_{50} + \phi_{84}}{3}$	$\sigma_1 = \frac{\phi_{84} - \phi_{16}}{4} + \frac{\phi_{95} - \phi_5}{6.6}$	$SK_1 = \frac{\phi_{16} + \phi_{84} - 2\phi_{50}}{2(\phi_{84} - \phi_{16})} + \frac{\phi_5 + \phi_{95} - 2\phi_{50}}{2(\phi_{95} - \phi_5)}$	$K_G = \frac{\phi_{95} - \phi_5}{2.44(\phi_{75} - \phi_{25})}$

(b) Descriptive terms applied to parameter values

Sorting (σ_1)		Skewness (SK_1)		Kurtosis (K_G)	
Very well sorted	<0.35	Very positively skewed	+0.3 to +1.0	Very platykurtic	<0.67
Well sorted	0.35–0.50	Positively skewed	+0.1 to +0.3	Platykurtic	0.67–0.90
Moderately well sorted	0.50–0.70	Symmetrical	+0.1 to −0.1	Mesokurtic	0.90–1.11
Moderately sorted	0.70–1.00	Negatively skewed	−0.1 to −0.3	Leptokurtic	1.11–1.50
Poorly sorted	1.00–2.00	Very negatively skewed	−0.3 to −1.0	Very leptokurtic	1.50–3.00
Very poorly sorted	2.00–4.00			Extremely leptokurtic	>3.00
Extremely poorly sorted	>4.00				

Fig. 11.4. Statistical formulae used in calculating the grain size parameters; a. Statistical measures in the metric and phi systems; b. Descriptive terms applied to parameters values, after Mcmanus, 1998

diameter is lower than 0,002 mm. The diffraction of X-rays one of the methods is used for the determination of argillaceous minerals (for detailed grain size analysis, see McManus 1998).

STRATIGRAPHIC CHARACTERS OF THE LOWER TAGUS TERRACES

This section describes the stratigraphical and lithostratigraphica description of fluvial terraces at the municipalities of Mação and Villa Nova da Barquinha district of the Alto Ribatejo province in the central Portugal. The description of each area is as follows:

Mação area

This area is comprised a groups of villages (Mação, Ortiga and Outeiro du Pedro) and situated in the central Portugal in a pine tree forest area. The Mação area is located in the northern edge of the Lower Tagus Sedimentary Basin, close to the southwest limit of the Portuguese Central Range. The relief of this area is comprised of tertiary planation surfaces, quartzite ridges (*inselberge*), push up compartments, tectonic depressions, and forms related with the incision of drainage. Throughout the region there are several pre-historic vestiges, including Antas like the one in Foz do Rio, as well as Palaeolithic period vestiges next to the Ribeira das Boas Eiras, or Rupestrian paintings nearby the Ribeira de Ocreza.

Sondages (test trenches) at Outeiro du Pedro

In order to see the potential of the fluvial terraces (lower terraces) of Tagus River in Mação region, five test trenches have been dug in the village of Outeiro da Pedro in Macao municipality. The test trenches have been dug in the altitude of 39°28'25.6"N and the longitude is 8°03'29.5" and the altitude is between 40 to 55 m.

Test trench 1

The first first test tench was dug on the alluvial plain of the Tagus river, around 50 m far from the actual river course. In order to measure the thicknees of each unit and layers, we took a refrence level and the reference level is 2.97 meters. This sondage is 3 m in height. The stratification of the sondage 1 gives a perfect example of principal of stratigraphy. The coarse material is deposited in the lower unit and the fine matterial is deposited above the coarse matterial. Seven sediment samples have been taken for the sedimentological analysis in laboratory. Stratigraphically, sondage 1 can be divided into five Units from bottom to top:

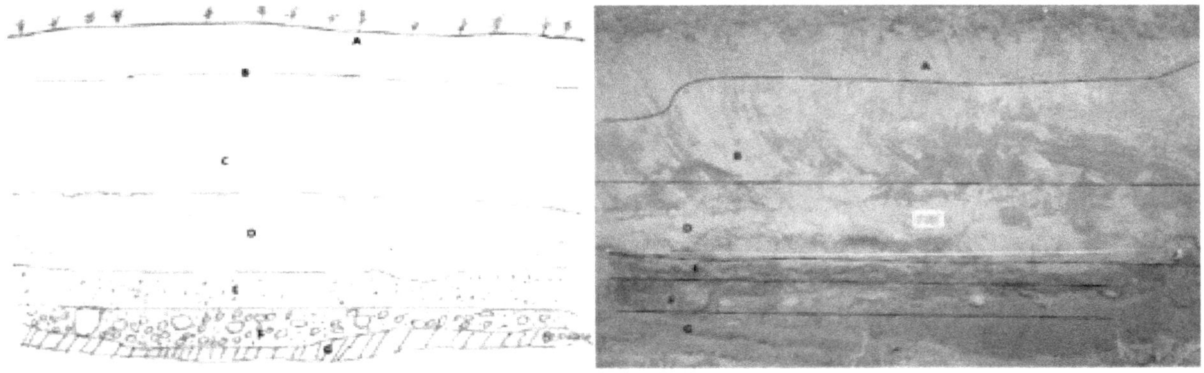

Fig. 11.5. Lithostratigraphy at test trench 1 at Outeiro du Pedro

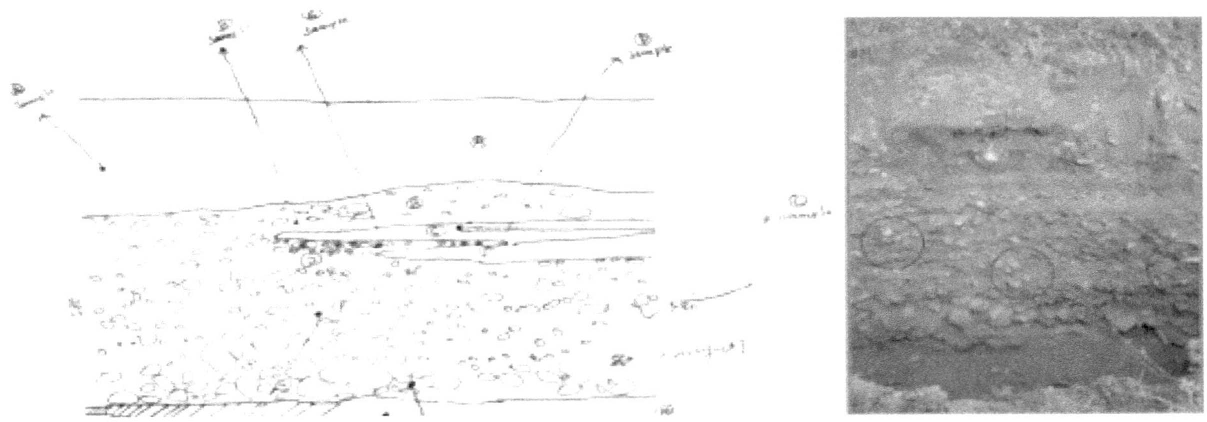

Fig. 11.6. Lithostratigraphy at test trench 2 at Outeiro du Pedro

Fig. 11.7. a. Bed rock and soil section at sondage 3; b. Lithostratigraphy at test trench

1) the first unit (G) is comprised of bed rock and the sample G is taken from here for the sedimentological analysis. 2) the second unit (F) is characterized by the boulders, cobbles, pebbles, granules and medium coarse sand. The sample F is collected from this unit and the thickness of this layer is 40 cm. The sorting size distribution of the clasts is supported by very poorly sorted matrix. The fabric is long axis. The stratification is horizontal and the grading is normal. It is unconsolidated in nature and the color of the unit is comprised of grey. The clasts matrix size ranges from 20 cm to 2 cm. It indicates that the hydrological energy was tremendous during the deposition of the sediments. The MPS is 16.8 cm. The stone clasts are made up of quartzite, quartz, milky quartz, sandstone and amphibolites. 3) the third unit (D) is comprised of medium to coarse sand and very few granules. It is 80 cm in thickness. It is very less consolidated in nature. The colour of the sediment is grey. When we are talking about the illustration of the concept of sorting, it is biomodal in sorting with well sorted sand and clay. Sample D and E is taken from this unit. 4) the fourth unit is comprised of granules and medium to fine coarse sand. In this layer the size of the granules are lesser than the previous layer. The thickness

of the layer is 90 cm. 5) the fifth unit is comprised of medium coarse sand to fine sand and silt. The thickness of this unit is almost 1 meter. The main characteristics of this layer alluvial sand. So far the consolidation of the sediments is concerned; this layer is also much unconsolidated in nature. It is poorly sorted sand. Sample B and C are taken from here. No artifacts have been found from this sondage. The fifth unit is comprised of actual soil and it is black and grayish in color.

Test trench 2

The second sondage has been dug around 40 to 50 m far from the sondage 1. It is almost four meter in thickness. The reference level of this sondage is 1.80 m. This sondage is also a good example of principal of stratigraphy. Three units have been identified and these units have been divided into layers. They have been explained from (bottom to top): 1) the first unit is comprised of bed rock and sample H has been taken from this unit, b) the second unit is characterized by the cobbles, pebbles, gravels and medium coarse sand, fine sand, and clay. The deposition of material shows that it is terrace. On the basis of altitude of the test trench, this terrace can be identified as T6. Thickness of this unit is 3 m. Several layers have been identified in this unit. This layer is named as G. In this layer, the biggest clasts are deposited at the bases of this layer and the size of the biggest clasts is ranging from 24 cm to 19 cm. The MPS is 18 cm. Some artifacts have been recovered from the pebbles and cobbles. The artifacts are comprised of flakes (side and end flakes) and different types of cores. The artifacts have been made on quartzite. The pebbles are rounded and elongated in shape. The sorting size distribution of the clasts is supported by the well sorted bimodal matrix. The fabric is long axis parallel flow. The stratification is horizontal. The grading is normal. Sample G is taken from this layer. The second layer is identified in the form of pebbles, gravels and coarse sand to fine sand. The thickness of this layer is around 1.40 m. The name of this layer is F. The pebbles of this layer are smaller than the pebbles of layer G. Some archaeological artifacts have also been found from this layer. They are comprised of few flakes and cores. The stratification of this layer is also horizontal. The clasts have been supported by the poorly sorted matrix and the grading of this layer is normal. The MPS is 17 cm. The layer F is superimposed by the layer E which is comprised of smaller pebbles as compared to the layer F. It is around 30 cm in thickness. The next unit (D) is comprised of lenticels of medium to fine sand with clasts that it 80 cm in thickness. Two samples have been taken from here (D). The next unit is named as Unit C that is comprised of fine sand with very few without clasts. It is consolidated in nature and the color of sediment is yellow; 3) the next layer is comprised of fine sand and it is grey in color with some pebbles. It is unconsolidated in nature. The MPS is 12 cm; 4) the third unit is comprised of fine sand and it is grey in color with some pebbles. It is unconsolidated in nature. The clasts size is ranging from 6 cm to 2 cm. The MPS is 5 cm. The layer number A and B could be a part of the alluvial plain.

Test Trench 3 and 4

The depth of sondage 3 is 70 cm and we found the bed rock at the depth of 20 cm. The bed rock is comprised of schist and it is covered by actual soil and cobbles and pebbles can be seen within the actual soil. A flake has been found from the surface. The depth of sondage 4 is 80 cm and the bed rock appears at the depth of 25 cm. No archaeological material has been found.

Test trench 5

Sondage 5 has been dug on the top of the river terraces. It is almost one meter in thickness. It can be divided into three units. The basal unit A is comprised of bed rock. The second unit B is made of clay, sand and with very small clasts. It can be observed that the clay is 75%. The origin of these sediments is alluvial. The third unit C is comprised of actual soil. Two samples have been collected from this section: A- Layer with organic matter B- Layer alluvial with water and clay. We did not find any artifacts here.

Test trench 6

This sondage has been dug between the sondage 1 and 2. The stratigraphical architecture of this sondage is divided in to several layers and it is as follows (bottom to top): 1) This layer is represented by the bedrock/rock (layer H); 2) this layer (G) is represented by the boulders with matrix of clay and sand. It is 50 cm in thickness. The MPS is 25 cm. It is identified as a river terrace (T6); 3) very fine sand and clay are the main constituents of this layer (f). The thickness of this layer is 30 cm. Some clasts and pebbles can also be observed. Some artifacts have been recorded from this layer. They are also observed very few end and side flakes. This layer also represents the same terraces (T6); 4) this layer (c) represents an alternation levels, which is represented clasts, pebbles and coarse sand matrix. The thickness of this layer is 20 cm. The lenticels are comprised of very fine sand; 5) this layer (d) is comprised of lenticels of lime, clay and clay-sand. It represents the lamination. The alluvial deposits starts from this layer; 6) the main constituent of layer C is coarse sand and it is 20 cm in thickness; 7) The second layer (b) is comprised of alluvial sand with clasts and thickness of this layer is 10-5 cm; 8) this layer (a) is comprised of alluvial sands with clasts on the base. It is 80 cm in thickness.

TERRACE SECTIONS AT ORTIGA VILLAGE (FABRIC FACTORY)

This section is located infront of the fabric factory in Ortiga village in Maćao munciplity. It is around 200 meter far from the railway track in the Ortiga village and around 2 km North-East of the Outeiro du Pedro Sondages. This section is located on the right bank of the Tagus river. The altitude is between 55 meter to 60 meter.

In this unit the substartum is not visible. It is a base of Quaternary terrace T4. The thicknees of this section is

Fig. 11.8. The lithostratigraphy of test trench 6

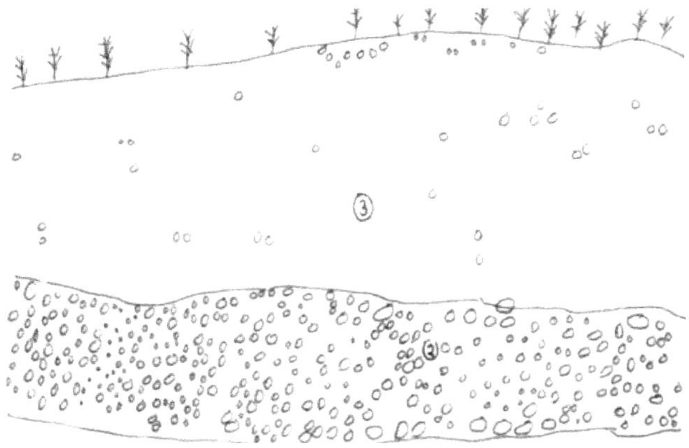

Fig. 11.9. Section at Ortiga village near Fabric factory

almost 4 meter. This section is divided into the following units (from bottom to top): a) this unit is characterised by the conglomerate of pebble, coarse sand and granules. The stone clasts are constituted of quartzite, quartz, milky quartz sand stone, imphibulite, granite. It is almost 1.50 meter in thickness. It is consolidated in nature. Sorting size distribution of the clasts is supported by very poorly sorted matrix. They are horizontaly stratified. The grading is normal. The colour of the sediments is red. The MPS is 13 cm. One sedimentological sample has been collected, from this unit, for the sedimentological analysis in the lab; b) the second unit is characterised by medium coarse sand, granules. It is also very consolidated and it is red in colour. It is 2.5 m in thickness. The gravels are very less in this unit. It is almost uniform in character; c) the third unit is characterised by the colluvial and natural soil. Archaeological this section is strile.

Section in the Centre of Ortiga Village

The altitude of this section is about 60-75 meter above the sea level. It is about 6 m in thickness and it can be

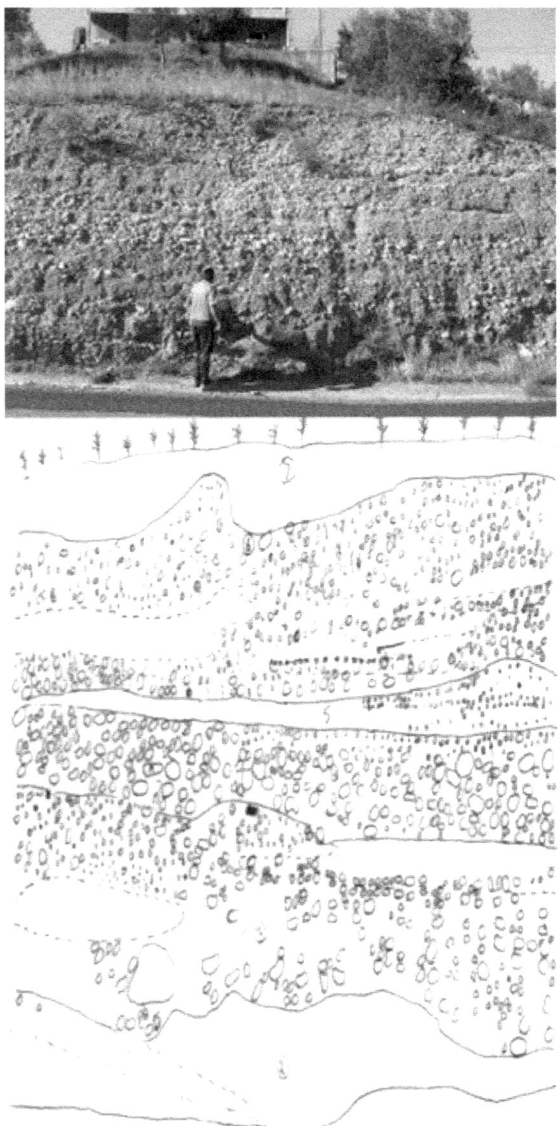

Fig. 11.10. River terrace T4 section in the centre of Ortiga village

divided into several units. It is a 6m height and ~30 m long a very big section divided into two quadrants (A and B).

The Quadrant A

It follows from the bottom to the top characterized by: 1) the basal rock unit that of the metamorphic rock, schist with the exposed thickness ~1 m. 2) the basal unit is superimposed by consolidated coarse sand, 20 cm in thickness, probably of the Miocene period. 3) unit is comprised of pebble, gravels, coarse sand forming several layers. The layer 2a is comprised of 60 cm very coarse sand, yellowish in colour. The energy of river was very low. This layer is succeeded by a consolidated layer of coarse sand and gravels. It is yellowish in color and 20 cm in thickness. In this unit (2b) the energy of water is more than the previous layer. The third layer 2c is comprised of pebbles (very small in size), gravels, and very coarse sand. It is also much consolidated in nature. It is 80 cm in thickness. In this movement of deposition of the sediments, the energy of water was higher than the previous depositions. The MPS is 13 cm, 3) the third unit could be channel of a course deposit of high energy because it is constituted of boulders, cobbles, pebbles, granules and coarse sand. This layer is Horizontal in character. This unit can also be divided in to several layers. The layer 3a is comprised of coarse sand. It is 60 cm in thickness. The MPS of this layer is 25 cm. The layer 3b is comprised of coarse sand with the granules. The layers 3a and 3b are superimposed by a layer which indicates a very high energy of river during the deposition of the sediments. It is also a very good example of the stratification where the coarser sediments are deposited at the base and the fine sediments are on the top. This layer 3c is comprised of boulders, cobbles, pebbles, granules and coarse sand. This layer contains two archaeological artifacts, one is side flake and other is core. It is around 1.5 m in thickness. The MPS is 18 cm. Small lenticels of coarse sand have also been observed in between the pebbles layers. The layer of coarse sand is 15 cm in thickness. The main constituents of the layer 3 d are pebbles, granules and coarse sand. The MPS is 12, 4) he fourth unit begins with the coarse sand indicates less energy of river water during the deposition of the sediments. It is yellowish in color. It is also consolidated in nature. The name of this layer is 4a. This layer (4a) is superimposed by a layer (4b) that is made of pebbles, gravels and coarse sand. Small lenticels of coarse sand can also be observed in between the pebbles, 5) the unit 5 is 20 cm thickness. It comprised of coarse sand and lenticels of pebbles, 6) the sixth unit is comprised of colluvial. It is around 1m in thickness.

This section is continuation of the previous section at the centre of Ortiga village. It can also divide into the following units from depositional points of view (bottom to top): 1) this unit is comprised of granules coarse sand clay. We can observe the granules at the base. There are also some lenticels (layer 1b) which are deposited at the angle of 15°. This unit is homogenous in character and it is consolidate and the color of this layer is grayish white. It is 90 cm in thickness; 2) the second unit could be recognized as river channel. It is also comprised of boulders, cobbles, pebbles, gravels and medium coarse sand. The base of this unit shows a picture of wavy line that is made of cobbles, pebbles, coarse sand and very few boulders. The name of this layer is 2a. It is around 1 m in thickness. There are also some lenticels of coarse sand can be observed in between the pebbles. This layer is nominated as 2b and 2c. The layer 2d is comprised of pebbles and coarse sand. In this layer it is an example of inverse stratigraphic deposition. Here the smaller pebbles have been deposited at the base and bigger at the top. The color of this layer is yellow; 3) the unit third is homogenous in character. It is around 1.5 meter in thickness. It is comprised of pebble, granules, medium coarse sand. The bigger pebbles are deposited at the base and smaller at the top. It is also yellowish in color. The clasts are made up of quartzite, quartz, milky quartz and sandstone. One side flake has been found from the section; 4) the fourth unit can be divided into two layers. The layer 4a is comprised of lenticels of pebbles and 4b is made of coarse sand. It is 30 cm in thickness. It is also consolidated in nature. The color of the sediments is yellow; 5) the fifth unit is 2 m in thickness. It is showing a various phase's depositional history of the river. The basal layer (5a) of this unit is made pebbles and gravels. Bigger pebbles are deposited at the base and the smaller pebbles have been deposited on the top. It is around 40 cm in thickness. This layer (5a) is superimposed by two horizontal layers (5b, 5d and 5f) that are made of the lenticels of coarse sand. The layer 5c is comprised of pebbles granules and coarse sand; 6) the sixth unit is comprised of colluvial and coarse sand.

Section along Macao-Ortega Road

This section is located on the same road that leads to Allege and Ortega. The longitude of this section is 8.01°W and the latitude is 39°50'N. The altitude of this section is 65 meter above the sea level. This section represents the top of T4. The deposition architecture of this section is as follows (bottom to top): 1) the first unit is comprised of very fine sand and clay. They are yellow in color and it is much consolidated in nature. It is 50 cm in thickness. These sediments seem to have been deposited in the Miocene period; 2) the second unit is constituted of granules and medium coarse sand. It is almost 70 cm in thickness. The size of granules is bigger at the top of the unit than the base of the unit. It is much consolidated in nature. It is red in color. The bigger size of the granules at the top of the unit indicates that the energy of river water was more during the deposition of upper portion of the granules. There are some lenticels of pebbles has also been observed; 3) the consolidated layer of the gravels is succeeded by a unit that is comprised of pebbles, gravels and medium coarse sand. The layer 3b is comprised of very coarse sand, whereas the layer 3b comprised of pebbles and granules. This part of the section is also much consolidated in nature. The color is yellowish and reddish and grayish. No archaeological material has been from this section; 4) the unit 4 is comprised of cobbles, pebbles, gravels and Medium coarse sand. The unit can also be divided into layers. The layer 4b is comprised of unsorted pebbles. The size of the pebbles is bigger at the base and smaller at the top. Some lenticels of coarse sand can also be observed in between the pebbles layers. The color of the sediment is yellow and grey. We could not find any artifacts from this region; 5) the top of the section is comprised of colluviums and actual soil.

Grain size and Sedimentlogical data from Macao region

The grain size analysis of the sedimentological samples that were collected from the sondages 1 at Outeiro du Pedro, in the alluvial plain of the Tejo, indicates that clasts are ranging from the fine sand to very fine sand and the grains size of the clasts ranging are from 78 mm to 163 mm. The sorting of the sediments are between poorly sorted to extremely poorly sorted. The value of skewness indicates that the sediments are very negatively skewed. The value of kurtosis of the samples shows that they are ranging from very platykurtic to extremely leptokurtic. The clasts distribution is supported by polymodal matrix.

The grain size analysis of the data from the sondage 2 indicates that the clasts size is ranging between 79 mm to 1030 mm. So, the sediments are ranging from fine sand to very coarse sand. The sorting of the sediments are ranging from the poorly sorted to extremely poorly sorted. So far the skewness of the sediments is concerned, it 50% of the sediments are positively skewed and rest of the sample are negatively skewed. The kurtosis range from leptokurtic to extremely leptokurtic; only one percentage of the sample is mesokurtoic.

The grain size analysis of the sondage 6 shows that the grain size ranging between 46 mm to 218 mm. So, these results explain that sediments are ranging from the coarse silt to fine sands. The sorting size of the sediments shows that they are poorly sorted to extremely sort. The sediments are very negatively skewed. The kurtosis of the sediments is very platykurtic to very leptokurtic. It means the sediments are very poorly sorted.

The grain size of the sediments of sondage five are between 84 to 233 mm. The gains are made of fine sand. They are extremely poorly sorted and they are negatively skewed. The kurtosis is very platykurtic.

The grain size analysis of the fabric section, at base of T4, Ortega reveals that the grain size is between 395 mm to 450 mm and the sedimentological texture is comprised of medium coarse sand. The sorting size distribution of the clasts is polymodal and supported by the poorly sorted matrix. They are positively skewed and the kurtosis is between the leptokurtic to extremely leptokurtic.

The grain size analysis of the river terrace (T4) at the centre of Ortiga village shows that the grain size ranges between 57 mm to 1000 mm. The sediments are comprised of coarse silt, fine sand, and coarse sand. The clasts are supported by polymodal matrix and very poorly sorted matrix. The clasts are very positively skewed and the clasts are leptokurtosis.

Studied outcrops in the Vila Nova da Barquinha

This area is located in the muncipality of Villa Nova da Barquinha. The main loacalitied of this area area are Barquinha village, Atalaia village, football field near Barquinha village and a Tejo River section in front of Railway station of Villa Nova da Barquinha. Ribeira da Atalaia.

The archaeological site of Ribeira de Atalaia is located on a small stream, named Ponte de Pedra, in the Villa Nova da Barquinha municipality in central Portugal. The altitude of this site is 44 to 45 m. This site is located on a slope of terrace Q3 (T4) terrace. The site was first excavated under the guidance of Prof Oosterbeek and Prof. Rosina in early 1999 under the framework of "Territory, Mobility and Settlement in the Alto Ribatejo" (TEMPOAR projects).

The main aim of this research is to describe the archaeological potential of Tagus fluvial terraces and provide chronological and stratigraphical sequences to the archaeological findings (Oosteerbeek et al. 2004). This site is very important from the earliest human occupation point of view because it provides us the *in situ* archaeological records which are very important for understanding the Palaeolithic occupation, chronological sequences, and hominid land relationship depositional environment of the sediments. Since the site is located on the slope of terrace Q3 (T4), the stratigraphical sequence of the site can be described under the supervision of Professor Peirluigi Rosina (my research supervisor) in the followings: 1) the basal unit of the stratigraphy belongs to the Miocene period. It is comprised of reddish-yellowish medium coarse sand and clay. It is much consolidated in nature and while the recent stratum shows very well sorted white sand; 2) the Miocene sand is superimposed by the Q3 (T4) terraces. The lower part of Q3 (T4) is comprised of four different deposition morphology (Rosina and Cura, 2010; Rosina et al. 2006), and "the first deposition morphology is characterised by A bar (formed by reddish coarse sand and pebbles); the second is a channel (filled with big pebbles and cobbles, up to 35 cm, and reddish coarse sand) that cuts the bar; the third is flood plain fine grain deposits (grey to yellow); and the fourth is transverse channels (filled with reddish sand and pebbles, up to 10 cm) with a very erosive contact with the flood plain deposit" (Rosina et al. 2010, p. 168). The Q3 (T4) terrace is followed by the Q 4a (T5) terrace, 4) all the units of the site have been covered by the colluviums.

These colluviums are the result of landslides, water activity on the slope and gravity process. The colluviums can be dated to Holocene Period. Recently, ten stratigraphic units are classified. The lithostartitygraphy

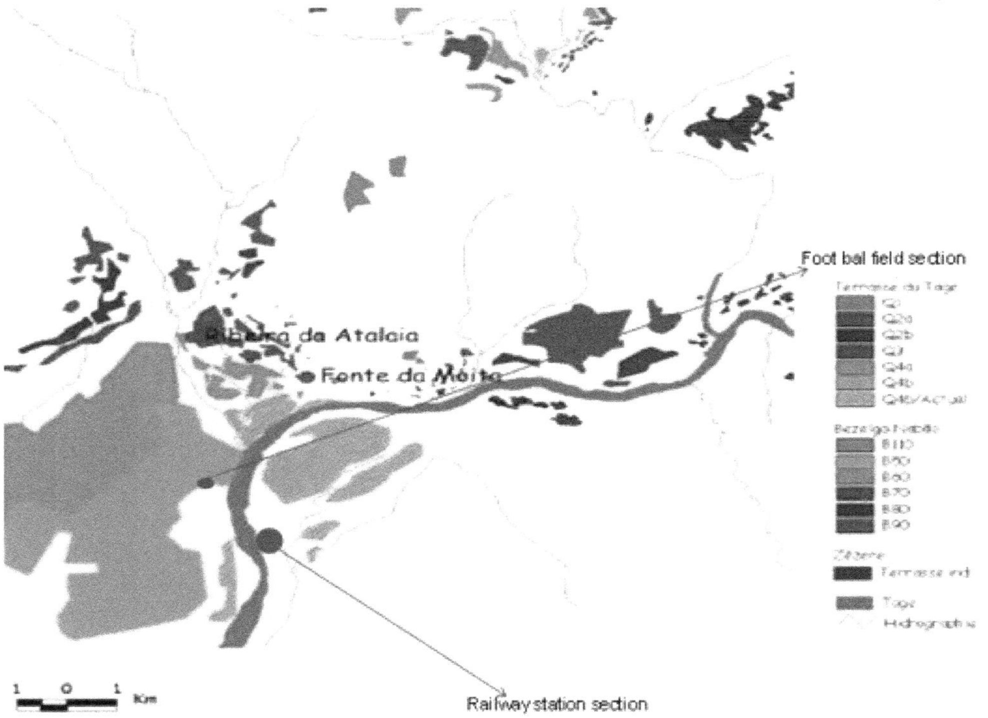

Fig. 11.11. Geomorphological map showing archaeological and river terrace sites in the Vila Nova De Barquinha (modified after Rosina, 2001, 2004)

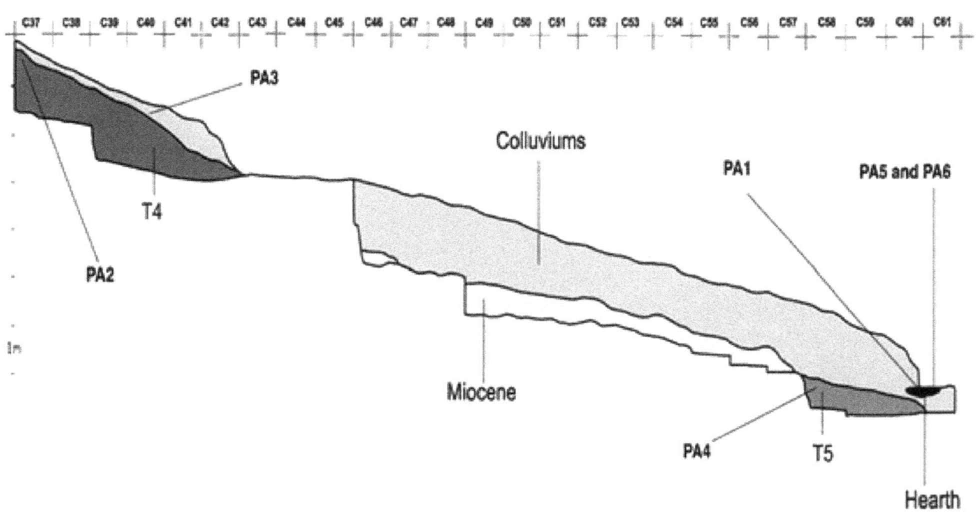

Fig. 11.12. Lithostartigraphic unit and dates for the archaeological sites of the Ribeira da Atalaia (after Rosina et al. 2004; Oosterbeek 2007; Martins et al. 2010; Rosina et al. 2010; Rosina et al. 2010)

of the archaeological site of Ribeira da Atalaia can be given in the following in the following manners (please see Rosina 2004; Cura and Rosina 2010; Rosina et al. 2006):

Unit 01 is constituted for a few developed soils, dark coloured, with a density which can vary between few centimetres until some ten of centimetres.

Unit 10 is characterized by colluviums. It seems that colluvial deposits (or colluviums) result from the gradual accumulation of weathered materials transported down slope by gravitational forces. Sediments of colluvium origins found down the trench (square B52-B63), constituted by middle coarse sand, little compact, and pebble of quartzite, with variable size. The density of this unit grows with the diminution quotation, since a few centimeters until 180 cm. In the base of this unit (squares B52-B56) come out to the surface/emerge a level of small rubbles and pebbles. At the end, the thickest one (square B60-B63), it seems possible to distinguish different colluvium occurrences.

Unit 20 and 99 formed of middle coarse sand, little compacted sand, with rubbles and pebbles of quartzite The pebbles are bigger in the lower part, with an average.

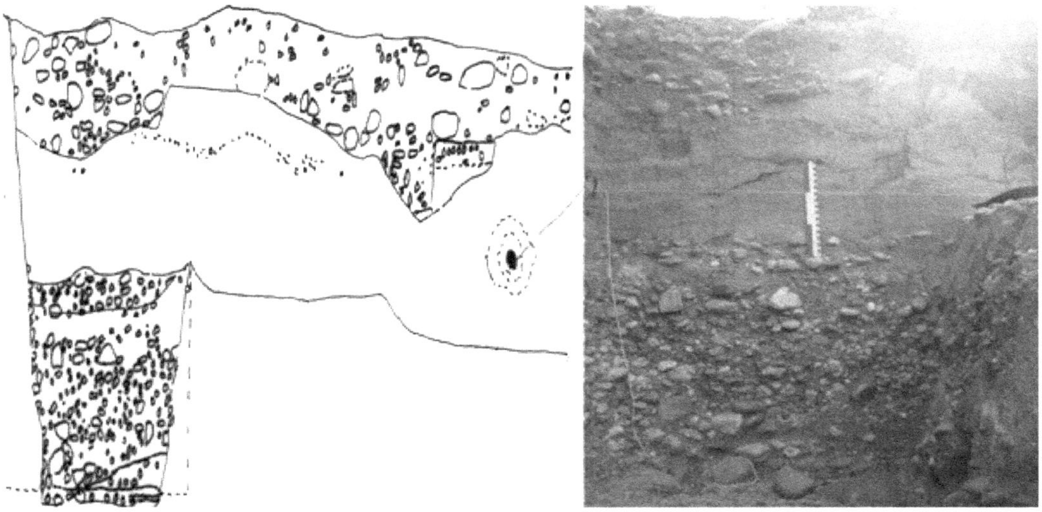

Fig. 11.13. Archaeological site at Q3 (T4) terrac of Ribeira da Atalaia: top of the section

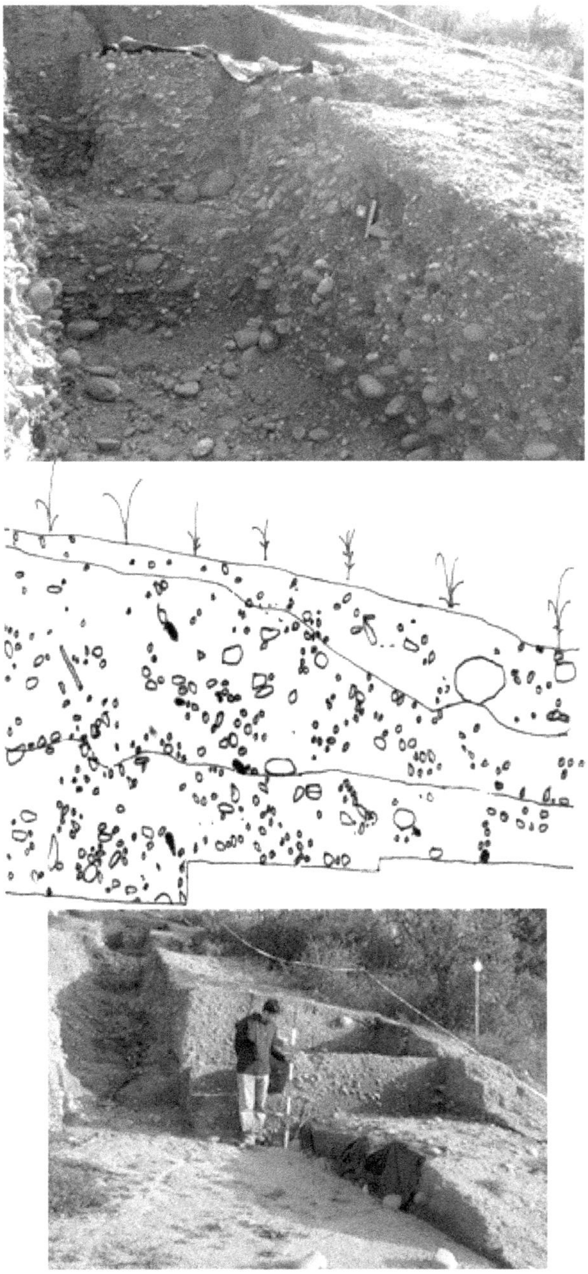

Fig. 11.14. Middle section showing colluvial deposits

Fig. 11.15. Basal part of the section showing Miocene sediments and Lower terraces

of 8-9 cm (comparing to an average of 5 cm in the superior part). The lithic artefacts are very rare and the most of them are in the dig's area. This area can also be interpreted as an accumulation of colluviums.

Unit 30 is constituted of middle sand with small rubbles and pebbles. On the base it's possible to observe a discontinuous increase of silt. The density is ~25 cm. They seem to form parallel and flat structures.

Unit 40 represents one of the most interesting units for archaeology, because of its concentration of lithic artefacts found in this sediment. This is middle-thin sand, with silt and rubbles, with pebbles of quartzite (max. 10 cm). The density varies between 50 cm and few centimeters. It seems possible to recognize different progressive colluviums fronts and overlapped. This unit can be interpreted as one constituted by old colluviums.

Unit 46 is a superficial piece constituted by middle-thin sand, rubble and silt, with dispersed pebbles of quartzite (average 5 cm; max. 10 cm). The highest density observed is 30 cm. This unit has an erosive contact on the

Fig. 11.16. The archaeological site of Ribeira da Atalaia showing locations of the samples collected for sedimentological analysis

Fig. 11.17. Base of T4 terrace

base, with the unit 47. In this deposit it was found several lithic artefacts. In spite spit of the small outcrop, it can be interpreted as a filling up channel episode/occurrence.

Unit 47 is thin sand and silt slightly cemented/consolidated; the highest density is ~50 cm. It's visible in its inside a band with few centimeters constituted by rubble and small pebbles (2-3 cm) as for unit 46, this deposit comes to the surface more affected by the erosion. It wasn't found any lithic artefacts.

Unit 48 is constituted by a subtle band of pebble (average of 3 cm, max. 5 cm) with a sandy matrix that, as the preceding units, ends directly at the surface. It was found lithic artefacts also with great dimensions. One of the possible interpretations, taking into account the stratigraphic sequence and the erosive contact over the unit 60, is that this unit represents an episode/occurrence of channel's closing or related with secondary drainage.

Unit 49 is characterized by thin sand, argil and slit, orange with reddish iron oxidations. The maximum density is ~30 cm, coming to the surface. This unit has also a lot of lithic artefacts. It can be observed the erosive contact of this unit over the unit 60, coinciding with the closing of this one. It is interpreted as an episode/occurrence of channel filling. It represents the most significant unit, from the archaeological point of view, weather for its density, whether for the abundance of lithic artefacts. These are conglomerate's levels with round pebbles or quartzite's sub-round, with a middle size of ~10 cm and maximum 25 cm; the pebbles are mainly directed according to a North-south axis, and a lot of imbrications are observed. The matrix is made of sand and silt, with greater concentrations of argil in the lower part.

Unit 50 has maximum density of ~90 cm (square B40); the basis of this unit wasn't still reached. It was interpreted as a channel's deposit. This unit is the main part of the T4 (Q3) terrace.

Unit 60 is mainly a deposit of silt, with small patches of thin orange sand and with structures correlated with phenomenon of paedogenesis. The density is ~20 cm. It's only possible to observer that the unit 49 has an erosive contact over this unit.

Unit 99 is represented by a set of levels of conglomerates with compact and gross sand, with silt, yellow coloured. The round or sub-round pebbles of quartzite frequently reach great dimensions (until 30 cm). In the upper part, the matrix is constituted by middle coarse sand, with rubbles and rests of argil and silt that increase in profoundness, surrounding the bigger pebbles. The oxidations are completely absent. The density is only visible in the squares B46-B52, where it can have 100

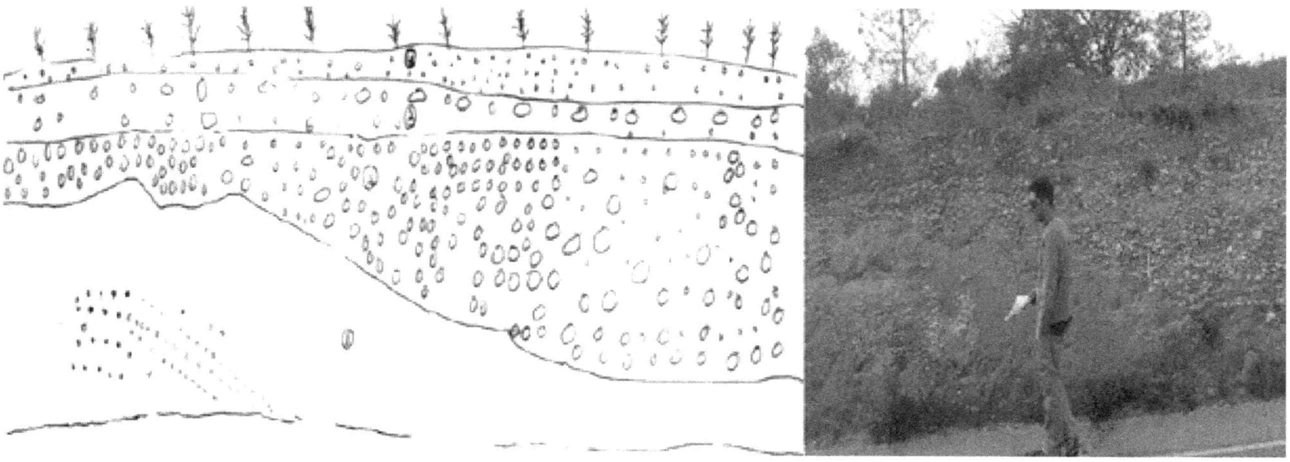

Fig. 11.18. T4 terrace on road opposite to the site Ribera da Atalaia

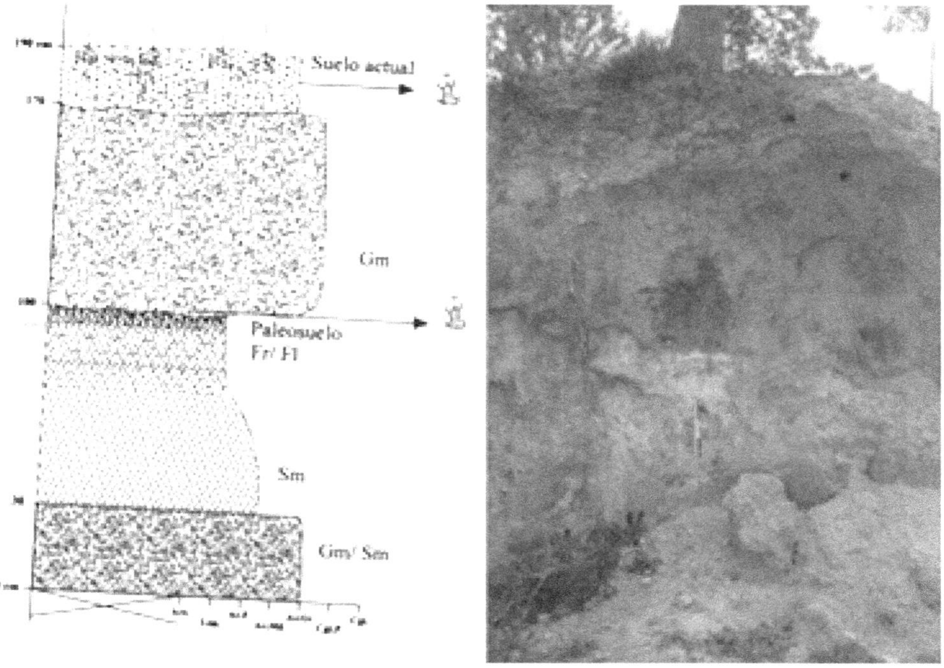

Fig. 11.19. Lithotratigphy at Football field at Atalaia near Villa Nova da Barquinha (Fernandez, 1998)

cm, decreasing slowly until disappear. This unit is archeologically totally sterile.

Unit 100 may become the most significant one for comprehension of the relations between the sedimentary processes, which determined the deposition of the terraces, and the other Quaternary deposits. This is coarse sand with gravels and silt, alternated with conglomeration's levels; the colour is white. This unit was found in a great part of the trench (square B46-B59) and isn't visible in the final part. There were any lithic artefacts. A possible hypothesis attributes this unit to the Miocene deposits observed in some natural cuts, present along the line of the transverse water to the Ribeira da Ponte da Pedra.

The square B37-B47 is the base of terrace Q3 (T4) at this site. Highest percentage of the artefacts has been recorded from the context of conglomerate level. This conglomerate level is the base of Q3 terrace. Few artefacts' have been recorded from the deposits of archaeological square of B.

SEDIMENTOLOGICAL SAMPLES

The sedimentological depositional history and palaeo-environment of the site of Ribeira da Atalaia is revealed from the ten samples collected were analyzed and described below:

Sample 1 (RPP1)

This sample is collected from the colluviums of the Q4a (T5) terrace. During the archaeological excavation, hearth has been discovered from this place. Oosterbeek (2007) provides dates for the colluviums of this terrace. Three

Fig. 11.20. T4 terrace section opposite to the railway station, Villa Nova da Barquinha

samples were collected near by the hearth for the TL dating. The first sample (ITN Lump -35, Sample PA1) was collected from the burned colluvial. The results of this sample are (as 24897±2194 y B.P., the second sample (sample PA4) was collected from the fluvial sand and the results of this are 90204±17776 y B.P. and finally, third was sample (sample PA5 and PA6) was collected from the burned quartzite and the results of samples are 24,897±2194 and 24.8810±2.184 y B.P. (Oosterbeek, 2007). Upper Palaeolithic artefacts have been recorded from this layer. The MPS (maximum pebble size) is 17 cm. The main type of the clasts is quartzite, granite with quartz, milky quartz, Amfibulite and sand stone. They are roundish in shape. The grain size analysis of the samples (RPP1) in the lab shows that the clasts size is 725 mm and they are comprised of coarse sand and the gain size is distribution is very poorly sorted.

Sample 2 (RPP2)

The sample 2 has been collected near the fire place. It is comprised of coarse sand, silt, clay, gravels and pebbles. Upper Palaeolithic artefacts have been recorded from this layer. Oosterbeek et al. (2007) dated this layer as 24897±2194 yr B.P.

The grain size analysis of the samples in the lab shows that the clasts size is 725 mm and they are comprised of coarse sand and the gain size is distribution is very poorly sorted.

Sample 3 (RPP3)

The sample number three is collected from the Miocene unit. This unit is comprised of reddish-yellowish sand. It is much consolidated in nature. The exposed Miocene sediments are 50 cm in thickness. The grain size analysis of the samples (RPP2) which collected from the Miocene sediments is 417 mm and they are comprised of medium coarse sand and the gain size is distribution is very poorly sorted.

Sample 4 & 5 (RPP4 & 5)

This unit is characterized by colluviums. It seems that colluvial deposits (or colluviums) result from the gradual accumulation of weathered materials transported down slope by gravitational forces. The sample 4 and 5 are collected from this unit. It represents the colluviums of Q4a and it is divided in to three different units. Rosina (personal communication) divided them three different layers; C1a, C1b, C3, respectively. These samples have been taken from the base and top of the colluviums, the place just above the Miocene sediments; it is around 1.5 meter in thickness. The MPS of the colluviums is 15 cm.

The grain size analysis of the samples (RPP 4 and 5) in the lab shows that the clasts size is 668 to 404 mm, respectively. They are comprised of coarse sand and medium coarse sand, respectively. The clasts size distribution is very poorly sorted. The clasts size distribution is supported by polymodal matrix.

Sample 6 (RPP6)

This unit represents the base of the terrace Q3 (T4). This unit is comprised of cobble, pebble, middle coarse sand and granules. It is almost 1m in thickness. The sample is collected from the base of this unit. The colour of the sediments is red. The MPS is 17 cm. The grain size analysis of the samples (RPP6) in the lab shows that the clasts size is 918 mm and they are comprised of coarse sand and the gain size is distribution is very poorly sorted.

Sample 7 & 8 (RPP7 & 8)

This unit is comprised of boulders, cobbles, pebbles, and gravels. It is much consolidated in nature. This unit represents the channel structure of the Q3 (T4 terrace). Most of the Lower Palaeolithic artefacts have been recorded from this unit. The thickness of this unit is almost 2 m. The maximum pebble size (MPS) is 25 cm. The grain size analysis of the samples (RPP7 & 8) in the lab shows that the clasts size is ranging between 1070 mm to 1109 mm and they are comprised of very coarse sand and the gain size is distribution is very poorly sorted.

Sample 9 (RPP9)

This sample is collected from the layer that is just super-imposed the channel structure. It is comprised of very few pebbles, granules, and coarse sand and some silt lenticels can also be observed. The colour of the sediments is grey. The thickness of this layer is 50 cm. The MPS is 12 cm.

The grain size analysis of the samples (RPP9) in the lab shows that the clasts size is 247 mm and they are comprised of coarse sand and the gain size is distribution is poorly sorted. The sorting size distribution of the clasts is supported by polumodal matrix.

Sample 10 (RPP10)

This unit is comprised of the fine sand and silt. It is consolidated in nature. The thickness of this unit is 50 cm. An OSL sample has been collected from this layer and Oosterbeek (2007) dated this unit as 302085±12635 y B.P. This unit is covered by the surface actual soil which is 25 to 50 cm in thickness.

The grain size analysis of the samples (RPP10) in the lab shows that the clasts size is 191 mm and they are comprised of coarse sand and the gain size is distribution is very poorly sorted.

Geological section next to the site Ribera da Atalaia

This geological section is situated just 500 meter next to the archaeological site of Ribeira da Atalaia. This section is 35 to 40 meter above the sea level. It is base of Q4a (T5). The stratigraphical architecture of this section is as follows (Bottom to top): 1) the first unit is constituted of cobble, pebble, gravels and medium to fine sand and clay. The thickness of the layer is 1.5 meter. It is consolidated in nature and the color of this section is red. The sorting size distribution of the clast is supported by the bimodal matrix or well sorted. It is horizontally stratified. The grading of the clasts is normal. The maximum pebble size (MPS) is 18 cm. The sedimentological sample (RPP0802-0803) is collected from this unit; 2) the second unit is comprised of medium reddish coarse sand with gravels. It is 80 cm in thickness. A channel structure can be observed in this layer. It is well sorted sand and very consolidated in nature. The sedimentological sample (RPP0804) is collected from this unit; 3) the third unit is also a much consolidated in nature. The thickness of this section is 60 cm. Medium to fine sand and clay are the main constituent of this section. It is yellowish in color. When we are taking about the concept of sorting, this layer is bimodal in nature: poorly sorted sand with silt. The sedimentological sample (RPP0805-0806) is collected from this unit.

Road Section opposite to Ponte de Pedra site

This geological section is also located on a road which is coming from Tomar Vila Nova da Barquinha and going to the football field. The altitude of this section is 43 m. The height of this section is 3,00 meters. The stratigraphic description of this section is as follows (From Bottom to Top): 1) the first unit is comprised of very coarse sandstone. It is much consolidated and it is red in color. The sand is coarse to medium in fraction. We can also observe the planner lamination of gravels and the angle of this lenticels is 45°. The lamination of gravels is named as layer 1a. In this unit the energy of the river water was very low. The sorting size distribution of the unit is very poorly sorted matrix; 2) the second unit is a channel structure and it is constituted of pebbles, gravels, granules, Medium to coarse sand and clay. The pebbles are made of quartzite, different types of quartz, sand stone and amphibolites. The color of the sand is yellow. The size of pebbles is bigger in eastern portion of the section and the concentration of the pebble is more in the western portion of the layer. The MPS is 16 cm. We could not find any artifacts from this section; 3) there is also a layer of 2 cm fine to medium coarse sand which is yellow in color. In this layer, small pebbles and gravels have also been observed. The sorting size distribution of this unit is supported by the very poor sorted matrix, where as it is unsatisfied in stratification. It is ungraded in nature; 4) the fourth unit is comprised of colluviums and actual soil.

It is very important to explain that we did an exploration near by the archaeological site of Ribeira da Atalaia. We have collected some cores and one flake (for the detail description, please see the chapter of discussion and conclusion).

The Football Field or Sports Ground site at Atalia

The Sports ground (Football Field) site is located near Villa Nova da Barquinha and 1.5 km from the archaeological site of Ribeira da Italia. The site of Sport-Ground represent the upper most part of the middle terrace level. The altitude of this site is 51 to 52 m above the sea level. The stratigraphic section of this site is described in the following way (from bottom to top: 1) the lowest stratum is comprised of reddish micro conglomerate. The thickness of the layer is 30 cm, the pebbles are also important component of this site and the size of these pebbles is not more than 1 cm and the medium size is 3 mm and the lowest part of this layer cannot be seen; 2) the micro conglomerate stratum is followed by a layer that is comprised of fine sand and pebbles. It is much consolidated in nature. This layer is 70 cm in thickness. The granulometry of the stone pebble are bigger on the base and small on the top. We have also observed lime and it is a principal component till the top. This level finishes with the beginning of next layer of Palaeosoil. Under the Palaeosoil, grey clay can be observed with the great amount of iron oxide. The main clast types are quartzite and quartz. Grimaldi et al. (1998) collected some lithic implements from here; 3) Palaeosoil is the main constituent of this layer. This layer is 30 cm in thickness. The palaeosoil contains the leach grey clay; 4) the previous unit of Palaeosoils is superimposed by the thick reddish micro-conglomerate which is 1.10 m in thickness. The maximum size of the pebbles is 8 cm whereas the minimum size is 2 cm. lithic industry has been recorded in the basal part of this unit. This layer is comprised of clay, sand and gravel and it is covered by the actual soil or present soil. The thickness of this unit is 20 cm.

Railway Station Section (Vila Nova de Barquinha)

This section is located in front of railway station at Vila Nova de Barquinha. The thickness of this section is around 6 m to eight meter. The altitude of this section is 42 to 46 meter above the sea level. The stratigraphic architecture of this section is as follows (from bottom to top): 1) the first unit is comprised of fine sand and silt. It is 3 meter in thickness. It belongs to Miocene period. The colour of the sediment is yellow and grey. These sediments are much consolidated in nature. There are number of gravels have been in the Miocene sediments. These Miocene sediments also contain the lenses of cobble, pebbles and gravels and it is nominated as layer number 1c, 1e, 1f. Layer 1a is characterized by the pebbly coarse sand. The layer number 1b and 1d is constituted of pebbles and cobbles. These pebbles are made of quartzite, quarts, amphibolites, sand stone, and conglomerate. The shape of the pebbles is ranging from rounded, sub rounded, angular, oblong etc. The sorting size distribution of the clasts is supported by poorly sorted matrix. The stratification of the clasts is horizontal, but there is an inclination in the stratification of the clasts at the lower part of this unit. The MPS is 21 cm.

The presence of pebble pavements within the Miocene sediments indicates a movement of higher energy. There are two main structures have been identified which is cut into the Miocene sediments. And these structures are just below the unit which is comprised of pebble unit. The shape of these structures is semi-circular. The semi-circular pebble structures within the Miocene sediments indicate episodes of river activity in which the water energy is very high. It is very important describes that few flakes have been recovered from the interface of Miocene and the channel structure, 2) the second unit is comprised of Palaeo-channel structure. The orientation of the pebbles indicates that the direction of the water course could be in the north-south direction. This unit can also be divided into two main layers. The layer 2a is marked by the big pebble size whereas the pebble size is smaller in the layer number 2b. Layer 2a is superimposed by a small layer of sand. The clasts composition is almost the same as in the previous unit. The thickness of this unit is around two meter. This unit is comprised of pebble, cobble, and gravels and very coarse to medium coarse sand, which is yellowish in color. In this unit, the stratification is uncertified and they are ungraded. The MPS is 23 cm. Archaeological materials from this section: There are six archaeological stone tools have been recorded from the channel structure that it situated just above the Miocene sediments. The archaeological material is comprised of end flakes, side flakes and cores.

Chronology of fluvial terrace in Villa Nova da Barquinha area

Earliest human occupation has been recorded from the T4 terraces at archaeological site of Ribera da Atalai and Fonte De Pedra. At Ribera da Atalai, two meter above the base of T4 terrace is dated by OSL dating method as 304±20 Kya (PA2;) the sandy matrix of T5 boulder as 90±13 Kya; the sandy matrix of colluvial is dated as 24.8±1.6 Kya; the hearth in the base of colluviaum is dated by TL dating as 25±2 Kya (PA1); the mean date for culluvium sand is 24.9±2.2 Kya (PA5 and PA6); the fine layer of sand within basal gravel of T4 terrace is dated as ≥175±16 Kya; top of T4 terrace at the football field of Atalai as ≥175±16 ka (Dias et al. 2009); and the top of T4 terrace at Vila Nova Birquinha chapel dated as ≥169±9 Kya. The recent luminescence dates for Tagus terraces are as follows: T3 older than 300 Kya, T4 dated as ~300-160 Kya, T5 as ~136-75 Kya, and T6 as 62-30 Kya; culluvium and aeolian context as ~26-14 Kya (Cunha et al. 2008; Martins et al. 2009). The new dating results for the Tagus river terraces are ascribed by Martins et al. (2010) in the following pattern: marine oxygen Isotope stage 9-6 (MIS) for T4, MIS 5 for T5, and MIS3.

THE GRAIN SIZE ANALYSIS AND INTERPRETATIONS

The grain size analysis of the sediments which are collected from the different localities at T4 in the Villa Nova da Barquinha shows that the clasts size is ranging between 32 mm to 1109 mm. The sediments are ranging between coarse silt to very coarse sand. The sorting size distribution of the sediments is supported by polymodal matrix and the sorting of clasts is poorly sorted to extremely poorly sorted. The skewness ranges between very negative skewness, negative skewness, positive skewness and very positive skewness. The kurtosis is ranging between very platykurtic, platykurtic, mesokurtic, leptokurtic, very leptokurtic, and extremely leptokurtic.

In order to interpret the energy of deposition and the environment of deposition, grain size analysis of 50 sediments samples (Table 3) were carried out at the sedimentological laboratory in the University of Coimbra. The sedimentological samples have been collected from the sondage at alluvial plain / T6 of the and T4 terraces of the river. In grain size analysis, two principle graphical forms of analysis are used. The grains size is entered into

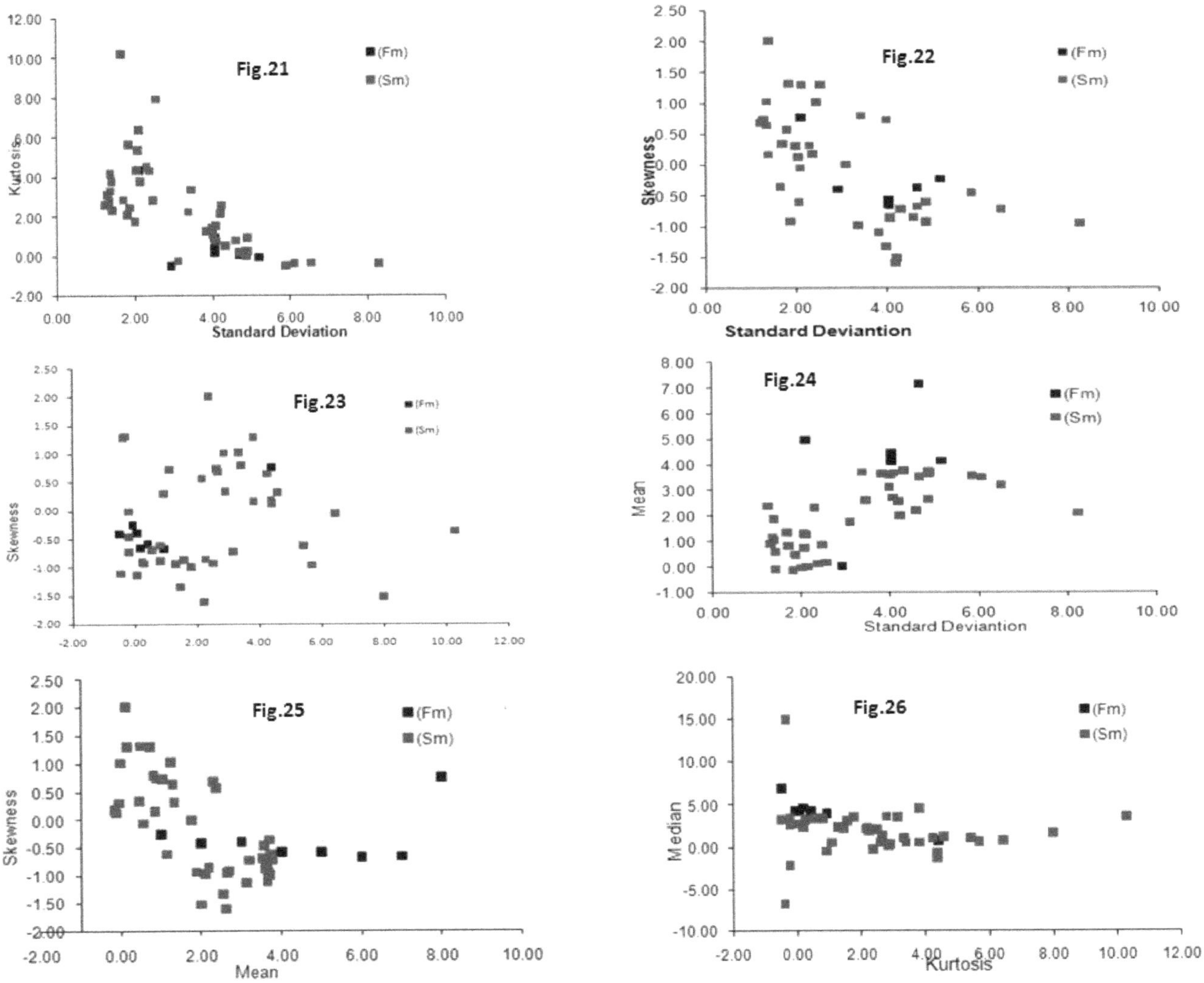

Fig. 11.21-26. Grain size interpretation of: Kurtosis/Standard deviation, skewness/standard deviation, Kurtosis/Kurtosis, Mean/Standard deviation, Skewness/Mean, and Medians/Kurtosis, respectively

a establish formulae and in which the value directly derived from the plotted cumulative curves. In these formulae, characteristics of every grain in the sedimentary deposit are used for computation. In all cases the number obtained serves to define the position of the distribution plot, its slope, and nature of any irregularity.

By using statistical method, following parameters have been derived: a) average size (Mode, Median, and Mean); b) spread size about the average sorting; c) preferential spread-skewness; kurtosis or degree of concentration of the grain to central size. We have produced six graphs for the grain size data, the description of the grain size analysis is as follows:

Standard deviation/kurtosis

The first graph has been produced on the basis of standard deviation and kurtosis. Standard deviation is been taken on the "X" axis and kurtosis on "Y-axis. Kurtosis is related both to depression and normality of the distribution. So, very flat curves are platykurtic. In this graph, the plots are ranging between -0.07 to 10. So, two main population of grain size can be identifies. The first population of the grains ranges between platykurtic-mesocurtic and represents 40% of the total percentage of the grains population. This population of the grains is related to the very poorly sorted sediments with biomodal frequency. The second type of the grains population can be identified as Leptokurtic to extremely leptokurtic. This population of the grains represents 60% of the total percentage and they are very exceptionally good sorting. This graph also indicates that the percentage of the coarse sediments is more than the fine sediments. On the basis of the distribution of the grain size, it can be also explained that the coarser grains are better sorted than the fine grains.

Standard deviation/Skewness

The skewness has a positive or negative value when more fine or coarse materials are present than in a normal distribution seen as tail to the right or left respectively on a frequency distribution plots. The skewness of the grains is ranging between -1.50 to 2. In this graph, two main population of the grains size can be identified. The first population is comprised of the grains are ranging between positively skewed (+0.1 to 0.3) to very positively skewed

(+0.3 to +1.0 or more). The second population is identified of the grains which are very negatively skewed and the size is ranging between -0.1 to -1.0 (or more).

Kurtosis/Skewness

The skewness/kurtosis interrelationship is used to recognize differing environmental energy levels in deposits of low energy environments to high energy depositional environment. On the basis Skewness, two main populations can be identified. The first population is comprised of samples which are ranging between -0.1 to -1.0 and these values indicates that they are very negatively skewed grains are coarser samples. The second population of the grains is comprised the samples which are ranging between +0.1 to +1 which indicates that this population of the grains is positively skewed. The percentage of the negatively skewed grains is more than the positively skewed grains. So, on this base we can explain that the population of coarse grains is more than the fine grains.

The kurtosis values indicate that the population of the grains is comprised of very platykurtic and the second population is extremely leptokurtic.

Standard deviation/Mean

Mean size has been used for determining the sorting nature of the sediments. This graph gives us an indication that the coarser sediments are more than the fine sediments.

Skewness/Mean

The combination of skewness and mean indicates the depositional energy of the river and depositional environment of the streams. The grain populations which have the values of skewness between -0.50 to +0.50 and the mean values between 0 to 2.0 give the representation of the high energy river. On the contrary, the grains population which has the values of skewness between +0.50 to +2.50 and the mean values between 3 to 7 gives the representation of the quit water or slow depositional environment.

Median/Kurtosis

The combination of kurtosis and median indicates that there is large population of samples which are ranging between very platycurtic to leptokurtic

The diffraction of X-rays is one of the methods used for the determination of clay minerals. The identification of the composition of the fraction <2 µm provide us very interesting results regarding the types of miners and environmental conditions during the deposition of the sediments. The present mineral analysis indicates that the different stratigraphical settings provide different types of clay mineralogy. We can produce the following type of clay mineralogy and associated depositional environments.

1) The results of clay mineral analysis of the sample BARQ1, collected from the base of the T4 at the railway station of the Vila Nova da Barquinha, indicates that there are three clay minerals are identified in the sample. The name of minerals is kaolinite, ilite and smectite. Kaolinite is found in the abundant quantity in the sample. Kaolinite is originated from the digenesis of feldspar. It indicates that there was a temperate climatic condition during the deposition of the sediments. A small percentage of geotite clay minerals is also be identified from this sample.

2) The sample OP2 A, 2D, 2H, collected from the T6/alluvial plain at Outeiro do Pedro, indicates that three clay minerals are identified. The name of the minerals is smectite, kaolinite and ilite. Smectite is identified in abundant quantity from these sediments. Smectite is a group of minerals. When the group of smectite mineral mixes with water, then it starts to expand and it breaks the structure of the sediments.

3) Three minerals have been identified from the sample SOND 5A (Sondage 5 – Outeiro do Pedro). The name of the identified minerals is vermilite, kaolinite and ilite. The vermilite is the most abundant mineral from these sediments. Vermilite is made of mafic minerals. The abundance of vermilite mineral indicated that there are temperate climatic conditions during the deposition of the sediments. The sample SOND 5B has abundance of ilite. Ilite could be derived from the weathering of kaolinite and muscovite. The abundance of ilite shows that there could be a low temperature and it could be also abundant in the hydrothermal conditions.

4) Vermilite, ilite and kaolinite are the three mineral which are indentified from the sedimentological sample (RPP1, RPP 2 and RPP3). The vermilite is most abundant mineral in the sediments of the sample (RPP1 and RPP2), whereas Kaolinite is abundant in the sample RPP3. The abundance of vermilite and kaolite indicates that there was temperate environment.

5) The sample RPP8 and RPP10 shows the presence of vermilite, kaolite and smectite. But, we can see the abundance of ilite. The abundance of ilite shows that there could be a low temperature and it could be also abundant in the hydrothermal conditions.

SUMMARY

The present research obtained interesting depositional and palaeoenvironmental data concerning the sedimentary record at Outeiro do Pedro, in Mação municipality and also in the areas of Ortiga and Vila Nova da Barquinha. On the basis of the six trenches dug out, I have described the stratigraphy, sedimentology and interpretation of the artifacts associated with the geological context of alluvial plain/T6 of the Tejo River in the Mação-Ortiga region. The stratigraphical units are the following:

i. the first unit is comprised of schist, a metamorphic rock; ii. the second unit is comprised of boulders, cobbles and pebbles; this indicates that fluvial energy was very strong during the deposition of these sediments. On the basis of this facies, observed in the trenches 6 and 2, in which very few artifacts have been recovered, it can be interpreted that the T6 terrace can be ascribed to the Upper Pleistocene; iii. The third unit comprises pebble, gravels and coarse sand; many artifacts have been recovered from this unit as well; iv. Pebbles and sand matrix; v. Alluvial sand. The composition of the <2 µm fraction (XRD) of selected samples show that in Sondage 2 smectite is abundant, Sondage 5 has abundance of vermiculite and SOND 5B has abundance of illite.

The Tejo River T4 (Q3) terrace in the Ortiga locality can be described and interpreted in the following way:

Unit 1. It contains basal rock, namely a metamorphic rock-schist.

Unit 2a. It is comprised of very consolidated coarse sand, which is yellow in colour. It could be interpreted as Miocene. The thickness of this yellow coarse sand is 40 cm. Layer 2b is represented by coarse sand with a bit of gravel. This unit is not horizontally represented through the length of the section and it is followed by the some pebbly gravelly coarse structure that could be identified as a river channel. This indicates a high energy deposition of the sediments. In this same level, the deposits of gravels are in an angle of 15°. This pebbly gravelly unit is followed by a small layer of coarse consolidated sand which is 40 cm thick and represents a very low energy of it deposition.

Unit 3a. It is comprised of boulders, cobbles and pebbles, gravels and separated by coarse sand at a regular interval. Flakes have been recovered from this unit. The presence of boulders and cobbles indicates very high energy river stream during this time period. Some erosive events can also be observed in this unit. In between the sand and pebbles discontinuity can be observed.

Unit 4. represented by pebbles and coarse sand.

Unit 5. marked by coarse sand.

Unit 6. comprised of colluvium and actual soil.

In Vila Nova da Barquinha area, the stratigraphical sequence of T4 (Q3) can be recorded in following way:

i. The first unit is comprised of Miocene sand, exposed at the section of the archaeological site of Ribeira da Atalaia at the Vila Nova da Barquinha sect. The colour of these Miocene sediments is yellow in Ribeira da Atalaia, and it is ranging between grey to white at the base T4 (Q3) section at Vila Nova da Barquinha railway station. The grain size analysis indicates that these Miocene sediments are comprised of medium sand, very fine sand and coarse silt. They are very poorly sorted, the skewness is positive and with platykurtosis. The X-ray diffraction of the <2 µm fraction indicates that the clay minerals are dominated by kaolinite.

ii. The second unit is identified as a channel structure. The Ribeira da Atalaia archaeological section is represented by boulders, cobbles and pebbles. The boulders are absent in the Q3 terrace section (located on the road coming from Tomar to the football field at Atalaia, and also absent in the Q3 section at Vila Nova da Barquinha railway station. It can be observed that the Q3 of Ribeira da Atalaia was deposited during a time period when the energy of the river was very high. But, it is also important to note that the size of boulders at Ribeiera da Atalaia is smaller than the boulder size at the Q3 section at the village of Ortiga.

iii. The top part of the T4 (Q3) terrace is represented by the archaeological site that is located at the Atalaia football field. It is very interesting that the all the lower Palaeolithic artifacts have been found at the base of T4 level at the Vila Nova da Barquinha and some artifacts have also been recovered from the channel structure of the T4 terrace base at the village of Ortiga.

The grain size analysis of the T4 terrace sediments indicates that they are comprised of medium to very coarse sand and they are very poorly sorted. The grains population is very positively skewed and the kurtosis is leptokurtosis. The X-ray diffraction of the <2 µm sediment fraction indicates abundance of vermilite and kaolinite.

On the basis of morphology and typology of the stone tools, they can be divided into several categories. The first category is comprises core tools. The cores assemblage is comprised of five specimens. They were collected from the railway station section at Vila Nova da Barquinha and Outeiro do Pedro. On the basis of length, breadth and thickness of the core tools, they can be classified into small and large cores. It seems that the large cores were used for the production of heavy duty flakes, whereas the small cores were used to produce light duty flakes. A striking platform can also be observed in all of them. Most of the cores show the negative bulb, indicating that flakes were removed from the core and it also indicates that the result of the strike could have been an end flake.

The second category of artefacts is comprised of unifacial and bifacial tools. We have also collected handaxes. The third category is comprised of flakes which are small in size. These flakes have been recorded from the various contexts. The third category of artifacts is comprised of small side and end flakes.

On the basis of the literature, these artefacts can be classified into three categories:

1) The first category is classified on the base of the geological context of the artefacts. There are some artefacts which have been collected from the river terrace

T4 (Q3) of the Tejo River (at Barquinha and Ortiga). This terrace, T4, has produced the earliest records of human occupation in central Portugal, which are dated as early as OIS 6-9 (300 ka, according to Oosterbeek, 2007) at the archaeological site of Ribeira da Atalaia in Vila Nova the Barquinha.

2) The second category is made of artifacts which have been found in the geological context of the alluvial plain/T6 of the Tejo River at Outeiro do Pedro trenches (Mação locality). The artifacts are a few flakes and cores.

3) The third category is for the artifacts which have been recorded from surface scatters. These artifacts have been found in surface contexts of the T6 terrace at Ribeira da Atalaia in Vila Nova da Barquinha and Outeiro do Pedro, at the surface level of the T4 at Carregueira vand Chamusca villages. The group contains bifaces, unifacial tools, cores and few flakes; they are extensively made on local river quartzite pebbles and cobbles.

The above described lithic assemblages recorded from surface are very common in Portugal and they are called "Languedocense". This term is frequently used in the present archaeological literature for describing the lithic assemblage which is morphological and typologically very similar to the artifacts which have recorded from the Middle/Lower Palaeolithic period to the Bronze Age. These archaeological findings are very similar to the lithic assemblage which is described by Grimaldi et al. (1997, 1998) and Raposo (1986) in the Portuguese Middle Tejo and Breuil (1937) described very similar artifacts in the region of Languedoc in southern France.

The origins of geoarchaeological investigations in the Middle Portuguese Tejo can be traced back to the scientific enquiry that began with the period of scientific reasoning of the 19th century. What has revolutionised the approach have been the advances in the field of Earth Sciences making the archaeology of the region of Middle Tejo far more challenging. The interpretation of the rich archaeological records of this region received great impetus with the findings of new sites in primary geoarchaeological context. Several goal-oriented case studies in the Alto Ribatejo region eventually led to realisation of the significant position that the Middle Tejo region has with respect to early human behaviour, migrations, ecological adaptations, cultural transformations, and environmental conditions in the formative period of human evolution, especially in the context of Iberian Prehistory.

Acknowledgements

I want to thank Prof. F. Semha and Prof. L. Oosterbeek for choosing me for Erasmus Mundus fellowship and provide me research opportunities in European prehistory. My special thanks go to Prof. Pedro P. Cunha, Prof. P. Rosina, and Prof. Martins for helping me stratigraphic and lithostratigraphic interpretation for the river terraces of Tagus River. I want to thank Prof. P.P. Cunha for teaching me minute details of sedimentology and grains size analysis. I must my heartiest thanks to Mr. Hugo, Mr. Pedro Cura who provides me unconditional help during the field and lab analysis. I also must mention the European Union and European tax payers who supports this fellowship and provide research opportunities to the researcher from third world country. The last, but not least, I will be failing in my duty if I do not thank Dr. A.R. Sankhyan for inviting my contribution and accepting my full report, besides taking much trouble to recompose the text and the figures.

References

BARBOSA, B.P. 1995. Alostratigrafiae Litostratigrafia das unida descontinentais da Bacia Terciária do Baixo Tejo. Relacoes com o eustatismoe a tectonica. PhD dissertation Univ. Lisboa, 253.

BARBOSA, B.P., PENA, DOS & REIS, R. 1989. Litoestratigraphiae modelo deposicional dos sedimentos aluviais do Neogenico superior da Baciado Tajo (Tomar-Lavvre), Portugal. *Comune Service of Geology of Portugal* 75, 89-97.

BREUIL, H. 1937. Terrasses et quartzitees tailles de la hautevallee de la Garonne, *Bulletin de la Societe Preistorique, Francaise* XXXIV (2): 104-128.

BREUIL, H. & GEORGES, Z. 1942. Contribution a l'itude des industries paliolithiques du Portugal et de leun rapports avec la giologie du Quaternaire. *Comunicq Besdus Servigos Geoligicos de Portugal* XXIII, Vol. 1. Lisbon.

BREUIL, H. & GEORGES, Z. 1945. Contribution a l'itude des industries paliolithiques du Portugal et de leun rapports avec la giologie du Quaternaire. *Comunicq Besdus Servigos Geoligicos de Portugal mTI,* Vol. 2. Lisbon.

CARVALHO, A.M. 1968. Contribuição para o conhecimento geológico da bacia terciária do Tejo. *Memórias dos Serviços Geologicos de Portugal* 15, 210.

CORRAL, I. 1998a. Depositos Cuaternarios en el area de Constancia-Barquinha-Entroncamentoy la Rib. del Bezelga. In: Cruz, A.R., Oosterbeek, L., Reis, R.P. (Eds.), *Arkeos 4, Quaternário e Pré-História do Alto Ribatejo (Portugal).* CEIPHAR, 59-144.

CORRAL, I. 1998b. Seccionescom material arqueologico en estrato en las proximidades de Atalaia. In: Cruz, A.R., Oosterbeek, L., Reis, R.P. (Eds.), *Arkeos 4, Quaternario e Pre-História do Alto Ribatejo (Portugal).* CEIPHAR, 227-250.

CUNHA, P.P. 1992. *Estratigrafia e sedimentologia dos depósitos do Cretácico Superiore Terciário de Portugal Central, a leste de Coimbra.* PhD Thesis. Univ. Coimbra, 262.

CUNHA, P.P., BARBOSA, B.P. & REIS, R.P. 1993. Synthesis of the Piacenzian onshore record, between the Aveiro and Setubal parallels (western Portuguese margin). *Ciências da Terra* 12, 35-43.

CUNHA, P.P., MARTINS, A.A. & DAVEAU, P.F. 2005. Tectonic control of the Tejo River fluvial incision during the Late cenozoic, in Radao—Central Portugal (Atlantic Iberian border). *Geomorphology* 64, 271-298.

CUNHA, P., MARTINS, A., HUOT, S., MURRAY, A., & RAPOSO, L. 2008. *Dating the Tejo River lower terraces (Ródão, Portugal) to assess the role of tectonics and uplift. Special Issue of "Geomorphology"* (Impact of active tectonics and uplift on fluvial landscapes and river valley development) P.G. Silva, F.A. Audemard & A.E. Mather (Editors).

CUNHA, P.P., ALMEIDA, N.A.C, AUBRY, T., MARTINS, A.A. et al. 2012. Records of human occupation from Pleistocene river terrace and aeolian sediments in the Arneiro depression (Lower Tejo River, central eastern Portugal). *Geomorphology* 165-166, 78-90.

CUNHA-RIBEIRO, J.P. 1995. Elementos para o estudo da cadeia operatoria de produçao de bifaces da indústria acheulense do Casal do Azemel. In: *Actas do 1_Congresso de Arqueologia Peninsular, vol. 8.* Vale do rio Lis, Distrito de Leiria, Centro de Portugal, 51-68.

CUNHA-RIBEIRO, J.P. 2000. A indústria lítica do Casaldo Azemel no contexto da evoluçao do Paleolítico Inferior na Iberia Ocidental. In: *Actas do 3_Congresso de Arqueologia Peninsular, vol. 2*, 137-167.

CURA, S. & GRIMALDI, S. 2009. The intensive quartzite exploitation in Midlle Tagus Valley Pleistocene open air sites – the example of Ribeira da Ponte da Pedra. In *Technological Analysis on Quartzite Exploitation*, Oxford: BAR Publishing, 2009, p. 49-56 (BAR S1998). Atti di: XV World Congress UISPP, Lisbon.

DAVEAU, S. 1993. Terraços fluviais elitorais. In: *O Quaternário em Portugal, Balanço e perspectivas*, Colibri, 17-28.

DELGADO, J.F.N. 1904-7. Contribucoes para estodu dos terrenos paleozicos. *Commune Service of Geology of Portugal*, 143.

DIAS, M.I., PRUDÊNCIO, M.I., FRANCO, D., CURA, S. et al. 2009. Luminescence dating of a fluvial deposit sequence: Ribeira da Ponte da Pedra – Middle Tagus Valley, Portugal. In: Dias, M.I., Prudêncio, M.I. (Eds.), Archaeometry. *Proceedings of the XV UISPP Congress*. Oxford, BAR Publishing, 103-113.

FERANDEZ, C. 1998. Depositos Cuaternarios en el area de contancia-Barquinha-entrocamneto y la Rbadel Bezelga. In: Cruz, A.R., Oosterbeek, L., Rui, P.R. (coord.), *Territórios, mobilidade e povoamento no Alto-Ribatejo. IV: Contextos macrolíticos*, Tomar: Arkeos 4, CEIPHAR – Centro Europeu de Investigação da Pré-História do Alto Ribatejo, 207-246.

GONCALVES, F. 1969-70. Contribuicoes para o estudo dos terrenos paleozoicos. *Commune Serviceof Geology of Portugal, Lisboa*, (1): 56-122.

GRIMALDI, S. & ROSINA, P. 2001. O Pleistoceno medio final no Alto Ribatejo (Portugal) in *Arkeos perspectivas em diálogo*, v., n. 11, 89-115.

GRIMALDI, S., ROSINA, P. & FERNANDEZ, I. 1997. Interpretazione Geo-Archeologica di Alcune Industrie Litiche "Languedocensi" del Medio Bacino del Tejo. In Cruz, A., Oosterbeek, L., Reis, P., (Ed.), Quaternário e Pré-Historia do Alto Ribatejo (Portugal). *Arkeos 4*, 145-226. Ceiphar, Centro de Pré-Históriado Instituto Politécnicode Tomar.

GRIMALDI, S., ROSINA, P. & CORRAL, I. 1998. Interpretazione Geo-Archeologica dial cune industrielitiche Languedocensi del Medio Bacino del Tejo. In: Cruz, A.R., Oosterbeek, L., Reis, R.P. (Eds.), *Arkeos 4* Quaternario e Pre-Historia do Alto Ribatejo (Portugal). CEIPHAR, 145-226.

GRIMALDI, S., ROSINA, P., CRUZ, A.R. & OOSTERBEEK, L. 1999a. A geo-archaeological interpretation of some "Languedocian" lithic collections of the Alto Ribatejo (Central Portugal). In: Cruz, A.R., Milliken, S., Oosterbeek, L., Peretto, C. (Eds.), *Arkeos* 5, Human Population Origins in the Circum-Mediterranean Area. CEIPHAR, 231-241.

GRIMALDI, S., ROSINA, P., BOTON, F. & GARCIA, F. 1999b. A behavioral perspective on "archaic" lithic morphologies in Portugal: the case of Fonte da Moita open air site. *Journal of Iberian Archaeology*. ADECAP 1, 33-57.

GRIMALDI, S., ROSINA, P. & BOTON, F. 2000. Um sítio ao ar livre do Plistoceno Médiono Alto Ribatejo (Portugal): Fonte da Moita. In: *Proceedings 3 Congresso de Arqueologia Peninsular, Porto, vol.* 2. ADECAP, 123-131.

GRIMALDI, S. & CURA, S. 2006. The quartzite exploration of in a middle Pliestocene open air site: Riberia da Ponte De Pedra (central Portuga). in *Proceedings of the International Workshop 15, Technology analysis on quartzite exploitation,* Oxford: BAR Publishing, In: corso di stampa, p. – (BAR). Atti di: XV UISPP Congress, Lisbon.

GUTIERREZ-ELORZA, M., GARCIA-RUIZ, J.M., GOY, J.L., GRACIA, F.J. et al. 2002. Quaternary. In: Gibbons, W., Moreno, T. (Eds.), The Geology of Spain. *The Geological Society*, 335-366.

LEMORINI, C., GRIMALDI, S. & ROSINA, P. 2001. Observacoes funcionais e tecnológicasnum habitat Paleolítico: Fonte da Moita (Portugal central). In: Cruz, A.R., Oosterbeek, L. (Eds.), *Arkeos* 11, Territórios, mobilidade e povoamento no Alto-Ribatejo. II: Santa Cita e o Quaternario da regiao. CEIPHAR, 117-140.

LOURANDOS, H. 1996. Change in Australian prehistory: Scale, trends and frameworks of interpretation. In S. Ulm, I. Lilley and A. Ross (eds), *Australian Archaeology '95. Proceedings of the 1995 Australian Archaeological Association Annual Conference*, pp. 15-21. Tempus 6. St Lucia, QLD: Anthropology Museum, University of Queensland.

LUSSU, T., ROSINA, P., OOSTERBEEK, L. & COSTA, F. 2001. O Musteriense de Santa Cita (Tomar, Alto Ribatejo, Portugal): investigacao e conservacao. In: Cruz, A.R., Oosterbeek, L. (Eds.), *Arkeos* 11,

Territórios, mobilidade e povoamento no Alto-Ribatejo. II: Santa Cita e o Quaternário da região. CEIPHAR, 13-70.

MARKS, A.E., MONIGAL, K. & CHABAI, V. 1999. Report on the initial excavations of Brechadas Lascas and Galeria Pesada (Almonda, Portuguese Estremadura). *Journal of Iberian Archaeology* 1, 237-250.

MARKS, A.E., BRUGAL, J.PH., CHABAI, V.P., MONIGAL, K. et al. 2002a. Le gisement Pleistocene moyen de Galeria Pesada, (Estrémadure, Portugal). *Premiers resultats. Paléo* 14, 77-100.

MARKS, A.E., MONIGAL, K., CHABAI, V.P., BRUGAL, J.PH. et al. 2002b. Excavations at the Middle Pleistocene Cave Site of Galeria Pesada, Portuguese Estremadura: 1997-1999. *O Arqueólogo Português, Série IV* 20, 7-38.

MENDEZ, Q.E., QUINTEIRO, V.R., GOMEZ, S.M., PEREZ-GONZALEZ, A. & LED O BERNARDEZ, M. 2006. Miscelanea en homenajea Victoria Cabrera. Zona Arqueologica 7, 184-193.

MARTINS, A.A. 1999. Caracterização morfotectonica e morfossedimentarda Bacia do Baixo Tejo (Pliocenico e Quaternario). PhD Thesis. University of Evora, 500.

MARTINS, A.A., CUNHA, P.P., HUOT, S., MURRAY, A. & BUYLAERT, J.P. 2009a. Geomorphological correlation of the tectonically displaced Tejo River terraces (Gavião-Chamusca area, Portugal) supported by luminescence dating. *Quaternary International* 199, 75-91.

MARTINS, A.A., CUNHA, P.P., BUYLAERT, J.P., HUOT, S. et al. 2010a. K-feldspar IRSL dating of a Pleistocene river terrace sequence of the Lower Tejo River (Portugal, western Iberia). *Quaternary Geochronology* 5, 176-180.

MARTINS, A.A., CUNHA, P.P., ROSINA, P., OOSTERBECK, L. et al. 2010. Geoarchaeology of Pleistocene open air sites in the Vila Nova da Barquinha – Santa Cita area (Lower Tejo River basin, central Portugal). *Proceedings of the Geologists Association* 121, 128-140.

MCMANUS, J. 1998. Grain size determination and interpretation. In: Maurice Tucker (Ed.) Technique in sedimentology. Blackwell publishers, London, 63-108.

MOZZI, P. 1998. Evoluzione Geomorfologica della Bassa Valle del Fiume Nabão. *Arkeos* 4, 37-58.

MOZZI, P., RAPOSO, L., CRUZ, A.R., OOSTERBEEK, L. & REIS, R.P. 1999. Morpho-stratigraphy of Quaternary deposits and the archaeological record: the case of the Tejoand Nabao valleys (Ribatejo, Portugal). In: Cruz, A.R., Milliken, S., Oosterbeek, L., Peretto, C. (Eds.), *Arkeos 5*, Human Population Origins in the Circum Mediterranean Area: Adaptations of the Hunter-gatherer Groups to Environmental Modifications. CEIPHAR, 63-84.

MOZZI, P., AZEVEDO, T., NUNES, E. & RAPOSO, L. 2000. Middle terrace deposits of the Tagus river in Alpiarça. Portugal, in relation to early human occupation. *Quaternary Research* 54, 359-371.

OOSTERBEEK, L., CRUZ, A.R., CURA, S., ROSINA, P., GRIMALDI, S. & GOMES, J. 2004. "Ribeirada Ponte da Pedra – Relatório da campanha de escavação de 2003". *Techne* 9, Arqueojovem-Tomar, 21-54.

OOSTERBEEK, L. CURA, S., ROSINA, P., GRIMALDI, P., CRUZ, A. & GOMES, J. 2007. Cadre géochronologique et stratégies de gestion du territoire des cultures u Pléistocène Inférieur et Moyen dans la vallée du Tage (Portugal) *Les cultures à bifaces du Pléistocène inférieur et moyen dans le monde. Emergence du sens de l'harmonie* Colloque International à Tautavel (25-30 Juin, 2007).

OOSTERBEEK, L., GRIMALDI, S., RO SINA, P., CURA, S., CUNHA, P. & MARTINS, A. 2010. The earliest Pleistocene archaeological sites in western Iberia: present evidence and research prospects. *Quaternary International*, v. 223-224, n. 10.1016/2010.01.024 (2010), 399-407.

PENCK, A. & BRUCKNER, E. 1909. *Die Alpenim Eiszeitalter*. Tauchnitz, Leipzig.

RAPOSO, L. 1986. Mustierense. Mustiero-Languedocence ou Languedocense? *Arqueologia* 14, 13-21.

RAPOSO, L. 1987. Os mais antigos vestıgios de ocupacao humana paleolítica na região de Rodao. In: Da Pré-História à História, homenagem a O. Veiga Ferreira, Editorial Delta, 153-178.

RAPOSO, L. 1995a. O Paleolıtico. In: Medina, J. (Ed.), *História de Portugal, vol. 1*. Clube Internacional do Livro, 23-85.

RAPOSO, L. 1995b. Ambientes, territórios y subsistencia en el Paleolıtico Medio de Portugal. *Complutum* 6, 57-77.

RAPOSO, L. 2000. The Middle-Upper Palaeolithic transition in Portugal. In: Stringer, C., Barton, R.N.E., Finlayson, C. (Eds.), *Neanderthals on the Edge: 150th Anniversary Conference of the Forbes'* Quarry Discovery, Gibraltar. Oxford Books, Oxford, 95-109.

RAPOSO, L. 2002. Um se culo de estudos no Paleolítico Médioem Portugal: balanço e perspectivas. *Arqueologia e Historia* 54, 25-39.

RAPAOSO, L. & CARDOSO, J.L. 1998. *Osito-palaeolitico da conceicao* CEMA-Alcochete.

RAPOSO, L., CARREIRA, J.R. & SALVADOR, M. 1985. Aestação Acheulense final de Milharós, Vale do Forno, Alpiarça. *Proceedings of I Reunião do Quaternário Ibérico 2*, 41-60.

RIBEIRO, J.C. 1886. *Descripcao do terreno quaternário das bacias dos rios Tajo e Sado*. Mem. Com. Geol. Piort., Lisboa, 166.

RIBEIRO, J.C. 1990. Os primeiros habitantes. In: Serrão, J., Marques, A. (Eds.), *Nova História de Portugal, vol. 1*. Editorial Presenca 15-74.

ROSINA, P. 2001. Stratigraphie et Geomorphologie des Terrasses Fluviatiles de La Moyenne Vallee du Tage (Haut Ribatejo-Portugal). (In Ana Rosa Cruz and Luiz Oosterbeekeds). Perspectivas em Dialogo: Territorios, Mobilidade e Povoamento no Alto Ribatejo II—Santa Cita e o Quaternario da Regio. *Arkeos* 13, 11-52.

ROSINA, P. 2002. Stratigraphie et Geomorphologie des Terrasses Fluviatiles de La Moyenne Vallee du Tage (Haut Ribatejo-Portugal). (In Ana Rosa Cruz and Luiz Oosterbeekeds). Perspectivas em Dialogo: Territorios, Mobilidade e Povoamento no Alto Ribatejo II—Santa Cita e o Quaternario da Regio. *Arkeos* 13, 11-52.

ROSINA, P. 2004. *I Depositi Quaternarinella Media Valle del Tago (Alto Ribatejo –Portugallo Centrale) e l industrie lithicheassociate. Dottorato Internazionale in "Dinamicheambiebrali, umane e comportamentali".* Universita di Ferrara.

ROSINA, P. & CURA, S. 2010. Interpretation of lithicremains in fluvial terracecontexts: an example from Central Portugal. *Tome XII, Numero* 1, 2010, p. 7-24.

ROSINA P., OOSTERBEEK, L., JAIME, A. & CURA, S. 2004. Archaeological sites associated with Tagus Middle valley deposits (Alto Riba Tejo-Portugal) *Geoarqueología y Conservación del Patrimonio* M. Santonja, A. Pérez-González, M.J. Machado (eds.) (20-26). 2004 I.S.B.N.

ROSINA, R., CURA, S. & GRIMALDI, S. 2006. The Middle Upper Pleistocene open air site of Ribeira da Ponte da Pedra (Midlle Tagus Basin, Central Portugal) in Proceedings of the International Colloquium 39 *Lower Paleolithic: short and long chronologies, lithic industries linear evolution models, transition to the middle Palaeolithic*, Oxford: BAR Publishing, In corso di stampa, p. 165-176. (BAR). Atti di: XV UISPP Congress, Lisbon.

SANTISTEBAN, J.I. & SCHULTE, L. 2007. *Fluvial networks of the Iberian Peninsula: a chronological framework, Quaternary Science Reviews* 26, p. 2738-2757.

SANTONJA, M. & PÉREZ-GONZALEZ, A. 2000-2001. El Paleolítico inferior en el interior de la Península Ibérica. Un punto de vista desde la geoarqueología. *Zephyrus* 53-54, 27-77.

SHACKLETON, N.J., BERGER, A. & PELTIER, W.R. 1990. An alternative astronomical calibration of the Lower Pleistocene timescale based on ODP site 677. *Transactions of the Royal Society of Edinburgh* 81, 252-261.

SILVA, P.G. 2003. El Cuaternario del valle inferior del Manzanares (Cuenca de Madrid, España). *Estudios Geológicos* 59, 107-131.

SILVA, P.G., ZAZO, C., BARDAJÍ, T., BAENA, J., LARIO, J. & ROSAS, A. 2009. Tabla Cronoestratigráfica del Cuaternario de la Península Ibérica. *AEQUA*, http://www.aequa.es/.

WATERS, M.R. & KUEHN, D.D. 1996. The geoarchaeology of place: the effect of geological processes on the preservation and interpretation of the archaeological record. *American Antiquity* 61, 483-497.

12. MORPHOMETRICS OF THE FRONTAL BONE: A NEW METHOD FOR MEASURING INTRACRANIAL PROFILES

Yannick KORPAL
Associated Researcher, Cresoi Lab., University of Reunion Island, France
yanndoct@gmail.com

Abstract: *From a flat conformation, still existing in greater apes, to a more straightened and rounded frontal characteristic of modern humans, an evolutionary trend in the frontal conformation is observed in the present study about the midsagittal frontal profile. Previous works described the variation in the external shape of the frontal bone useful for taxonomic differentiation. But, Bookstein et al. (1999) introduced an endocranial approach which inferred that the shape of the endocranial frontal surface remains very stable. This study advanced this approach and carried out a geometrical procruste analysis of the midsagittal cranial profile introducing new semi-landmarks to describe shapes and the frontal medium sagittal profiles. It is observed that the external shape varies greatly across taxa, but the internal (endocranial) shape remains very much the same right from the apes (Pan & Gorilla) through prehumans, Homo erectus and archaic Homo sapiens to the modern humans. However, the present study introduces a new ellipse method which enables differentiation of the endocranial profiles across taxa permitting the intergroup diagnosis. Hence, we can distinguish the endocranial conformations of the frontal midsagittal profiles of man from the ape, and now can use them for specific diagnosis.*

Keywords: *frontal cranial profile, midsagittal, geometry, ellipse, endocranial, architecture, basicranial, landmarks*

INTRODUCTION

With the discoveries of Engis, Forbes' Quarry, Gibraltar and the Neanderthal, the frontal bone is an object since the very early years of Human Paleontology (Schmerling, 1833; Busk, 1861; Schaafhausen, 1861 (1858); Huxley, 1863; Lyell, 1863; Dubois, 1896). Brodmann (1912) and Weidenreich (1947) also considered the frontal to be distinctive between apes and men, and between Middle and Late Pleistocene archaic *Homo*, in particular Neanderthals.

The frontal bone is the subject of recent scientific work (Tillier, 1977; Van Hoesen, 1997; Bookstein et al., 1999; Nambiar et al., 1999; White & Folkens, 2000; Sladek et al., 2002; Oostra et al., 200 5; Guipert & Mafart, 2005; Kumar, 2009; Athreya, 2012 (2013)). This frontal bone represents both a cranium and a facial bone. It is involved in the mechanical protection of the brain, complementing the meninges. The ancient Greeks used to see highest mental attributes associated with frontal lobes (Hublin et al., 2010, Lewin, 1984, Habib et al., 2000). The frontal bone so has two surfaces with much contrasted functions, and so adaptive constraints.

Thus exocranial morphometrics cannot look like the same than intracranial conformation. Indeed, Bookstein et al. (1999), using Procruste methods, demonstrated intracranial shape stability among the whole human lineage, including very old fossil species. Though, the evolution of the front is part of the dynamics of the general evolution of the human skull, and so is a full component of it, with great contribution even if analyzed as an isolated bone.

The continuous development of analytical techniques includes three-dimensional slices from medical imaging (Zollikofer, 2002; Mafart, 2002; Mafart et al., 2004). Associated with methods like Procrustes superimposition, it becomes the basis of new approaches in human paleontology. Such methods permit spatial locations of la ndmarks and morphometric analyses using virtual images. Previous works used dimensional analysis methods, which are based on a small number of landmarks. In this way, analyzing the frontal bone equates to consider it in the sagittal plane; and from only three craniological points (bregma, glabella, nasion) associated with three other extra-frontal sagittal points: opisthion, inion, lambda (Millet, 2000; Penin et al., 2002; plus Penin & Baylac, 1999 for non-human primates).

Modeling the curvature of the frontal squama necessitate more points (Fig. 12.1). The number of studied skulls remains small, increasing the effect of the lack of fossils and their dispersion in time and space: *Australopithecines*, *Homo habilis*, *Homo ergaster / erectus* and *H. Neandertalensis / H. sapiens* cover more than 3 Mya (Bookstein et al., 1999; Detroit, 2000; MacLeod et al., 2003). The morphological characteristics of the frontal bone in *Homo neanderthalensis* were analyzed in numerous studies (Heim, 1986). Ontogenetic development of the skull and therefore frontal bones of fossil men benefited from the contribution of 3D imaging (Ponce de Leon 2001, Zollikofer, 2001), but little was undertaken considering evolutionary dynamics.

Though very focused compared to a full cranial morphometric analysis, an analysis of a single bone

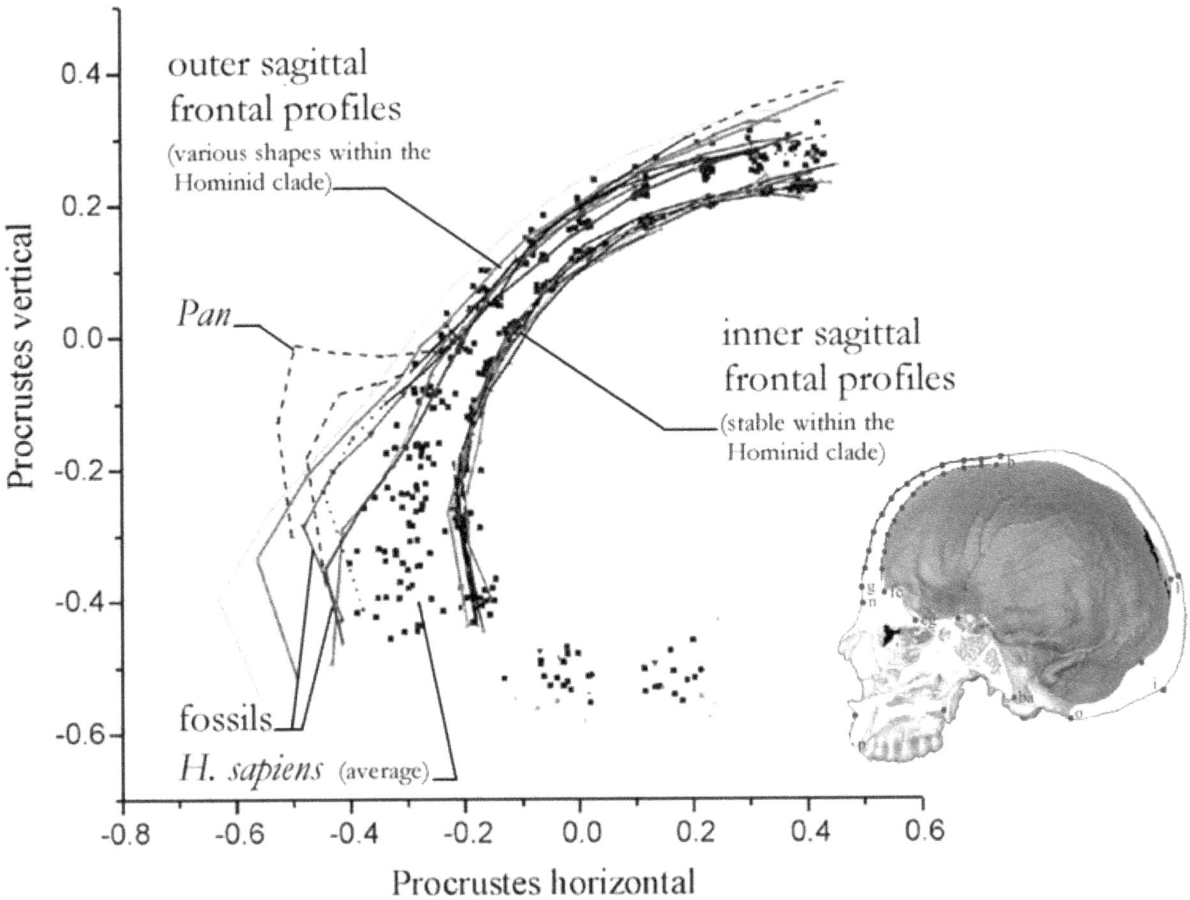

Fig. 12.1. Procrustes fits of 24 specimens of modern apes, human, archaic hominids and australopiths: The archaic and modern samples are indistinguishable by using Procrustes method. a: Modern forms only. b: Archaic hominids, australopith and Pan. Color, archaic Homo; short dashes, Sts 5; long dashes, Pan troglodytes. From Bookstein et al., 1999

shows certainly some advantages (Harvati, 2002). O ne can find with single boned studies a solution to taphonomic issues. The number of complete skulls of Pleistocene fossils is limited. Then most suffered fractures and deformations in sediments deposits and have to be restored or even reconstructed. The endurance of the frontal bone turns it into one of the best preserved bones. And considering that this frontal bone protects regions of the brain whose functions have played a key role in the cognitive evolution of the genus *Homo*; the study of the morphological evolution of sole frontal bone in the fossil collection is therefore of great phylogenetic interest.

Following the midsagittal plane, the frontal cranial profile have been studied before appearance of modern Morphometrics methods (Boule, 1913; Heim, 1976; Conroy et al., 1978; Stringer et al., 1979; Arsuaga et al., 1997). The endocranial frontal midsagittal shape have been demonstrated quite recently to be very stable in shape from apes to modern Human. And that despite one of the oldest remarks of the discipline, the observation in the frontal bone of fore-tilting and rounding trends during evolution. First notices were made during the 19th century. This encouraged us to choose the frontal bone as a subject.

The aim of the present paper is to introduce a new way for measurements for inner cranial profiles and surfaces. This path uses geometrical properties of sagittal profiles for bones observed on radiographs. Modern anatomical reference sample is compared with some typical fossils, such as Neanderthal, *Homo heidelbergensis* or *Homo ergaster*. Metric and Procrustes investigations can now be completed by another non -destructive intracranial method. The study was performed without neither scale nor orientation modifications.

MATERIAL

The metric and angular data were already collected in order to conduct a comprehensive study in morphometric while using 2D supports. The protocol was then applied to x-rays taken in the three planes of the skull, following conventional orthodonctic standard (Korpal, 2005), and for each individual of the sample. This protocol compares endocranial and exocranial measurements and was applied to different series of dry skulls from the collections of both the Department of Comparative Anatomy and Department of Prehistory of the National Museum of Natural History (NMNH) and the Institute of Human Paleontology of Paris (IPH). The frontal bone is considered as part of an integrated cranial architecture.

These series included extant Hominoid species (26 *Pan* and *Gorilla*, including three young individuals, two for *Pan* and one for *Gorilla*), 107 skulls of recent Anatomically Modern Human. Modern skulls are from Europe, Africa, Asia and America. It ranges from midst 19th to midst 20th centuries. A small number of children and women are included.

Original fossils of the early upper Pleistocene (Qafzeh 6, Cro-Magnon, La Chapelle-aux-Saints, Oberkassel) are completed by earlier *Australopithecus africanus* (Stw 505, Sts 5) thanks to Witwatersrand University, Johannesburg (Dambricourt Malassé 2005). Archeological remains are comprised of 23 fossils from Europe and Africa, with addition of series from the medieval cemetery of Saint Nicolas de la Queue en Brie (Cho and Hadjouis, 2005).

METHOD

The sagittal projection view is the most informative about the frontal conformation. Metric and angular data were gathered to support evidences of a 3D study of the whole skull (Korpal, 2005) under the inspiration of orthodontic standards. Though the use of sagittal x-rays is still an issue due to the loss of 3D information (Simpson, 2005), 2D cephalometric radiographs are widely used and tested. Our present analysis relies upon lateral radiographs and is thus willingly restricted to a study of the mid-sagittal profile.

The present method uses the Frankfurt plane (Martin, 1928) as a reference. The first reason is because it is a classical and easy to define reference plane. Because of this it enjoys a large data volume and many studies. The choice was made also in a concern for comparability.

On a second hand, the Frankfurt plane may actually retain some of its biomechanical significance as a biomechanical orientation, namely the average head port in erected bipedal posture. Reference plane such as the vestibular reference plane (Delattre A. & Fenart R., 1960), corresponding to the position of semi-circular canals of the inner ear, reflects the relation between balance sensors and cranial architecture too.

Though, standardised Frankfurt plane is widely used in biodynamical cephalometric analysis associated with Dentomaxillofacial Orthopaedic. On a living subject, and regarding easier cephalometrics, the plane defined by Martin is best and easier to define.

Before any measurements, the work starts with the curves tracing. First, classical landmarks are located on the x-ray along the mid-sagittal plane that naturally appear on radiographs due to standardized alignment. In clockwise order we plotted: nasion (highest point of the nasal bones roots, in the midline); glabella (most anterior point of the frontal); bregma (external intersection point of coronal and sagittal sutures); lambda (external intersection point of interparietal and occipital suture); opist hocranion (most posterior point of the skull, on the occipital squama); inion (spiking point at the junction of the two nuchal lines); basion (most anterior point of the foramen magnum).

The focus of the present study being the frontal, we needed to plot also 'inner bregma' (the external bregma corresponding point on the inside of the braincase); foramen caecum (as an equivalent of endo-nasion). Projections of the optic canals and crista galli were used in Bookstein et al. (1999). This protocol prefer s the use of Spheno-Ethmoidal (SEth: midline spheno-ethmoidal suture) and Sellae (deeper point of sella turcica, in the midline).

Like in previous studies, up to ten semi-landmarks were also added from foramen caecum to endo-bregma, to complete the inner frontal profile and allow the conical curve test. As for Fourier type analyses, the inner frontal profile is simplified into a single curve (Fig. 12.1). A simple and traditional test in Geometry determines which type of curve is drawn. The Hexagrammon of Pascal leads to the determination of an ellipse.

The method resembles that of Fourier: the geometrical construction is divided into eight steps: (1) plotting the useful profile for the frontal bone, (2) verification of the conical type using the Hexagrammon of Pascal, (3) curve drawing (three possible methods), (4) Layout of the centre O, (5) tracing the combined diameters, (6) obtaining greater axis and minor axis, (7) determining focuses, (8) equation setting. The parametric equation is:

p=e x FH0 with e: excentricity. If $e<1$, it is an ellipse.
If $e=1$, it is a parabola. If $e>1$, it is a hyperbola.

The frontal sagittal ellipse is set to describe the frontal endocranial medial profile. Drawing the profile is effective by plotting semi-landmarks using the X-ray reading in sagittal projection: an endocranial arc. Measurements were taken with appropriate software. In geometry conical family includes four types of curves: circle, ellipse, parabola, and hyperbola. In the case of endocranial frontal profile, it is best described by an ellipse, confirmed by using the parametric equation.

An ellipse is a geometric construction. The first two parameters are identified linear dimensions. The greater (i.e. long) axis of the ellipse is the measure of length. The minor (i.e. short) axis of the ellipse is the measurement of height. These parameters are size measurements.

As the Frankfurt plane (FP) is the reference plane, the angle between the greater axis and the Fankfurt plane can be measured. It is the third direct metric traits studied. A final fourth parameter is determined by the method. As an ellipse, the curve on which are based previous measurements is itself a parameter: the shape (Fig. 12.2). In addition to frontal parameters, a few non-frontal measurements were taken: cranium overall length; cranium height; basicranial angle (the Lantzert type was the choice of the present protocol) [see also Tab. 12.1]. Statistical analysis involves the comparison of statistical

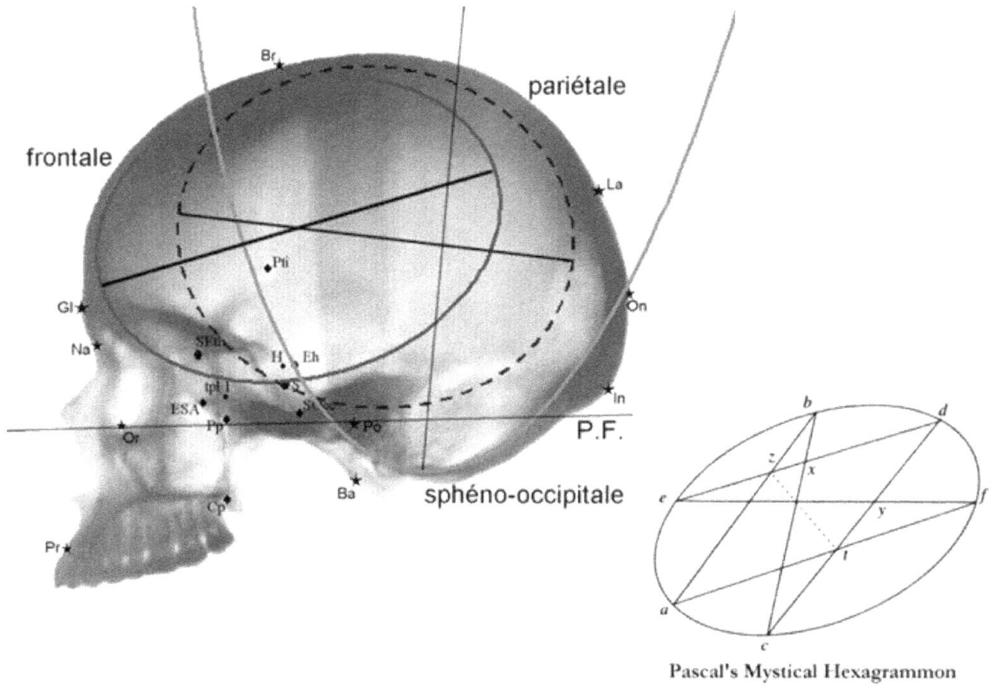

Fig. 12.2. Ellipses are drawn following endocranial mid-sagittal profile. The Frankfurt horizontal is the reference plane. Pascal's tool verifies the nature of the curve

average and analysis of variance test. The table I is a summary of the results. Simple one-to-one comparisons between traits values are applied in the present paper for the purpose of explanation. Extended analyses were done in previous work (Korpal 2005).

RESULTS

Drawing the ellipse is done by using Euclidean simple tools. The method can be commented as similar to that of Fourier type analyses. The Ellipse method proposes four parameters, each measured on the endocranial medial profile of the frontal bone in sagittal view. There are two linear measuring for size, one for shape and one for orientation. Data obtained for all four parameters concerning modern human are compared to African great apes *Pan* sp. and *Gorilla* sp., then to fossil hominids (Fig. 12.3).

Tab. 12.1 summarizes metric and statistical measurements with indication of standard deviation: three metric traits directly related to the frontal ellipse, plus maximum length and maximum height of the cranium. In addition the basicranial angle is also in the Tab. 12.2 because one observation made when drawing the endocranial ellipses was that the Sellae point (S) was always on the curve (Korpal 2005). The shape is so always described by an ellipse from apes to sapiens. Mathematically it is one of the conical curve, distinguishable from parabolic or hyperbolic curves. It is therefore one result similar to that obtained when using methods such as Procrustes analysis. Then the two axis of the ellipse are compared. The shift, from small ellipse found in apes and early hominins to large ellipses found in modern *Homo sapiens* and some from the Cro-magnon group, is consistent with what is expected in the evolution of the forehead.

Mean values for the frontal ellipse major axis inclination (Fig. 12.4) are 0,8° for *Gorilla* sp. and 1,1° for *Pan* sp.. The frontal ellipse is said to be horizontal in apes. The inclination is 0,5° for measured australopithecines which correspond to an apelike conformation. Archaic *Homo* generi shows an inclination around 7,7°, including the young *H. ergaster* WT15000. The consensus group for modern *Homo sapiens* is 14,5° of frontal ellipse inclination, a value found for earlier *H. sapiens* from Europe (Cro-Magnon) too. We may then conclude that the trend lead to a straighter frontal bone, an anterior-superior rotation of this osseous segment within the human lineage. Actually Neandertal as a group display a flatter frontal bone, even compared with *H. erectus*, with an inclination average of 3,4° regarding the Frankfurt Plane.

Classical Neandertal fossils show an even more horizontal ellipse at about 2° of inclination. The specificity of these European fossil men is confirmed using our new method. *Homo* heidelbergensis fossils considered on the Neandertal lineage displays less horizontal ellipse inclination than so called classical Neandertal. Such a phenomenon should correspond to a specific basicranial conformation: The neandertal evolutionary trend is on a basicranial longitudinal extension dynamic, more than a contraction one like in *Homo sapiens*. So the reverse behaviour of the frontal squama.

One notable case in this sample is A-1943-36. This skull belongs to a young hydrocephalic chimpanzee (* on Fig.

Tab. 12.1. The values of the cranial length and height with those of the frontal ellipses for modern H. sapiens, Cro-magnoid, H. neaderthalensis, H. erectus, Australopithecus, and Pan an d Gorilla

	ell-ft-inc	ell-ft-L	ell-ft-H	γ 1
STw_505	0,0	73,9	59,0	116,0
STS_5	1,0	64,5	55,9	120,0
KMN ER406	2,0			
WT_15000	7,0	87,2	66,0	
Atapuerca 5	15,0	96,6	69,6	111,0
Bodo	8,0	91,2	69,6	98,0
Kabwe	15,0	90,5	71,0	117,0
Gibraltar	2,0	94,0	76,7	120,0
la Chapelle aux Saints	2,0	91,4	63,5	
la Ferrassie 1	2,0	101,6	71,2	
Monte Circeo	2,0	106,7	72,7	133,0
Petralona	4,0	88,8	66,4	
Quina H18	12,0	103,1	76,6	
Saccopastore 1	5,0	111,5	73,0	
Steinheim	2,0	99,1	68,3	
Qafzeh 6	1,0	88,8	67,8	
Abri Pataud	15,0	111,8	78,8	125,0
Chancelade	15,0	103,3	82,8	
Cro Magnon 2	15,0	108,4	73,7	117,0
ES_11693	17,0	104,9	69,1	80,0
Oberkassel (f)	10,0	119,3	94,0	
Oberkassel (m)	18,0	101,6	70,5	106,0
Rond du Barry	13,0	98,8	75,4	119,0

Tab. 12.2. Fossil hominins table with some frontal ellipse values. Note Qafzeh 6 as a H. sapiens though flat frontal (i.e. neandertalian) ellipse

group		L. max	H.	ell-ft-incl	ell-ft-L	ell-ft-H	γ 1
AMHS	modern Hss	184,4 7,2	138,5 9,6	14,4 7,7	126,8 1,9	95,1 4,4	110,9 5,0
Hs	cro-magnoid	203,0 7,9	143,2 12,0	14,3 1,9	107,2 2,7	79,2 7,6	116,8 8,4
Hn	H neanderthal	206,0 9,2	114,7 2,2	2,7 19,5	99,0 1,3	70,3 8,3	126,5 4,5
He	H erectus/sp.	196,96 20,0	120,47 13,4	13,67 3,7	95,75 7,9	69,47 7,4	107,80 1,9
Aus	Australopith.	142,90	107,19	0,50	69,20	57,45	118,00
apes	Pan	137,82 7,0	91,82 8,4	1,07 8,6	94,14 1,5	64,39 7,6	152,29 9,1
apes	Gorilla	174,71 7,7	103,67 17,3	-0,77 5,1	94,52 2,4	62,20 5,8	151,67 5,5

σ is the small number for each

Abbreviations used: AMHS: anatomically modern Homo sapiens, Hs: Homo sapiens, Hn: Homo neandertalensis group, He: Homo erectus group (includes ergaster and heidelbergensis for simplification), Aus: australopithecine group, Pan: chimpanzee, Gorilla: gorilla (African great Apes; the Asian apes, Pongo and Hylobates not considered); L.max: glabella-opisthocranion length, maximum cranial length, H.: Basion Bregma height, Ell-Ft-incl: frontal ellipse major axis angle with Frankfurt plane; Ell-Ft-L: frontal ellipse major axis length; Ell-Ft-H: frontal ellipse minor axis length, i.e. height of the ellipse; Gamma 1: Lantzert basicranial angle (planum/clivus).

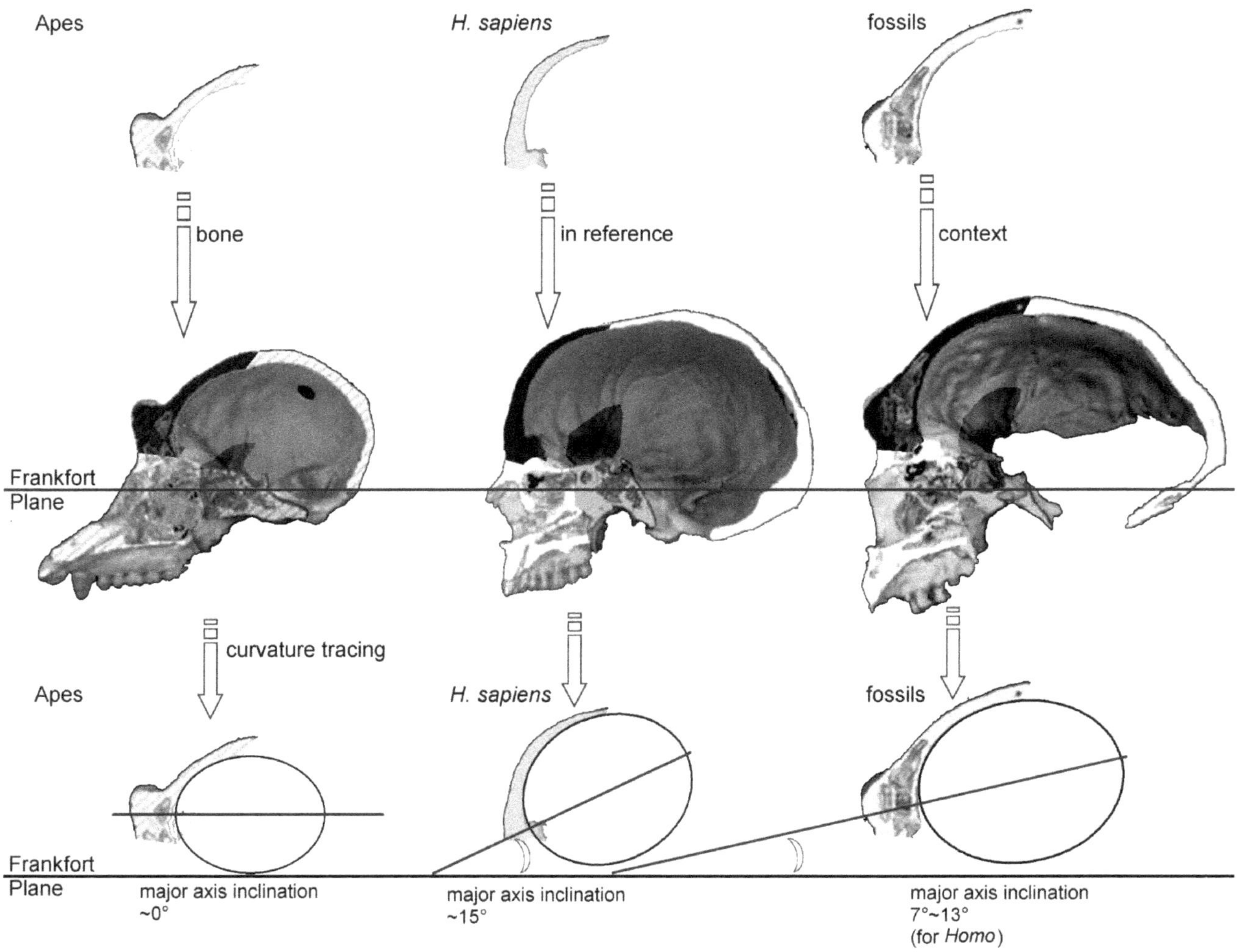

Fig. 12.3. Isolated frontal bones gives an endocranial ellipse. The angle between the Frankfurt plane and the greater (major) axis of the ellipse shows a pattern in human evolution; related to forehead bulging

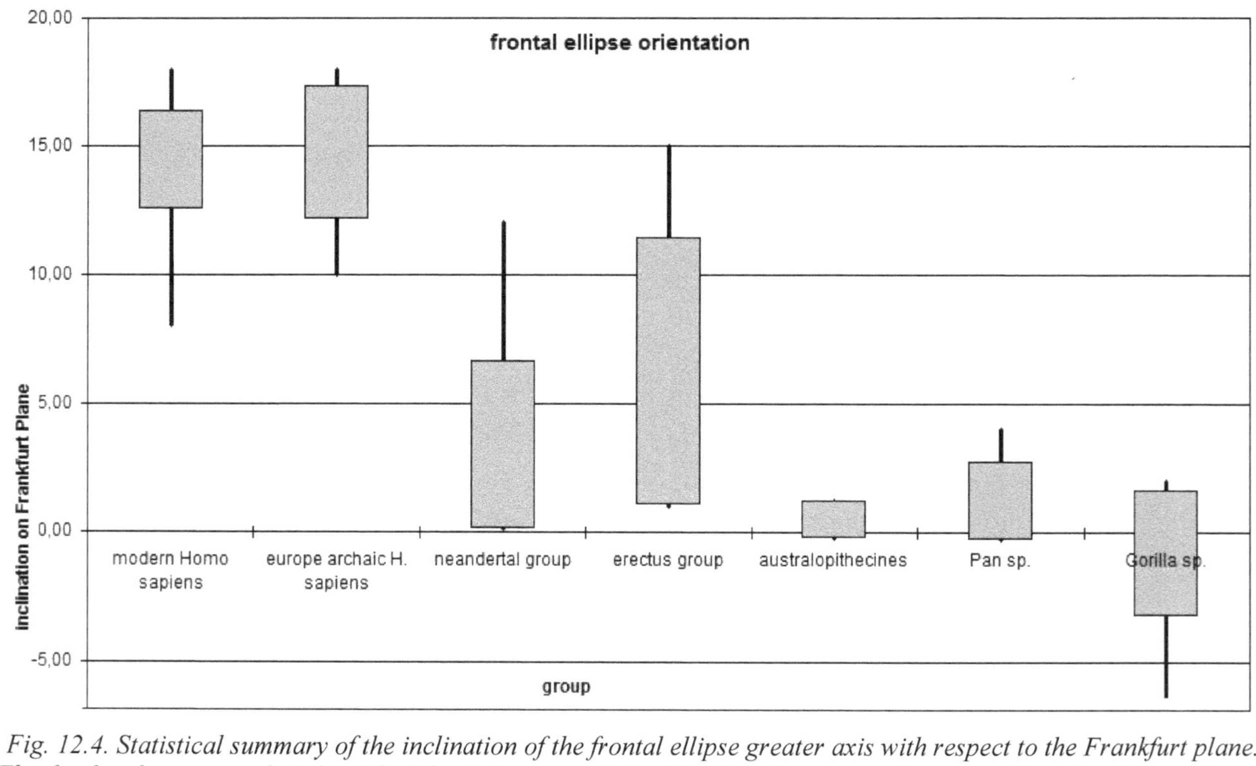

Fig. 12.4. Statistical summary of the inclination of the frontal ellipse greater axis with respect to the Frankfurt plane. Flat forehead corresponds to less tilted frontal squama. Bulged forehead corresponds to more fore-tilted frontal bones. Gorillas show negative angles (Boxes for 2. □)

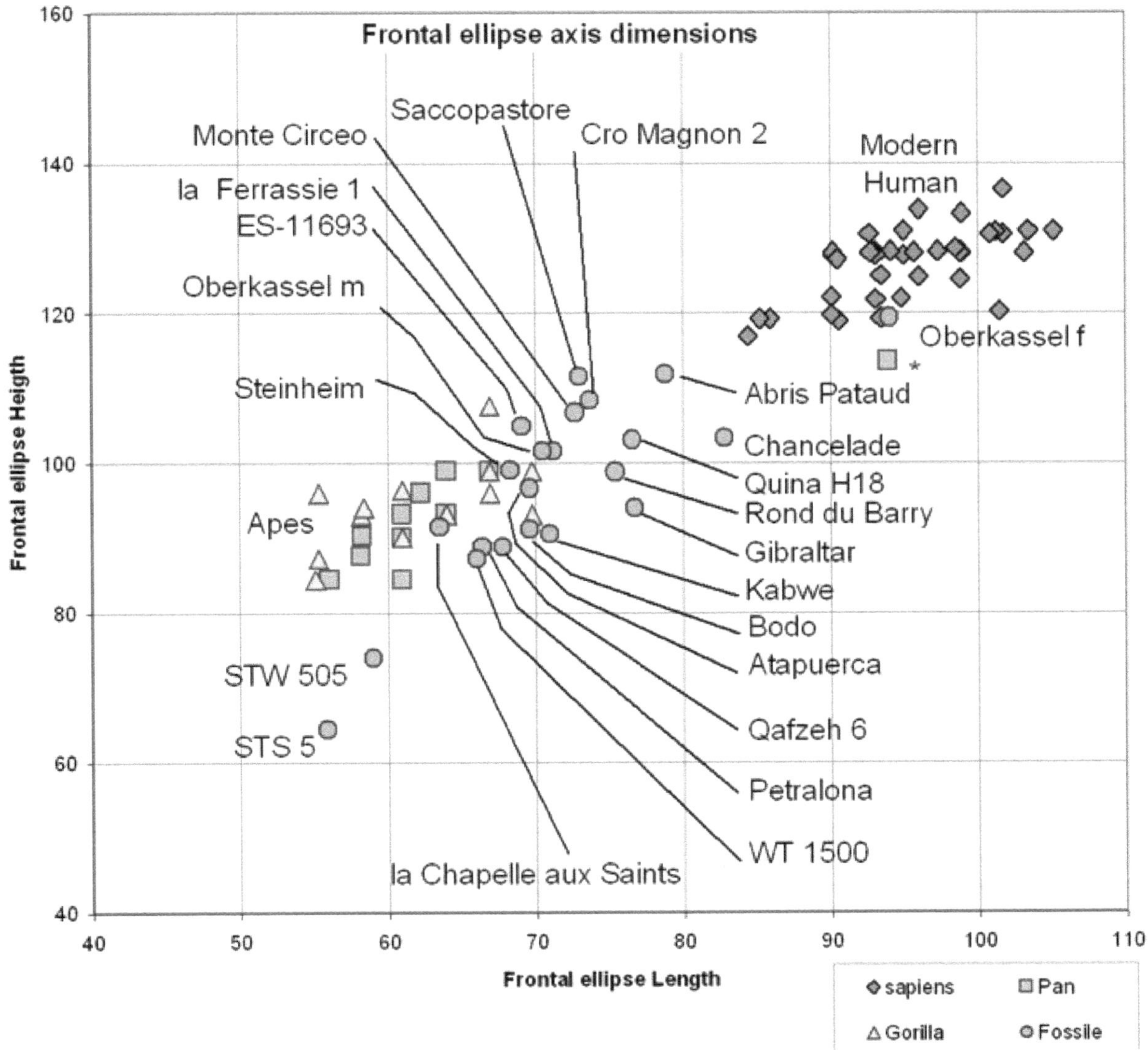

*Fig. 12.5. Metric traits of the frontal summarized in the figure: frontal ellipse length tends to increase slightly more than frontal ellipse height. Apes (Pan: & Gorilla:) and modern man () are clearly separated, with apes well regrouped. Pleistocene Homo () are standing between these two reference groups. More archaic are closer to apes, while early sapiens occupy the lower margin of the modern distribution. * A-1943-36 is a young hydrocephalic chimpanzee*

12.5). Endocranial curvatures type and orientation are corresponding to the Ape architecture despite highly increased cranial volume. Linear dimensions so reach the modern human area, however without structural changes: the frontal ellipse is still horizontal (Ell-Ft-incl: 4°) like a non-pathologic sub-adult (here 1974-15).

DISCUSSION

The aim of this study was to assess the morphological affinities found between human fossils when comparing frontal bones conformations. Bookstein et al. (1999) and Bruner et al. (2004) used Procrustes geometric morphometrics to evaluate variation in midsagittal cranial profiles of modern humans and mid-Pleistocene hominins. Similarly, Seidler et al. (1995, 1997) analyzed the human medial sagittal plane using trigonometric curve-fitting analyses of fifty modern humans compared to four fossil casts. Their results too show considerable divergences between patterns of variation when comparing external morphology of the frontal bone with endocranial corresponding parts. Relationships should then be made with the anterior brain (Fig. 12.6). Previous research generally described Pleistocene human species as not being significantly different in most descriptive traits, such as supraorbital torus projection, frontal chord or frontal subtense. According to the review by Athreya (2009), only anatomically modern human displays a really distinctive anatomy. Right in this line is the recent study of Freidline et al. (2012) about Zuttiyeh. They conclude that "Neanderthals largely retained a generalized [i.e. plesiomorphic] morphology, while the modern human frontal bone represents a significant departure from this presumably ancestral morphology." Frontal flatness is well associated with extended eyebrows (Rightmire, 2007) for any Pleistocene human.

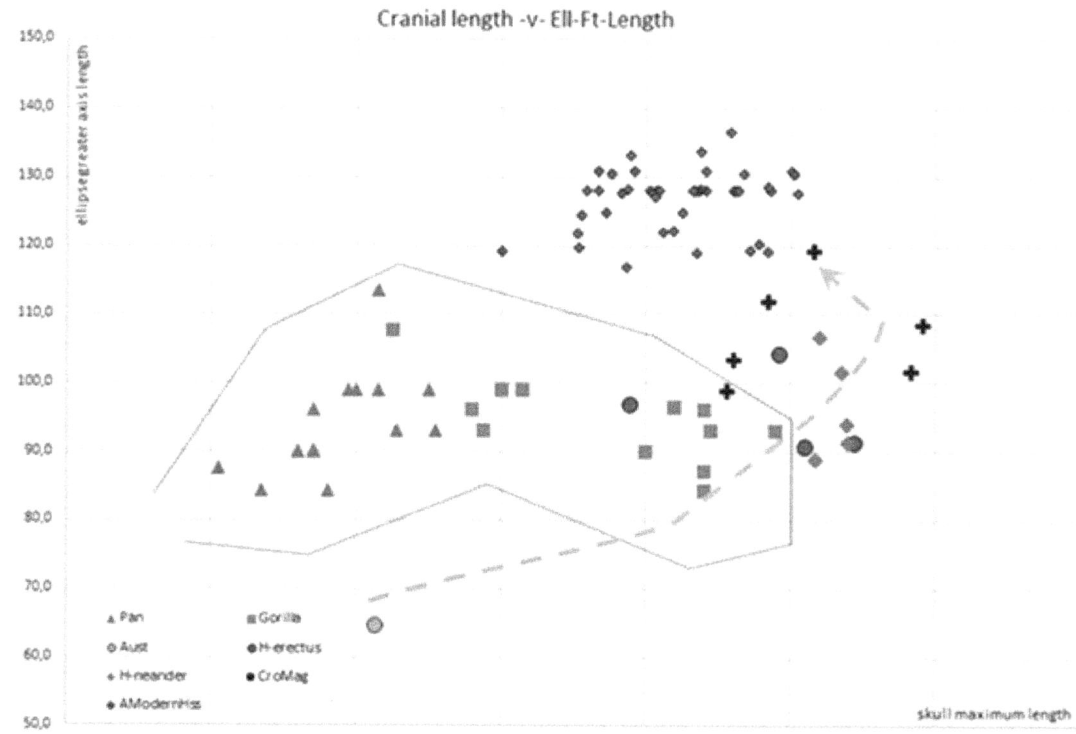

Fig. 12.6. Frontal ellipse angle with the Frankfurt plane separates very well modern man (dashed line) and both Pan and Gorilla. Human fossil species are distributed all along the trajectory seeing the overall cranial length increasing first during evolution. Only on a second time the ellipse get a more fore-tilted orientation. Consequently australopithecines lays within the apes distribution. Neanderthals, though far more recent, are in the same level of inclination

The present method introduces some contradictions. Analysis of differences in average shows a neat pattern from apes to modern humans. The frontal ellipse inclination tends to be more and more for-tilting during evolution. With the notable exception of the Neanderthal group, more recent but flatter. Behaving as a deformable structure on the long term ages of human evolution, the general pattern we observe in our results is an upswing of the whole squama. And, increasing this fore-tilting, anatomically modern *Homo sapiens* shows some continuity with earlier *H. erectus / heidelbergensis*. On the contrary, the whole group leading to neanderthal displays a reverse pattern: the ellipse is clearly less inclined (average angle with F.P. drops to a very low 2°).

Here, it is Neanderthal that represents a significant departure from the presumed ancestral morphology. We think that the ellipse method brings here a real tool for completing diagnosis between the two lineages. Bruner & Manzi (2005, 2007) evaluated Ceprano, confirming that “among archaic Homo, the frontal lobes are posteriorly positioned relative to the orbits instead of resting above them”. Further investigation (Bruner, 2007; Bruner & Holloway, 2010) demonstrated the frontal expansion having a more or less direct relation with neural anterior increase: frontal lobes are more developed. The frontal of anatomically modern *Homo sapiens* is considered as very different in its architecture: the reduction of such large structures like orbital arches appears to be related to modification of the encephalic dynamics (Lieberman et al., 2002). Our species is unique among hominids to wherein the frontal lobes are placed well above orbital cavities (Bookstein et al., 1999). The present method introduce a starting point for an explanation. The upswing of the frontal squama brings the main part of the structure more above the orbital region. The ancestral conformation sees it well behind. This aspect diminishes post-orbital constriction: relationships between frontal and zygomatic bones are modified, reducing the structural gap between the orbito-frontal block and the root of the frontal squama. Hence the orbito-frontal block is narrower than the shell of frontal bone. The modern forehead is bulging above the eyes.

CONCLUSIONS

Our objective is to develop a better understanding of the significance of frontal bone variation for modern human origins. The method presented in this paper analyzes the morphological and metrical characteristics of intracranial surface of human skulls, either fossils or modern, with a focus on the medio-sagittal profile of the frontal bone. The endocranial frontal midsagittal shape have been demonstrated quite recently to be very stable in shape from apes to modern Human. Differences are noticeable in supraorbital superstructures (Weidenreich, 1937, 1943;

Le Gros Clark, 1955; Rightmire, 1990; Antón, 2002). In the same time analysis shows the contrary concerning endocranial frontal surfaces. In previous studies, frontal midsagittal profile displays astonishing shape stability from apes to modern men, including fossil genera. That in spite of one the oldest remarks of the discipline: the observation in the frontal bone for fore-tilting and rounding/bulging trends during our evolution (Dubois, 1896; Cunningham, 1908, Boule, 1913).

Several authors have suggested that Neanderthals and archaic *Homo* share a conserved neurocranial architecture that is different from modern humans (Lieberman et al., 2002; Bruner et al., 2003; Trinkaus, 2007; Gunz et al., 2009). Our results also confirm this observation in the frontal bone.

The frontal shape is seen in this work too as similar in shape in every specific group, present or past. This similarity is translated in the geometrical nature of the curve used for the description of the inner profile: in any case the tracing is an ellipse. It is one type of the conical curves family. Beyond the shape comparison, the present geometrical model of the 'ellipse method' gives not only a pure morphological trait. As it is not a procruste method it gives quantitative parameters for orientation and linear dimensions, in addition to shape. Analysed as a deformable structure on the long time span of human evolution, we observe a trend in the frontal anatomical evolution.

The present results assume that a general tendency in *Homo* is in increasing endocranial volume, translated in a larger frontal sagittal ellipse. The refore, the reference group should be here the primitive australopithecines and not the present-day apes The differences in size are also separating the archaic Cro-Magnon from the modern *H. sapiens,* such that the general endocranial architecture has likely changed from the archaic to the modern *H. sapiens.*

By introducing some contradictions with previous studies, we think that the ellipse method brings here a real tool for completing diagnosis between the two lineages, one leading to *H. sapiens* (i.e. AMHS), and the other to *H. neandertalensis*.

The "ellipse method" is a new model. As a biomechanical method it puts in balance the skull base and the calvarium. As geometry is involved, all curves are ellipses among Apes. We think it is a translation of a compact and small endocranium. *Homo sapiens* display a more open architecture, with sagittal spheno-occipital and transversal spheno-parietal curves turning from ellipses to parabola (Korpal, 2005). Basicranial dynamics is now related to vault aspects.

The midsagittal frontal ellipse is one of the parameters from this method. We here focus on its translation of the differences between small brained apes and encephalized modern human. The evolutionary trajectory revealed by morphological exocranial observations is for once confirmed by endocranial measurements. The "ellipse method" gives not only morphological traits. As it is not a procruste method (no size, no orientation) the so-called 'ellipse method' gives quantitative parameters for orientation and linear dimensions, in addition to shape: more information is extracted for the morphometrical study.

The "ellipse method" would also help to reconstruct a fragmented or deformed skull from its own architectural references. The method is based on geometric tracing. The S (sellae) point is also a determinable central point. It enables us to place the remaining parts of the sella turcica. By applying then the principle of the "average figure for the basicranium" or "specific consensus bone", one can figure out an entire theoretical basicranial complex for the specimen.

Acknowledgments

I am grateful to Pr. Anek Sankhyan for the invitation to submit this article. A special thank is to Pr. Anne Dambricourt-Malasse for her advices in the writing operations.

References

ANTÓN, S.C. 2002. Evolutionary significance of cranial variation in Asian Homo erectus. *American Journal of Physical Anthropology* 118(4), 301-323.

ARSUAGA, J-L., MARTÍNEZ, I., GRACIA, A. & LORENZO, C. 1997. The Sima de los Huesos crania Sierra de Atapuerca, Spain). A comparative study. *J. Hum. Evol.* 33: 219-281.

ATHREYA, S. 2012. The frontal bone in the genus Homo: a survey of functional and phylogenetic sources of variation. *Journal of Anthropological Sciences* 90: 59-80.

BOOKSTEIN, F., SCHÄFER, K., PROSSINGER, H., SEIDLER, H. et al. 1999. Comparing frontal cranial profiles in archaic and modern Homo by morphometric analysis. *Anat. Rec,* 257: 217-224.

BOULE, M. 1913. L'Homme Fossile de La Chapelle-aux-Saints. Masson, Paris.

BRODMANN, K. 1912. Neue Ergebnisseüber die vergleichendehistologische Lokalisation der Grosshirnrinde mit besonderer Berücksichtigung des Stirnhirns. *Anatomischer Anzeiger* 41: 157-216.

BRUNER, E., MANZI, G. & ARSUAGA, J.L. 2003. Encephalization and allometric trajectories in the genus Homo: evidence from the Neandertal and modern lineages. *PNAS* USA. 100: 15335-15340.

BRUNER, E. & MANZI, G. 2005. CT-based descrip-tion and phyletic evaluation of the archaic human calvarium from Ceprano, Italy. *Anat. Rec. A*, 285: 643-657.

BRUNER, E. & MANZI, G. 2007. Landmark-based shape analysis of the archaic Homo calvarium from Ceprano (Italy). *Am. J. Phys. Anthropol.* 132: 355-366.

BRUNER, E. & HOLLOWAY, R.L. 2010. A bivariate approach to the widening of the frontal lobes in the genus Homo. *J. Hum. Evol.* 58:138-146.

BRUNER, E. & ATHREYA, S. 2013. The frontal bone in the genus Homo: a survey of functional and phylogenetic sources of variation. *J. Anthropol. Sci.* 90: 59-80.

BUSK, G. 1861. Translation with comments of "On the Crania of the Most Ancient Races of Man" by D. Schaafhausen. *Natural History Review*, pp. 155-175.

CHO, K-H. & HADJOUIS, D. 2005. Les asymétries cranio-faciales de la nécropole médiévale de la Queue-En-Brie (Ile-de-France, Val-de-Marne). Biom. *Hum. et Anthrop.*, Paris. 23 (1-2): 95-104.

CONROY, G., JOLLY, C.J., CRAMER, D. & KALB, J.E. 1978. Newly discovered fossil hominid skull from the Afar Depression, Ethiopia. *Nature* 276: 67-70.

CUNNINGHAM, D.J. 1909. XII.-The Evolution of the Eyebrow Region of the Forehead, with Special Reference to the Excessive Supraorbital Development in the Neanderthal Race. *Transactions of the Royal Society of Edinburgh* 46:283-311.

DAMBRICOURT-MALASSÉ, A. 2005. Un nouvel angle de la base pour une meilleure compréhension de la morphogenèse basi-crânienne des hominidés et des modalités évolutives. In: Biométrie et anthropologie de la tête, de la face et du cou, *Biométrie Humaine et Anthropologie* 23: (1-2): 53-62.

DELATTRE, A. & FENART, R. 1960. L'hominisation du crâne, CNRS, Edition.

DETROIT, F. 2002. Origine et évolution des Homo sapiens en Asie du Sud-Est. Descriptions et analyses morphométriques de nouveaux fossiles. Thèse de Doctorat du Muséum National d'Histoire Naturelle, Académie de Paris, pp. 444.

DUBOIS, E. 1896. On Pithecanthropus erectus: A Transitional form Between Man and the Apes. *The J. Anthropol. Instit. Great Britain and Ireland* 25:240-255.

FREIDLINE, S.E., GUNZ, P., JANKOVIĆ, I., HARVATI, K. & HUBLIN, J-J. 2012. A comprehensive morphometric analysis of the frontal and zygomatic bone of the Zuttiyeh fossil from Israel. *J. Hum. Evol* 62:225-241.

GUIPERT, G., MAFART, B. 2005. Evolution de l'os frontal d'Homo heidelbergensis et Homo neanderthalensis et comparaison avec l'homme moderne, *Anthropologie* XLIII (2-3): 159-167.

GUNZ, P., MITTEROECKER, P., NEUBAUER, S., WEBER, G.W. & BOOKSTEIN, F.L. 2009b. Principles for the virtual reconstruction of hominin crania. *J. Hum. Evol.* 57: 48-62.

HABIB, M., JOANETTE, Y., ROCH LECOURS, A. 2000. Le cerveau humain et les origines du langage. *Médecine/sciences* 16: 171-180.

HARVATI, K. 2002. Models of shape variation between and within species and the Neanderthal taxonomic position: A 3D geometric morphometrics approach based on temporal bone morphology. In: B. Mafart, H. Delingette (Eds.): Three-Dimensional Imaging in Paleoanthropology and Prehistoric Archaeology. pp. 25-30. BAR Publishing, Oxford.

HEIM, J-L. 1976. Les Hommes Fossiles de La Ferrassie. Masson, Paris, New York.

HUBLIN, J-J, GUNZ P., NEUBAUER S. 2010. Endocranial shape changes during growth in chimpanzees and human: a morphometric analysis of unique and shared aspects. *J. Hum. Evol.* 59:555-566.

HUXLEY, T.H. 1863. On Some Fossil Remains of Man. Evidence as to Man's Place in Nature. Williams and Norgate, London and Edinburgh.

KORPAL, Y. 2005. L'os sphénoïde dans la lignée humaine. Thèse de Doctorat du Muséum National d'Histoire Naturelle, Paris.

KUMAR, N.V.N. 2009. A new evolution time scale based on vertebrate frontal cranial slope osteometry for dating. *Evolution World Journal of Zoology* 4 (4): 270-276.

LE GROS CLARK, W.E. 1955. The Fossil Evidence for Human Evolution. University of Chicago Press, Chicago.

LEWIN, R. 1984 (rep. 2009). Human evolution: an illustrated introduction. Blackwell Publishing, Oxford.

LIEBERMAN, D.E., MCBRATNEY, B.M. & KROVITZ, G. 2002. The evolution and development of cranial form in Homo sapiens. PNAS U.S.A. 99:1134-1139.

LYELL, C. 1863. Post-Pliocene Period: Fossil Human Skulls of the Neanderthal and Engis Caves. The Geological Evidences of the Antiquity of Man with Remarks on Theories of the Origin of Species by Variation, pp. 75-80. John Murray, London.

MACLEOD, C.E., ZILLES, K., SCHLEICHER, A., RILLING, J.K., GIBSON, K.R. 2003. Expansion of the neocerbellum in Hominoidea, *J. Hum. Evol.* 44: 401-429.

MAFART, B., GUIPERT, G., LUMLEY, M.-A. DE, SUBSOL, G. 2004. Three-dimensional computer imaging of hominid fossils: a new step in human evolution studies. *Canadian Association of Radiologists* J. 55 (4): 264-270.

MARTIN, R. 1928. Lehrbuch der Anthropologie, Gustav Fischer, Jena.

MARTIN, R. & SALLER, K. 1957. Lehrbuch der Anthropologie in Systematischer Darstellung mit Besonderer Berücksichtigung der Anthropologischen Methoden. Abt 1. pp. 661, Gustav Fischer Verlag, Stuttgart.

MILLET, J.-J. 2000. Analyse du champ morphologique infantile Pan et Gorilla et comparaison avec l'architecture crânienne de *Homo*. In: P. Andrieux, D. Hadjouis, A. Dambricourt-Malasse (Eds.): L'identité humaine en question. Nouvelles problématiques et nouvelles technologies en Paléontologie humaine et en Paléoanthropologie. Pp. 67-78. Artcom, Paris.

NAMBIAR, P., NAIDU, M.D.K., SUBRAMANIAM, K. 1999. Anatomical variability of the frontal sinuses and their application in forensic identification. *Clinical Anatomy* 12: 16-19.

OOSTRA, R.J., VANDER WOLK, S., MAAS, M., HENNEKAM, R.C. 2005. Malformation of the axial skeleton in the Museum Vrolik II: craniosynostoses and suture-related condition. *Am. J. Med. Genet.* A136A: 327-420.

PENIN, X., BAYLAC, M. 1999. Comparaison tridimensionnelle des crânes de Pan et Pongo par superpositions procrustéennes. Comptes Rendus de l'Académie des Sciences, Paris, Sciences de la vie 322: 1099-1104.

PENIN, X., BERGE, C., BAYLAC, M. 2002. Ontog enetic study of the skull in modern humans and the common chimpanzees: Neotenic hypothesis reconsidered with a tridimensional Procrustes analysis. *Am.. J. of Phys. Anthrop.* 118: 50-62.

PONCE DE LEON, M.S., ZOLLIKOFER, C.P.E. 2001. Neanderthal cranial ontogeny and its implications for late hominid diversity. *Nature* 412: 534-538.

RIGHTMIRE, G.P. 1990. The Evolution of Homo erectus: Comparative Anatomical Studies of an Extinct Human Species. Cambridge University Press, Cambridge.

SCHAAFHAUSEN, D. 1861 (1858). On the crania of the most ancient races of man (from Müller's Archiv, 1858:453) with remarks on original figures, taken from a cast of the Neanderthal cranium. Translated with comments by G. Busk. Natural History Review, pp. 155-175.

SCHMERLING, P. 1833. Recherches sur des ossements fossiles découverts dans les cavernes de la province de Liège. L'Université de Liège, P-J. Collardin, Liège.

SEIDLER, H., FALK, D. STRINGER, C., WILFING, H. et al. 1997. A comparative study of stereo-lithographically modelled skulls of Petralona and Broken Hill: Implications for future studies of middle Pleistocene hominid evolution. *J. Hum. Evol.* 33: 691-703.

SIMPSON, E.K. 2005. Variation in cranial base flexion and craniofacial morphology in modern humans, Thesis at the University of Adelaide.

SLÁDEK, V., TRINKAUS, E., ŠEFČÁKOVÁ, A. & HALOUZKA, R. 2002. Morphological affinities of the Šala 1 frontal bone. *J. Hum. Evol.* 43: 787-815.

STRINGER, C.B., HOWELL F.C. & MELENTIS J.K. 1979. The significance of the fossil hominid skull from Petralona, Greece. J. Archaeol. Sci. 6: 235-253.

TILLIER, A-M. 1977. La pneumatisation du massif craniofacial chez les hommes actuels et fossiles. *Bulletins et Memoires de la Société d'Anthropologie de Paris* 4: 177-189, 287-316.

TRINKAUS, E. 2007. European early modern humans and the fate of the Neandertals. PNAS USA. 104: 7367e7372.

VAN HOESEN, G.W., SEMENDEFERI, K. DAMASIO, H. & FRANK, R. 1997. The evolution of the frontal lobes: a volumetric analysis based on three-dimensional reconstructions of magnetic resonance scans of human and ape brains, Journal of Human Evolution 32: 375-388.

WEIDENREICH, F. 1937. The Relation of Sinanthropus pekinensis to Pithecanthropus, Javanthropus and Rhodesian Man. *J. Royal Anthropol. Instit. of Great Britain and Ireland* 67: 51-65.

WEIDENREICH, F. 1943. The Skull of Sinanthropus pekinensis: A Comparative Study on a Primitive Hominind Skull. Geological Survey of China, Pehpei, Chungking.

WEIDENREICH, F. 1947. Facts and speculations concerning the origin of Homo sapiens. *Amer. Anthrop.* 49: 187-203.

WHITE, T.D. & FOLKENS, P.A. 2000. Human Osteology, 2nd Edn. Academic Press, San Diego.

ZOLLIKOFER, C.P.E. 2002. A computational approach to paleoanthropology. *Evol. Anthrop.* 11: 64-67.

ZOLLIKOFER, C.P.E. 2012. Evolution of hominin cranial ontogeny. *Progress in Brain Research, Evolution of the Primate Brai*n 195 (13): 273-292.

13.
DISCOVERY OF TWO PREHISTORIC SITES AT GALUDIH IN EAST SINGBHUM, JHARKHAND: A STUDY IN TYPO-TECHNOLOGY AND GEOMORPHOLOGY

Ratna BHATTACHARYA
Department of Anthropology, Bangabasi College, Kolkata – 700 009, India
ratnaardhendu@gmail.com

Abstract: *Surface and partly on excavated findings were discovered at Galudih-Mahulia and Chandrarekha in Jharkhand in very rich stratigraphic section. The two sites are located in a natural geographical setting and have yielded different types of tools made on different rock raw material, mainly quartzite pebbles, but quartz, mica, quartzite and basalt as well. We can note a change in the nature and pattern of the soil with stratigraphic variations due to the erosional activities of the Subarnarekha River. Primary laterite was present on the hill tops and due to erosion; it is converted into secondary laterite. A succession of dry and wet periods during the Pleistocene can be recognized from the sequence of deposits. Step-like formations are observed throughout and stratigraphy shows layers of calcified bed, kankar, hard sandy soil and lateritic soil from bottom to top. The findings prove that there was continuity in tool industry from chopper-biface to flake to flake-blade industry, followed by a rich microlithic industry. The successions reveal a relationship between geomorphology and typo-technology of the artifacts and climatic fluctuations during Middle to Late Pleistocene, whereas evolving tool types and techniques indicate continuous intellectual development and human evolution in this part of eastern India.*

Keywords: *Galudih, Jharkhand, archaeology, geomorphology, typo-technology*

INTRODUCTION

The remains of the prehistoric cultures are found all over the globe, in every part of the land which was suitable for human habitation in the Pleistocene period. In some places the remains are found open on the surface and in other places they are embedded down into the earth. So, to find such remains one has to explore and at some places excavate areas with such potentials.

In the present context two sites are explored where not only tools are collected, but the relevant basic geology, especially Quaternary geology and stratigraphy geomorphology are also studied to point out the palaeo-climatic states. Stone tools have been collected from the two sites and have been categorized in terms of the nature of raw materials, technology of fabrication, tool types and others. Observation of natural sections and stratigraphy have been done in those sites, the stratigraphic section of artificial trenches have also been studied. Other associated samples like soil, rocks have been collected. The work was carried out by a team of undergraduate students and Gopal Mondal, earlier initiated by eminent scholars like V. Bal, V.D. Krishnaswami and A.K. Ghosh in the area.

METHODS

For exploration of a limited place in this area following methods and techniques were of great help. These are – Toposheet Consultation, Surface Collection, study of topography and soil conditions, trial excavation, analysis of data on the background of geomorphology and stratigraphic evidences, Collection of in situ tools, core debitages, study of various stones and rocks for understanding the nature of raw materials used for tool making. Also geomorphological information raw material transportation and palaeoclimatic relation for tool making are studied. Lastly, after analysis, the reconstruction of complete stratigraphic sequence is done for better understanding of cultural succession.

THE AREA AND LOCALITY

East Singbhum district is situated at the southeast corner of Jharkhand. It has been formed after isolating nine blocks from old Singbhum on 16th January 1990. From the industrial growth and mining quarrying point of view this district has leading position in Jharkhand. Before independence the entire area of east Singbhum district was part of old Manbhum district and old Dhalbhum estate. After independence it has been merged with greater Singbhum. The total geographical area of district is about 3533 Sq. km, which is about 2.03% of the whole state. About 53% of the total area of district is covered by residual mountains and hills consisting of granite, gneiss, schist. It is a part of Chotanagpur plateau. The Dalma range extends from west to east covered by dense forest on the northern side. The Subarnarekha River flows from west to south-east direction. The district is rich in minerals and these are found abundantly. From the administrative point of view this district has been divided into two Sub-Division – Dhalbhum and Ghatshila. The district consists of eleven blocks namely Golmuri-cum-Jugsalai (Jamshedpur), Potka, Patamda and Boram in Dhalbhum Sub-Division and Ghatshila, Musabani, Dumaria, Baharagora, Dhalbhumgarh, Chakulia and Gurabandha in Ghatshila Sub-Division. There are 231 panchayat and about 1810 revenue villages out of which 1669 revenue villages are inhabited and rest 141 revenue villages are unhabitated. The Head Quarter of this district is Jamshedpur.

CLIMATE & RAINFALL

The climate of the district is tropical. Its minimum temperature in winter ranges from 12 degree centigrade to 18 degree centigrade. In summer maximum temperature goes up to 46 degree centigrade and minimum at 24 degree centigrade. Average rain fall is 1310 mm per year. The climate of the district is temperate. Annual rain fall is 1200 mm to 1400 mm. This area comes under the path of south-west monsoon, so sometimes it receives heavy rain during July to September. During the summer season maximum temperature goes up to 40-45 degree centigrade whereas in winter it has recorded a minimum 8 degree centigrade. The district is rich in minerals and is found abundantly. Iron Ore, Copper, Uranium, Gold Kynite are the main minerals.

SOIL

Soil content of Jharkhand mainly consists of soil formed from disintegration of stones & rocks, and the soil composition is further divided into:

Red soil: found mostly in Damodar valley & Rajmahal area.

Micaceous soil (containing particles of mica): found in Koderma, Jhumeritilaiya, Barkagon and areas under Mandar hill.

Sandy soil: generally found in Hazaribagh & Dhanbad.

Black soil: found in Rajmahal area.

Laterite soil: found in western part of Ranchi, Palamou, and part of Santhal Parganas & Singbhum.

ROCKS

About 53% of the total area of E-Singbhum district is covered by residual mountain and hill consisting granite, gneiss, schist and basalt rocks. It is a part of Chota Nagpur plateau igneous, sedimentary and metamorphosed rocks of Dharwarion period are found everywhere. The Dalma Range is as main hill has been extended from west to east covered by dense forest.

RIVER

Subarnarekha River & other small tributaries pass through the district. Swarnarekha Multi Purpose Dam is being constructed at its upper ridge. SMP right bank canal as well as left bank canal will irrigate about 45,000 ha of land in the district if completed. Apart from the main dam a Barrage has been constructed in Galudih which will provide irrigation to the nearby states of Orissa and West Bengal. Swarnarekha supplies water to industrial city of Jamshedpur and its suburban areas.

GEOMORPHOLOGY

Geomorphology is the scientific study of landforms and the processes that shape them. Geomorphologists seek to understand why landscapes look the way they do, history and dynamics of landforms & to predict future changes through a combination of field observations, physical experiments and numerical modeling. Geomorphology is practiced within geography, geology, geodesy, engineering geology, archaeology & geotechnical engineering. This broad base contributes to a wide variety of research styles and interests within the field.

At our present fieldwork which was held in Galudih, East Singbhum, Jharkhand we observed the natural setting and the geographical formation of the land under our concerned supervisors. Two sites i.e., GLD (site 1) & CRH (site 2) were stu died respectively which yielded different rocks and tools. The primary laterite may be from Pleistocene or early Pleistocene & the low level secondary laterite is a criteria of Pleistocene and recent" (A.K. Ghosh, 1961).

A succession of dry and wet periods during the Pleistocene can be recognized from the sequence of deposits. The gravel on detrital laterite indicates a wet phase and silt of clay indicates a dry phase.

Upper Pleistocene – Upper clay, upper loose gravel

Middle Pleistocene – Lower Silty Clay, Lower gravel

Lower Pleistocene – Clay, compact secondary laterite.

The exact location of Galudih is 86025'. Geographically the region appears to be partly covered with dense forests. The settlements are in scattered pattern. Physiographically the region is highly dissected plateau, terrain, undulated lands with river Subarnarekha flowing on the western side through the area. Practically the river divides the region into two units. The north eastern part is covered with forests, with highest altitude of 351 ft. There is a slope south west wards with erosional rolling plains. The Jadugora range lies in the west and the Dalma range in the east.

The river Subarnarekha emerged from Pisca/Nagri near Ranchi (230°18'N: 850°11'E) of Chota Nagpur plateau. The elevation of the river is 610 m or 2,001 ft from the sea level. The total length of the river is 470 km. It traverses a long distance and finally drains into the Bay of Bengal near Talsiri.

The surroundings of the area can be easily identified from the topographic sheets. The map shows that the river is passing through highlands and a deep river bed is formed due to river erosion. The river across the barrage on the north eastern part meets with the river Gauri. The map also shows important terrace like formations at an average height of 200-350 ft. The entire land is gradually declining towards the east. The topographic sheet shows the elevation of the land through contour lines. The land surface shows direct evidences of climatic factors such as erosion, weathering & lateritic formation.

STRATIGRAPHY

The area has been searched thoroughly and surface collection has been yielded from each side. 'GLD' (site – 1) and 'CHR' (site – 2) yielded different rocks and tools. The sites seemed to be well accumulated with suitable materials for tool making. The change in nature and pattern of the soil, the stratigraphic variations were noticed. The erosional activity of the river was taken into account and the river bed was exposed. Primary lateritic present on the hill tops and due to erosion it is converted into secondary laterite. From excavation we found a lot of in situ lateritic pebbles, quartz, mica, quartzite & basalt. The sites are very much accumulated with suitable materials for tool making. Geomorphological variations also express the climatic change from Paleolithic times. Nature and pattern of erosion has also changed from time to time. Stratigraphic variations denote the climatic fluctuations & the river banks show erosional activity, so the site is entirely exposed. Primary laterite is present on the hill tops and due to erosion it is converted into secondary laterite. From excavation we found a lot of in situ lateritic pebbles, quartz, mica, quartzite & basalt. In India laterite is a weathered product of various rocks; mainly basalt & other aluminous rocks under a warm, humid, tropical monsoon climate (Pascoe, 1962). Stratigraphy was done on the bank of river Subarnarekha, Mahulia, Galudih, East Singbhum (near Barrage). The area was located in 22°38'N: 86°24'E at an altitude of 292 ft above sea level.

Site 1 (Galudih, Mahulia, Subarnarekha River Bank)

Step like formation were observed with layers of sandy river bed, calcified bed, kankar soil, hand sandy soil and lateritic soil from bottom to top. Granite and basalt rocks encrusted with mica, quartz & quartzite were found. The 1st and 2nd Erosions were observed at a height of 112 cm and 118 cm respectively.

Site 2 (Chandralekha Village along Subarnarekha)

The site is located at 22°39'N: 86°22'E: 332 ft above sea level. We observed the land formation & the topology of the concerned area. We took the account of the stratigraphy and also collected stones, rocks, including few tools & pebbles. Some prehistoric tools such as hand axe, crizer, chopper & cleaver were observed and collected. The latitude, longitude and altitude of the site were jot down according to the data received with the help of GPS machine. The course of the river was also observed within and from the particular site concerned was quite broad with respect to the main flow.

FINDINGS

Rocks and stones from two sites– *GLD* (site 1) & CRH (site 2)

Quartz, granite, gneiss, mica schist, quartz with phyllite banding, hornblende, pyroxene and mica; we found quartz banded with phyllite, and calcified laterite, marble and *gutting*.

Gneiss: the metamorphosed granite with bands. We got one piece of gneiss with reddish banding.

Granite: contained minerals of quartz, feldspar, hornblende, pyroxene & mica.

Schist: was found along with mica.

The relative chronology and evolutionary succession is observed from the typo-technology of the tools. The hand axes, cleaver s of the Lower, the scraper s and flakes of the Middle to the points, flakes and flake blades of the Upper Palaeolithic, and finally the microliths on the top, indicating change s with time on the tool typology and technology.

CONCLUSIONS

Pleistocene stratigraphy, climate and cultures in Singbhum are correlated. These sites were a meeting ground for the "biface" industries, "flake" industries and "flake-blade" industries during Palaeolithic period. Later on the area was inhabited by Mesolithic folks as reported from the microliths. In the present study, only the selected areas are taken into consideration where tools are mainly found from secondary laterite. These findings prove that there was a positive continuity in tool industry from "biface" to "flake" industry to "microliths". Our work was mainly exploratory. But from the finds it can be concluded that both the sites showed very high potential in terms of cultural remains and should be exposed to a more sophisticated and specialized long term excavation work in near future.

References

DISTRICT ADMINISTRATION EAST SINGBHUM, Jamshedpur (n.d.) East Singbhum Jamshedpur. Retrieved from http://jamshedpur.nic.in/

GHOSH, A.K. 1967. Laterite & associated artifact bearing strata in Eastern India, Quaternary Soil, edited by R.B. Morrison and H.B. Wright.

GHOSH, A.K. 1961. Prehistoric studies in Eastern India. *Science and Culture* 27 (8): 369-375.89.

HUGGET, R.J. 2007. Fundamentals of Geomorphology. Routledge Taylor & Francis group: USA; 2nd Edition. http://www.whereinciti.com/India/jharkhand/

NATIONAL ATLAS & THEMATIC MAPPING (n.d.). ORGANIZATION, Govt. of India (Dept. of Science & Technology) C.G.O. Complex, 7th Floor, DF block, Bidhannagar, Kolkata – 700064.

WIKIPEDIA (n.d.). East Singbhum district. Retrieved from: en.wikipedia.org/wiki/East_Singbhum_district

14.
UNBALANCED ENDEMIC ISLAND FAUNA: ARE HOMININS THE EXCEPTION?

Anneke H. VAN HETEREN
Universität Bonn, Steinmann Institut Bereich Paläontologie, Nußallee 8, 53115 Bonn, Germany
Anneke.vanHeteren@uni-bonn.de

Abstract: *Unbalanced endemic island faunas lack mammalian predators, and, as a result, evolution takes courses which fundamentally differ from those on the mainland. Key insular dwarf adaptations include size reduction, most pronounced in the limbs, adaptations for low gear locomotion, and heterochrony. Additionally, observed changes often include reversals to ancestral states.* Homo floresiensis *displays adaptations to low gear locomotion in the form of short legs and broad hips. Furthermore,* H. floresiensis *displays archaic wrist morphology and heterochronic features in the skull and postcrania. These typical island-endemic features indicate that* H. floresiensis *dwarfed from* H. erectus *in an insular setting on Flores.*

Keywords: *Island, fauna, hominin,* H. floresiensis, *heterochrony*

INTRODUCTION

Several types of differences between endemic island mammals and their mainland ancestors occur so frequently that they can be considered patterns. The most common and obvious change in large mammals is a reduction in size (van der Geer et al., 2010). When dwarfism occurs in endemic island animals, the limbs often show relatively more shortening (sometimes accompanied by an increase in robusticity of the limbs) than the rest of the body, such as in *Myotragus* from Mallorca, Pleistocene cervids from Greece, Japan and the Philippines, and Candiacervus from Crete (van der Geer et al., 2010). Islands are generally relatively small and no mammalian predators are present, high gear locomotion is unnecessary for survival and low gear locomotion provides valuable energy savings. The adaptation to low gear locomotion is frequently seen in insular mammals, such as in *Hoplitomeryx*, Cervidae and *Myotragus* (van der Geer et al., 2010).

Many adaptations of insular vertebrates can be recognized as the result of heterochrony. For example, juveniles of any species tend to have short stubby limbs compared to adults. While the femur of island species *Candiacervus ropalophorus* is relatively massive, like the femora of newborns from the mainland species *Cervus elaphus*, *Dama dama* and *Capreolus capreolus*, adult specimens of these mainland species have more slender bones. The short snout of *Sus nanus* can also be classified as heterochronic, because juveniles mammals tend to have shorter snouts than adults of the same species. Van Heteren and de Vos (2007) give more elaborate examples of heterochronic island adaptations in vertebrates.

Island animals regularly show reversals to a more primitive condition often associated with their decrease in size. Primitive ruminants are very small compared to the derived forms. Island ruminants are also very small and have developed some features that are similar to the primitive ruminants for functional reasons. *Phanourios*, a pygmy hippopotamus developed a more lophodont dentition than its ancestor, which is a reversal to the more primitive condition. *Archidiskodon (=Elephas) celebensis*, a dwarfed descendent of *A. planifrons*, is also more primitive than its ancestor based on the dentition. The Japanese *Hemimastodon (=Gomphotherium) annectens* has molars which resemble those of the primitive genus *Phioma*. *Gymnesicolagus aff. gelaberti* has a short diastema, a primitive character shared with continental ochotonids and insular leporids. Another good example is *Hoplitomeryx*. The orbital position of the horns of *Hoplitomeryx* is a reversal to a more primitive configuration. The absence of 'pogo-stick' locomotion (i.e., the absence of high gear locomotion) is characteristic of more primitive ruminants, but is also found in both *Hoplitomeryx* and *Myotragus*. Additionally, *Hoplitomeryx* has patellae which have returned to the primitive condition.

Homo floresiensis is a small-bodied hominin from the Indonesian island Flores. The type specimen, LB1, is approximately 1 metre (m) or a bit more than 3 feet tall with a cranial capacity of around 400 cubic centimetre (cc) (Brown et al., 2004; Morwood et al., 2005). While there is still no agreement on the cause of the small stature and small cranial capacity of LB1 and associated individuals, one current hypothesis states that *Homo floresiensis* is a dwarfed descendent of *Homo erectus*.

Because the features described above for insular faunas form a general pattern, the presence of a reduction in size, adaptations to low gear locomotion, the presence of heterochronic features and reversals to an ancestral state are all indications of island adaptations. The next step is to discuss the morphology of *Homo floresiensis* and assess whether, relative to *H. erectus*, this species displays morphological changes typical of insular dwarfism.

OBSERVATIONS

Elaborate descriptions of the fossil material recovered from Liang Bua and its stratigraphical context have been published in a special issue of the Journal of Human Evolution (Morwood and Jungers, 2009). The skeleton is depicted here (Fig. 14.1) and a brief summary is given below.

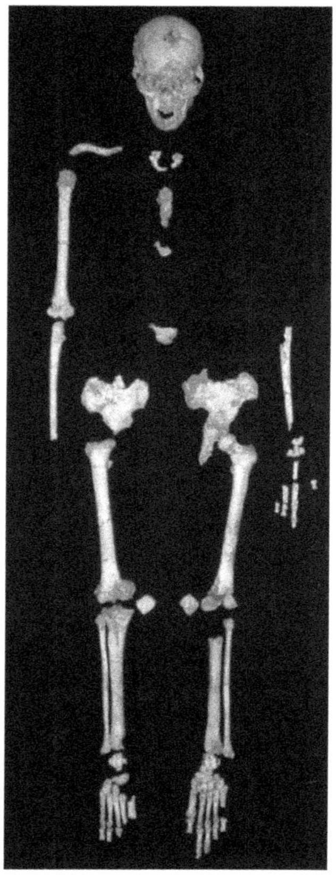

Fig. 14.1. The skeleton of LB1, type specimen of H. floresiensis (courtesy of W. Jungers)

Only one cranium has been found in Liang Bua cave so far, belonging to LB1. According to its craniodental morphology it is most likely a member of the genus *Homo*. LB1 has a short and relatively orthognathous face with small cheek teeth compared to *Australopithecus*. LB1 lacks both upper third molars, the right congenitally and the left ante-mortem, and shows rotation of the fourth upper premolars on both sides of the jaw. Additionally, LB1 has a long and low cranial profile, mound-like occipital torus, mastoid fissure, and its maximum cranial breadth is in the supramastoid region. The frontal bone is rather narrow relatively constricted across the frontotemporale. The cranial vault is thickened posteriorly and in areas of pneumatisation in the lateral cranial base. LB1 has a narrow foramen magnum. LB1 has a projecting occipital, but no occipital bun. The canine fossae of LB1 are relatively deep with large infraorbital foramina, and the canine juga are pronounced. LB1 has a sloping forehead and brow ridges which arch over the orbit. The orbits have a particularly arched superior border.

Tab. 14.1 presents several measurements on a stereolithographic cast of the cranium and published measurements taken directly on the cranium of LB1 with regards to the orbits. The masticatory apparatus is similar to that of other members of the genus *Homo*. Based on the LB1 endocast, the cranial capacity and relative brain size of LB1 are similar to those found in apes and australopiths. The shape of the endocast resembles that of *H. erectus* with respect to its relative height, the relative widths of its caudal and ventral surfaces and its long, low lateral profile, however. LB1's endocast is extremely brachycephalic because of lateral expansion of the caudal part of the temporal lobes. The occipital lobe extends caudally relative to the cerebellum, but without the presence of an occipital expansion.

Tab. 14.1. Measures taken by Brown et al. (2004) on the cast of LB1's cranium and measurements taken by the author on the left and right sides of a lithographic cast

MEASURES / INDICES	MEASURES from Brown et al. (2004)	LEFT	RIGHT
Cranial base length[1]	-	71,5 mm	Not possible
Orbital breadth	32 mm	32,0 mm	31,7 mm
Orbital height	31 mm	Not possible	30,7 mm
Orbital index[2]	96.9	-	96,9

Two *H. floresiensis* mandibles have been found to date, belonging to LB1 and LB6. The gonial angle ranges from 109° to 113°. The average gonial angle for LB1 is 111°, whereas LB6 has a gonial angle of 112°. The mandibular angle is more comparable to juvenile *H. erectus* than adult *H. erectus* or *H. sapiens* (Tab. 14.2 & 14.3). Both mandibles lack a mental tuberosity and are relatively robust. The morphology of the premolars is primitive.

The LB1 and LB6 finds include various bones of the upper limb. The skeleton of LB1 includes a partial clavicle and a nearly complete humerus. The clavicle is relatively short and displays a primitive double curvature in posterior view. The right humerus of LB1 has a relatively limited degree of humeral torsion. Estimates of the humeral torsion vary between 110° and 120°. The scapula of LB6 is also available. The scapula might have been more protracted than in modern humans, resulting in an anteriorly facing glenoid fossa. A low humeral torsion in this configuration would have resulted in a 'normal' position of the arm. Portions of both the left and the right ulnae of LB1 are preserved. The right ulna is nearly complete. The left ulna of LB6 is missing the distal end. A radius of LB6 has also been recovered from the sediments. The brachial index of *H. floresiensis* is similar to that of a child at birth or at the age of 1 year (Fig. 14.2). Five carpals belonging to the type specimen have been recovered, all from the left wrist. A three-dimensional quantitative comparative analysis of the scaphoid, trapezoid and capitate has shown that these

Tab.14.2. Mandibular angles of LB1 and LB6 (after Van Heteren and Sankhyan, 2009)

ANGLE (in °)[1]	MEASURED FROM CAST	MEASURED FROM PHOTOGRAPH
α (LB1)	112	113
β (LB1)	100	100
γ (LB1)	110	109
δ (LB6)	No cast available	112

Tab. 14.3. Ontogeny of the mandibular angle of H. sapiens and H. erectus s.l. For the latter individuals were only specified as juvenile or adult. For Homo sapiens the ranges indicate the 25th and the 75th percentile of the total sample (N=76)

Age	Angle (in °) of *Homo sapiens*	Angle (in °) of *Pan troglodytes*[1]	Angle (in °) of *Homo erectus s.l.*
Early embryo	130	No data	No data
Birth	140-145	No data	No data
Juvenile	128-133	121-129	107-113
Adult	120-127	108-120	97-108
Old age	117-124	No data	No data
After loss of teeth	124-131	No data	No data

carpals display a symplesiomorphic or primitive pattern of features similar to African apes, *H. habilis* and australopiths (Morwood and Jungers, 2009).

The left os coxae of LB1 is relatively complete; the right is more fragmentary. The ilium exhibits substantial lateral flare relative to the position of the acetabulum, similar to the pelvic morphology of australopiths. The width of the greater sciatic notch falls in the range of modern human females, but is relatively narrow compared to australopiths. The iliac blade is relatively short and wide, and the ischial spine is not very pronounced. Both the left and right femora of LB1 have been preserved, the left being more complete. The shafts of the femora are relatively robust. The collodiaphysial angle is approximately 128-130°. The femoral bicondylar angle is 14°, which is within the range of australopiths. Tibial torsion in *H. floresiensis* is rather low compared to adult modern humans, but not necessarily outside the range. The tibial shafts are oval in cross section, which also suggests a very small torsion angle. The skeleton of LB1 includes a relatively complete left foot and parts of a right foot. The foot of LB1 is clearly very different from a modern foot. The feet are approximately 70% of the femur length. The first metatarsal is relatively short compared to the other metatarsals and without sesamoid grooves. The metatarsals of LB1 are long relative to tarsal length, and the proximal phalanges are long relative to the metatarsals. The shape of the navicular is primitive and the medial longitudinal arch is weak or absent, like in australopiths.

DISCUSSION

Several authors have provided evidence for close affinities between *H. floresiensis* and *H. erectus s.l.*. Zeitoun (2007) provides a cladogram in which LB1 belongs to a *H. erectus*-clade, including the type specimen for *H. erectus*. Tocheri et al. (2007) discusses the morphology of the wrist and supports the hypothesis

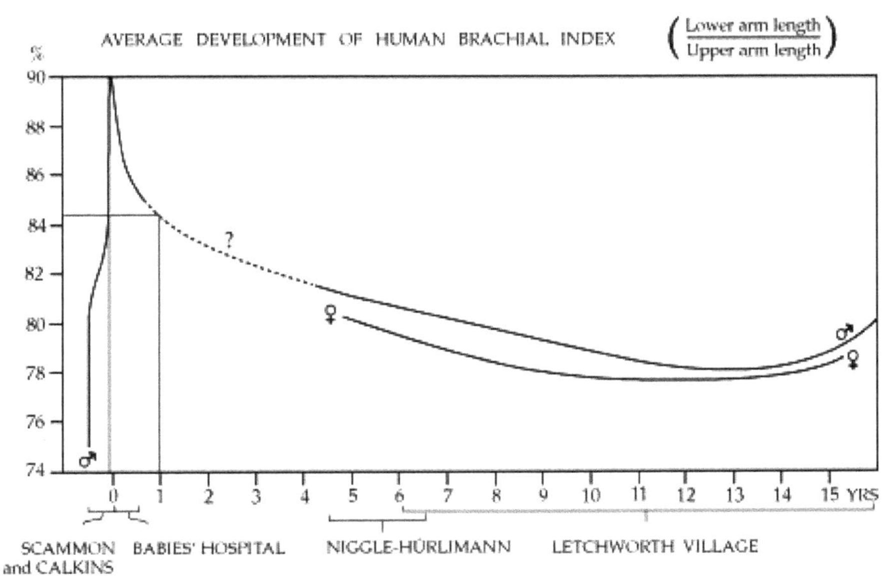

*Fig. 14.2. Development curve of brachial index from intrauterine life to puberty. Data have been taken from various sources as displayed. 0.5 to 5 years, data sparse. The grey lines indicate the brachial index of LB1, which is similar to the brachial index of babies. Modified after Davenport (1934). Using the data of Morwood et al. (2005) the brachial index of H. floresiensis was calculated as follows: (length ulna*100)/length humerus = (205*100)/243 = 84.4*

of a new species. In the absence of preserved *H. erectus s.l.* wrist fossils, it cannot be excluded that *H. floresiensis* and *H. erectus s.l.* had similar wrist morphology. An interpretation of the cranium of LB1 and postcranial remains in terms of size and scaling results in LB1 appearing to have been closely related to *H. erectus s.l.* (van Heteren and de Vos, 2007). Brown et al. (2004) write that the most likely explanation for *H. floresiensis* "is long-term isolation, with subsequent endemic dwarfing, of an ancestral *H. erectus* population." This is a very plausible idea. It has been shown that the environment on Flores during the Pleistocene was very suitable for insular dwarfs and giants to develop (Meijer et al., 2010). *H. floresiensis* may have been one.

Reduction in Size

The most obvious change in the skeleton of *H. floresiensis* compared to *H. erectus* is the decrease in body size. Whereas *H. erectus* had a body size comparable to modern *H. sapiens, H. floresiensis* only reached a height of around 1 m. Morwood et al. (2005) state that the humero-femoral index is primitive and outside the range of adult *H. sapiens*, which means that the legs of *H. floresiensis* are relatively short compared to its arms. This indicates that the weight bearing extremities have decreased in size more than the rest of the body, just like in other island dwarfs. Burness et al. (2001) have made estimations of the body mass of endothermic and ectothermic top herbivores and top carnivores based on island size. The top carnivore on Flores at the time of *H. floresiensis* was the 70 kg ectothermic *Varanus komodoensis* (Komodo dragon). The top herbivore was the small endothermic *Stegodon florensis*. The possible body mass of a top omnivore, such as *H. floresiensis*, is not dealt with directly in this paper, but it can be derived by interpolation. To determine the expected body mass of *H. floresiensis*, it is assumed that *H. floresiensis* is omnivorous and endothermic. The expected body mass of an endothermic top omnivore can logically be predicted to lie somewhere around the average of the values for endothermic top herbivores and endothermic top carnivores. The size of Flores as given by Burness et al. (2001) is 14154 km^2. Using this value in the regression equations of Burness et al. (2001), an endothermic omnivore such as *H. floresiensis* is predicted to have a weight of approximately 36 kg, possibly slightly less, because the top niches on Flores are already occupied by *V. komodoensis* and *S. florensis*. This value closely agrees with that estimated for LB1 (36 kg based on the cross-sectional area of the femur) by Brown et al. (2004). Thus, the body weight of LB1 is appropriate for an endothermic omnivore adapted to an island the size of Flores.

Low Gear Locomotion

The relatively wide pelvis and the short legs of LB1 are very similar to those of *Australopithecus afarensis* AL 288-1, aka Lucy. The large iliac flare would have led to an increase of the abductor moment arm about the hip and as a consequence a reduction in the abductor force necessary. Although the distance between the acetabulae of LB1 cannot be assessed quantitatively without an associated sacrum, it appears from the general morphology and resemblance with australopith pelves that the acetabulae were likely relatively far apart. This is supported by the high bicondylar angle, which is associated with a large interacetabular distance, particularly in a short lower limb. Interacetabular distance has the potential to contribute to stride length through rotation of the pelvis. Furthermore, pelvic rotation has relatively greater influence on stride length in individuals with short limbs. Additionally, short legs decrease the energy required for walking. Therefore, the lower body configuration of *H. floresiensis* and australopiths likely made them more efficient walkers at low speeds than are modern humans. The adaptation to low gear locomotion of *H. floresiensis* is also visible in the long bones of the lower body (van Heteren, 2012). The femora and tibiae of *H. floresiensis* have been found to be relatively robust. The increased robusticity and shortening of the limb elements compared to its likely ancestor *H. erectus* resulted in the centre of gravity being closer to the ground, which increases stability during stance and locomotion.

An additional characteristic of shortened limbs among island fauna is that the distal limb segments are shortened relative to overall limb length. The intermembral index of *H. floresiensis* resembles that of australopiths more than that of *H. sapiens*, and LB1 is sometimes said to have relatively long arms. In fact, the arms (and feet) of *H. floresiensis* are not relatively long, but the legs are relatively short. Other island fauna exhibit the same feature with dwarfing being due for the most part to a shortening of the locomotory limbs. For example, changes in body proportions are also observed in endemic insular pigs. Short-leggedness is seen in the extant Sulawesi warty pig (*Sus celebensis*) and the two Pleistocene Sulawesi pigs (*Celebochoerus heekereni, C.* sp.), and the Sardinian island pig (*Hyotherium insularis*) also had shortened phalanges. As *H. floresiensis* is not quadrupedal, shortening is expected mostly in the legs, and indeed this is the case. Although it may seem that the distal limb elements in *H. floresiensis* are elongated and not shortened, because it has such large feet, from a functional perspective this should be viewed as an increase in robusticity of the most distal part of the limb rather than an elongation (van Heteren, 2012).

Heterochrony

The orbital index of adult *H. floresiensis* clearly is more similar to a *H. erectus* juvenile than an adult. A comparison can be made between the right orbital index of *H. floresiensis* (Tab. 14.1) and the orbital indices of adults and children. The right orbital index of *H. floresiensis* falls within the middle 50% of *H. erectus* children, but outside the range of *H. erectus* adults for this sample (Fig. 14.3). When *H. floresiensis* is compared with *H. sapiens*, it falls outside the range of adults and in the upper quartile of children. *H. floresiensis'* orbital index is also closer to the median orbital index of children of

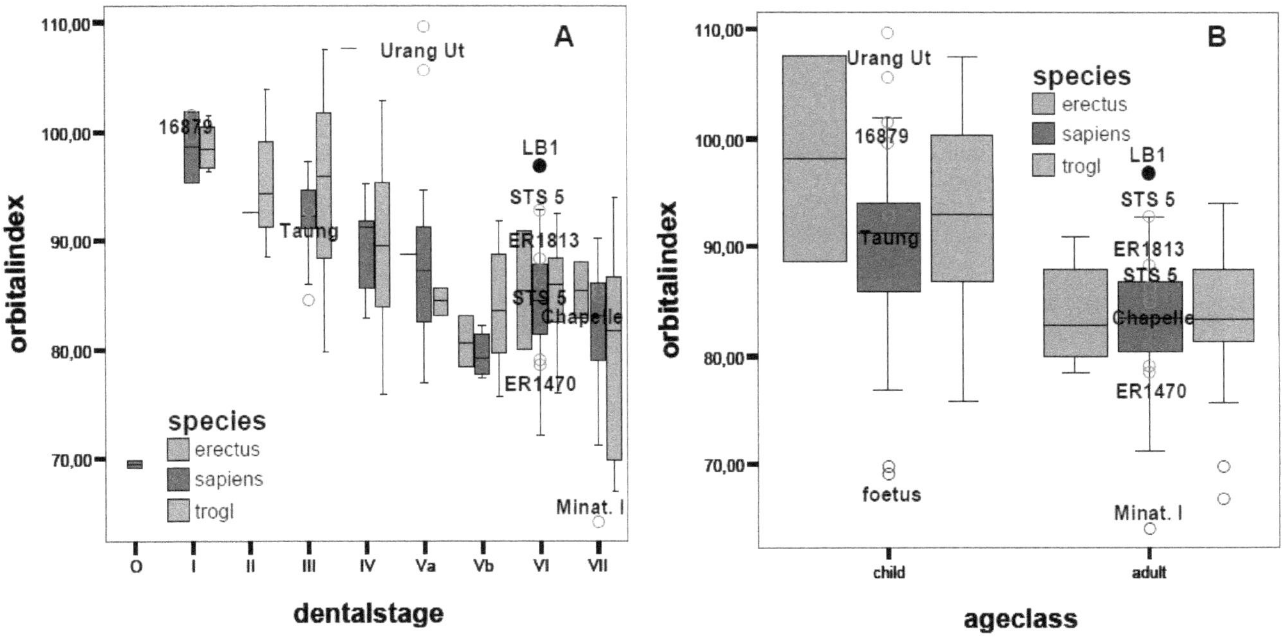

*Fig. 14.3. Boxplots of the orbital index (A and B) of LB1 compared to other hominids. H. sapiens is indicated in blue (sapiens), H. erectus in green (erectus), P. troglodytes in orange (trogl), all other species in red and the black dot is LB1. Minat. I is Minatogawa I Man from Japan. The crania and casts are stored in the National Museum of Natural History Naturalis in Leiden, The Netherlands. The orbital index is calculated as: Orbital index = orbital height * 100 / orbital breadth. A: Boxplot of the orbital index classified by dental stage. LB1 has a dental index of VI and falls within the ranges of the children of dental stage I, II, III, IV and Va when all species are taken into account, but outside the range of all adults. B: Boxplot of the orbital index for adults and Juveniles. LB1 falls within the middle 50% of H. erectus and P. troglodytes juveniles and the upper fourth quartile of H. sapiens children, but outside the range of adults of any species. Modified after Van Heteren & de Vos (2007)*

both *H. erectus* and *H. sapiens*, than to the median orbital index of adults for of species in this sample. The mandibular angle of *H. floresiensis* ranges between 109° and 113°. This is much more comparable to juvenile *H. erectus* than to either adult *H. erectus* or juvenile or adult *H. sapiens* (Tab. 14.2 & 3).

Comparisons can be made the endocast of *H. floresiensis* and the endocasts of both healthy and microcephalic *H. sapiens, P. an troglodytes* and *H. erectus*, as depicted in Falk et al. (2005, Fig. 14.1B) and with Mojokerto child (*H. erectus* juvenile), as depicted in Balzeau et al. (2005, figure 7). LB1 lacks the occipital expansion over the cerebellum of the adult *H. erectus*, and its endocast is more brachycephalic. Mojokerto child also misses this occipital expansion. The endocast of LB1 is more similar to *H. erectus* than to healthy or microcephalic *H. sapiens* or to *P. troglodytes*. Further, the endocast of LB1 looks more like juvenile *H. erectus* than adult *H. erectus* with respect to the occipital expansion. Thus, the endocast of LB1 resembles the endocast of juvenile *H. erectus* more than adult *H. erectus* or the other species.

In island animals, such as *Elephas falconeri*, there is a lack of humeral and femoral torsion present in mainland species. This is exactly the same adaptation as *H. floresiensis* displays, indicating the same evolutionary pattern of heterochrony, as torsion increases during development and a lack of it indicates an developmental changes. The right humerus of LB1 also has relatively limited degree of humeral torsion according to Morwood et al. (2005). Its humeral torsion is approximately 110°, which is less than in large bodied apes and humans (141-178°) and other known hominins. The tibial torsion of *H. floresiensis* is also rather low for an adult (van Heteren, 2008; Morwood and Jungers, 2009).

Finally, the relative length of arms and legs in *H. floresiensis* is characteristically heterochronic. The brachial index of *H. floresiensis* is similar to that of a *H. sapiens* child at birth or at the age of 1 year (Fig. 14.2). Morwood et al. (2005) reported that the humero-femoral index is outside the range of adult *H. sapiens*. In *H. sapiens*, during foetal life the arms are longer than the legs; at birth the arms and legs have approximately the same length; between the sixth and the ninth year the legs become longer than the arms; and during adulthood the relation between arms and legs is 4:5. The intermembral index of *H. floresiensis* is similar to *H. sapiens* children.

Reversals to an Ancestral State

H. floresiensis displays primitive features in the cranium, mandible, brain, shoulder, wrist, pelvis, limb bones, and foot, and in its general body proportions (van Heteren, 2012). Australopiths are small compared to later members of the genus *Homo*, and *H. floresiensis*, similar in size to australopiths, has secondarily returned to this primitive state. The presence of reversals to a more primitive condition should be viewed as supportive of the insular dwarfing theory.

CONCLUSIONS

The low body weight of *H. floresiensis* is congruent with expectations based on an omnivorous endotherm adapted to an island the size of Flores. The legs of *H. floresiensis* are relatively short and robust and the pelvis relatively broad as an adaptation to low gear locomotion and to increased stability on uneven terrain. Several features of *H. floresiensis* display differences from adult *H. erectus* in ways that are strongly suggestive of heterochrony. In addition, *H. floresiensis*, similar in size to australopiths, has secondarily developed primitive features. In conclusion, the data presented here support the hypothesis originally put forward by Brown et al. (2004) that the distinctive features of *H. floresiensis* were caused by island dwarfing. Consequently, hominins are not the exception that prove the rule, but rather comply with the rules of island evolution.

Acknowledgements

I would like to express my gratitude to Tony Djubiantono for providing me with a cast of the cranium and mandibula of LB1. I would also like to thank W. Jungers for giving permission to use one of his photographs. Further, I would like to thank D. Kaz for comments and suggestions.

References

BALZEAU, A., GRIMAUD-HERVE, D. & JACOB, T. 2005. Internal cranial features of the Mojokerto child fossil (East Java, Indonesia). *J. Hum. Evol.* 48: 535-553.

BROWN, P., SUTIKNA, T., MORWOOD, M.J., SOEJONO, R.P. et al. 2004. A new small-bodied hominin from the Late Pleistocene of Flores, Indonesia. *Nature* 431: 1055-1061.

BURNESS, G.P., DIAMOND, J. & FALANNERY, T. 2001. Dinosaurs, dragons, and dwarfs: the evolution of maximal body size. *PNAS* USA 98: 14518-14523.

DAVENPORT, C.B. 1934. Ontogeny and phylogeny of Man's appendages. *PNAS* USA 20: 359-364.

FALK, D., HILDEBOLT, C., SMITH, K., MORWOOD, M.J. et al. 2005. The brain of LB1, *Homo floresiensis*. *Science* 308: 242-245.

MEIJER, H.J.M., VAN DEN HOEK OSTENDE, L.W., VAN DEN BERGH, G.D. & DE VOS, J. 2010. The fellowship of the hobbit: the fauna surrounding *Homo floresiensis*. *Journal of Biogeography* 37: 995-1006.

MORWOOD, M. & JUNGERS, L. 2009. Paleoanthropological research at Liang Bua, Flores, Indonesia. *J. Hum. Evol.* 57(5): 437-648.

MORWOOD, M.J., BROWN, P., JATMIKO, SUTIKNA, T. et al. 2005. Further evidence for small-bodied hominins from the Late Pleistocene of Flores, Indonesia. *Nature* 437: 1012-1017.

VAN DER GEER, A.A.E., LYRAS, G., DE VOS, J. & DERMITZAKIS, M. 2010. *Evolution of island mammals: Adaptation and extinction of placental mammals on islands*, Wiley-Blackwell, Chichester, pp. 461.

VAN HETEREN, A.H. 2008. *Homo floresiensis* as an island form. *PalArch's J. Vertebrate Palaeontology* 5: 1-12.

VAN HETEREN, A.H. 2012. The hominin of Flores: insular adaptations of the lower body. *Comptes Rendus Palevol* 11: 169-179.

VAN HETEREN, A.H. & DE VOS, J. 2007. Heterochrony as a typical island adaptation in *Homo floresiensis*. In: Indriati, E. (Ed.) *Recent advances on Southeast Asian Paleoanthropology and Archaeology*. Faculty of Medicine, Gadja Mada University, Yogyakarta, pp. 95-106.

VAN HETEREN, A.H. & SANKHYAN, A.R. 2009. Hobbits and pygmies: Trends in evolution. In: Sankhyan, A.R. (Ed.), *Asian Perspectives on Human Evolution*. Serials Publications, New Delhi, pp. 172-187.

15.
IMAGING OLDOWAN-ACHEULIAN KNAPPERS: SCOPE & LIMITATIONS

Tanusree PANDIT
Senior Research Fellow
tanusree2@gmail.com

Anek R. SANKHYAN
Ex-Anthropologist at Anthropological Survey of India, 27, Jawaharlal Nehru Road, Kolkata-700016, and President at Palaeo Research Society, Ghumarwin (H.P.)-174021, India
arsankhyan@gmail.com

Abstract: *Scholars in recent years have used fMRI, PET, MRI scanning techniques in experimental research conducted on naïve and expert knappers in order to understand differences in the activated brain parts performing Oldowan and Acheulian stone tool knapping. There are observed differences in the activated parts but the inferences remain provisional and inconclusive without having control on the raw materials, diet, manual dexterity, and a complexity of many influencing factors that activated early hominin brain during its evolution.*

Keywords: *Broadmann Area (BA) 19, 44; fMRI, PET, MRI; Oldowan, Acheulian*

INTRODUCTION

Homo sapiens is an accomplished species with advanced technological skills as a precursor to its brain development. Substantial mental prowess, afforded by a large, slow-developing brain, affords both time to acquire complex social and technological skills of human groups and the plasticity to modify behaviour in response to environmental change which gradually emerged to human social intelligence. Behaviour is the quickest, simplest, least costly way of responding to environmental pressures. Therefore, cumulative knowledge of stone tool production has changed not only the physical attributes but also helped in carrying out the production tasks which can be examined through the artefacts. Dietrich Stout (2005: 273) conclusion is worth quoting, "Conventional archaeological, ethological and psychological approaches to understanding the origins and evolution of human tool use employ a representational perspective that emphasizes underlying cognitive mechanisms rather than overt performance".

The skills of making stone tools of varied shapes and sizes indicates advancement in technology accountable to utilization of the raw materials, changes in dietary regime, the morphological change in human hand and social and genetic transmission. This paper critically evaluates the complexity of the brain by analysing the Oldowan and Acheulian stone tool techniques. For analysing these technologies we found that the scanning techniques a useful tool for recording the activated brain parts. The scanning came into use since the 18th century for medical purpose, but soon extended to the study of fossil skulls for imaging the endocasts and exocasts.

AIMS & OBJECTIVES

Present study aims to: **a.** present a brief overview of the new analytical techniques applied in the analysis and understanding of the activated brain parts with the Oldowan-Acheulian archaeological traditions; **b.** to judge the merits and limitations of the such experiments; **c.** to discuss a multiple of factors other than stone tool technology influencing human brain development and evolution.

OBSERVATIONS

The New Techniques

Palaeoanthropologists employed two types of scanning techniques, the structural (CT and MRI) and the functional (PET and fMRI) (Fig. 15.1). The CT scan is 'non-invasive' and exhibits the extent of soft tissues in fossilised bone. The 3D imaging technique transforms the virtual reality to real virtuality, but it leads to destruction of certain parts. To overcome its shortcoming SEM technique emerged, which has the ability to replicate and enable speculation of successive changes. Both the techniques are used to analyse exocast and endo-casts.

There is addition of new technique called TMS (Trans-cranial Magnetic Stimulation), which helps in an spotting the neural configuration. But, PET technique can perform neuroanatomical correlations of the brain function and helps in speculating the neuronal functions when the brain parts are activated. However, its drawback is that the injected radiotracer can affect blood groups.

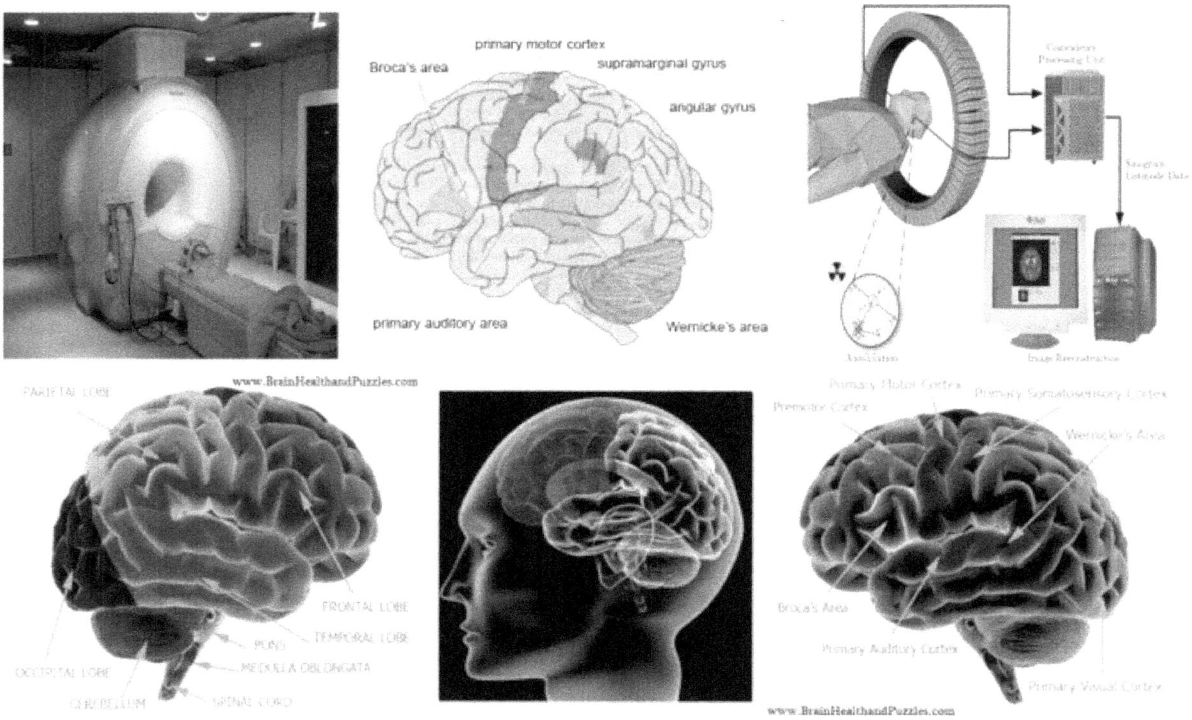

Fig. 15.1. Scanning techniques and their application in understanding brain activation during stone tool knapping and major parts and scans of modern human brain (assembled from various web sources)

The shortcoming is overcome by a new technique called fMRI (meaning magnetic resonance imaging). Like the CT the fMRI determines the soft tissues of the brain without X-Rays and with magnetic non-ionizing radiation of protons and electrons. Hence, it is the only safe method and can be applied several times on the patient without harm.

Frey, 2007; pp. 368 observed: "simple tool use behaviours involve using objects to implement motor-to-mechanical transformations that amplify movements of the upper limbs (e.g., using a rock to increase pounding force, or a stick to extend reach). By contrast, complex tool use involves objects to implement transformations that convert movements of the hands into qualitatively different mechanical actions".

RESULTS OF RECENT EXPERIMENTS ON STONE TOOL TECHNIQUES

Stout (2005) noted that the results drawn from knapping-related activations were found in the primary motor and somatosensory cortices on the central sulcus, the superior parietal lobule, the cerebellum and the fusiform gyrus of the right inferior temporal lobe of the brain (Fig. 15.2). The PET scanning techniques over the stone knapping reveals that the Mode-I technology is associated with certain activated areas mentioned below:

Result 1: The Pericentral cortex, which is surrounded by the central sulcus, procures the primary motor cortex (M1) of the precentral gyrus and the primary somatosensory cortex (S1) of the postcentral gyrus. The M1 receives information from not only the afferent regions of the brain but also from the S1 (motor-planning areas), thalamus and cerebellum. Therefore M1 is related to the limbic movements and replied with the perception action movement which is ecologically valid.

The Superior parietal lobule consists of the polymodal association cortex and has mappings between diverse visual, somatosensory, auditory, vestibular and motivational signals. This area supports the sensory perception and motor action and the Mode-I activation clearly reflects on the perception-through-action of polymodal mappings in knapping.

The occipital lobe is composed of the primary (V1) and secondary (V2, V3, V4 and V5) lobules which reflects the sensitive dependence with visual environment.

The Cerebellum is divided into three parts: vestibulocerebellum, spinocerebellum and cerebrocerebellum which are responsible for the temporal coordination on the perception motor mappings.

Result 2: Bilateral activation on the primary somatosensory and motor cortex on the central sulcus and cerebellar hemispheres is found. The knapping control is found to be activated in both the hemispheres while in the movement of the limbs. The main difference in the results from the pilot study revealed that it has the robust and extensive occipital visual cortex which appeared through the test. The novice results generally reflected on the visual processing areas as well as inferior parietal, while the Broadmann area 19 is involved in the secondary

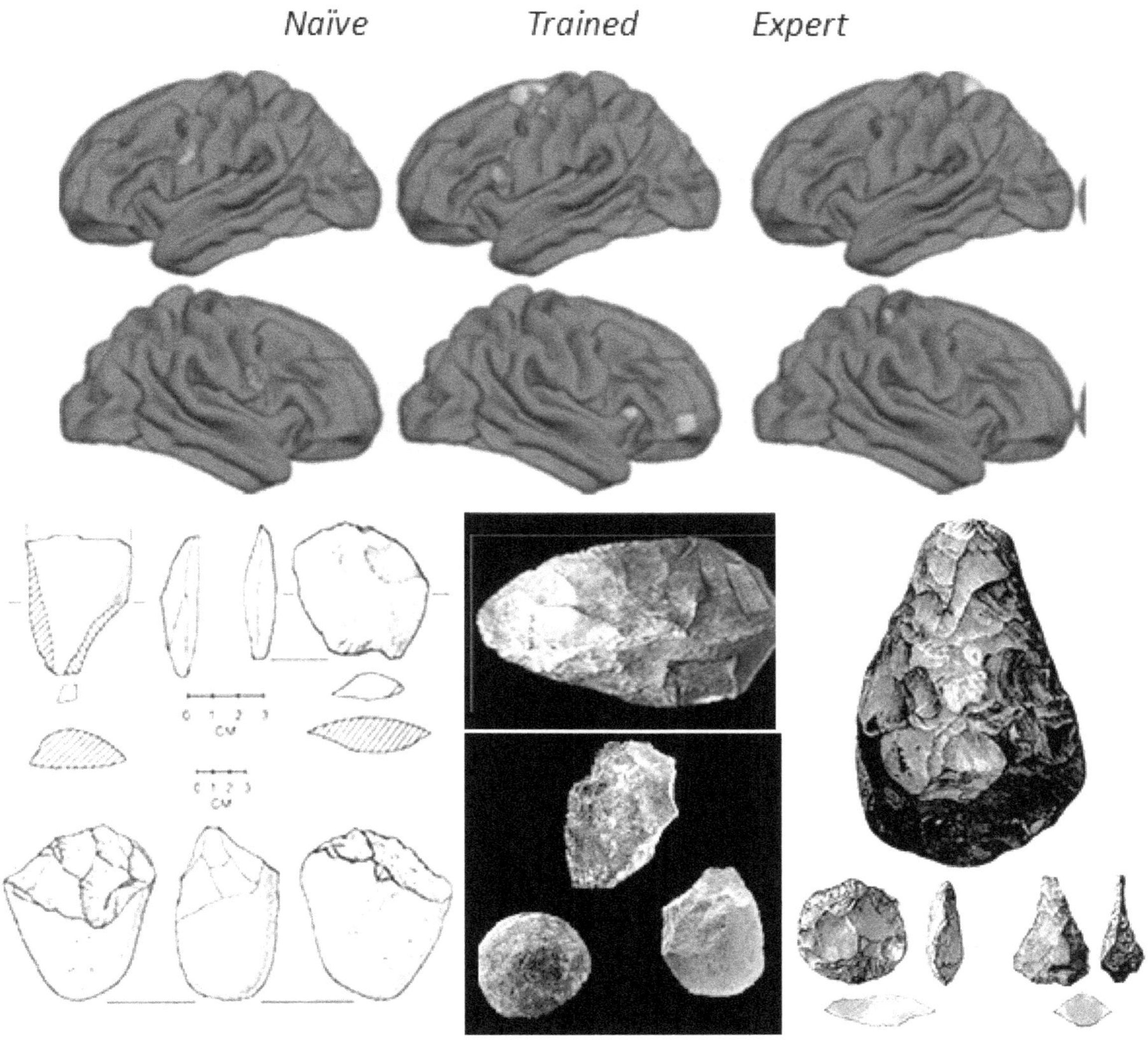

Fig. 15.2. Brain activation in during Oldowan & Acheulean Stone tool knapping experiments (FDR P <0.05, extent k >20; modified from Stout et. al., 2011, pp. 5)

visual processing but the prefrontal activation was not observed (Stout, 2007).

Result 3: Two contrasts of interest were applied where the brain activation was analysed before the tool practice and after tool practice and contrasts between the two. The results inferred were seen in the left dorsal primary motor cortex and right central sulcus regions with the premotor, primary motor and somatosensory cortices and some regions including the parietal cortex and intraparietal sulcus. When the knapping skills judged with practice, the results were reflected on the two ventral primary motor cortex, also in the ventral, lateral and dorsal visual areas within broadmann area 19 after practice. Thus it reveals that the controlled action of knapping skills gathered after stone tool practice (Stout, 2008). More recently, Stout (2011) and Stout et al. (2011) conducted experiments with 11 naïve, 10 trained and 5 expert Oldowan and Acheulian knappers and did statistical analysis as: 1. (Acheulian – control) + (Oldowan – control); 2. Acheulian – control; 3. Oldowan – control; 4. Acheulian – Oldowan. The findings are summarised in the Tab. 15.1 and Fig. 15.1 & Fig. 15.2.

DISCUSSION

Firstly, construction of layers in sequences is often highlighted by the memorization and apprenticeship which can also be aimed at the chain imitation like the tradition passed on through generation to generation. Secondly, it is the constellations of knowledge where planning is required before any action. When we try to organize the holistic approach of archaeological neuroscience based on the cognitive capacity we basically blame the individual neurons, acting as a precursor to the quantitative succession of the stone tool manufacture (Grove et al., 2008). Both the previous and present research reveals that the knappers are right handed and the left hemisphere is much more activated.

The experts who have experience of 2-10 years of knapping infer the activated areas of primary motor and somatosensory cortices, parietal lobule, cerebellum and fusiform gyrus whereas the naïve have the BA 19, ventral precentral gyrus and left occipital gyrus, left frontal cortex and dorsal superior frontal gyrus and ventral BA 44 of inferior frontal gyrus. Thus, previous research lacks

Tab. 15.1. Results of the Oldowan and Acheulian knapping experiments by Stout et al. (2011)

	Expertise Response	Non-Expertise Response
Tool making control	**Cortical regions** including primary visual and lateral occipital cortices and inferior temporal cortices, intraparietal sulci, inferior parietal cortices and post central gyrii. **Frontal cortex** includes BA 44, bilateral BA 45 of the inferior frontal gyrus and dorsal precentral gyrii.	**Observation of tool making (naïve)**: bilateral ventral precentral gyrus and left middle occipital gyrus. **Observation of tool making (experts)**: left superior parietal and right post central gyrus. **Observation of tool making (trained)**: medial frontal cortex, right par orbitalis, left BA 45, bilateral BA 44, right anterior insula, left posterior middle frontal gyrus and left recentral gyrus, left middle temporal gyrus and right inferior temporal gyrus.
Acheulian vs. Oldowan stone tool making	Anterior part of the left intraparietal sulcus, left prefrontal sulcus with inferior frontal sulcus.	**Naïve**: left frontal cortex, dorsal superior frontal gyrus and ventral BA 44 of inferior frontal gyrus. **Experts**: right medial frontal and parietal cortices and inferior parietal lobule, anterior intraparietal sulcus.
Tool making with control condition	**Cortical regions** including inferior frontal gyrus, dorsal premotor cortex, intrapa-rietal sulcus and inferior parietal lobule, anterior intraparietal sulcus and inferior frontal sulcus on the left hemisphere.	**Naïve**: motor resonance activation with ventral prefrontal cortex. **Trained**: strong response of motivational set with training. **Expert**: medial prefrontal cortex inferring high level of intentional reading.

the number of experimentees and concentration should also be given to the lefthanders, females and apes. Further research can be proceeded through the analysis of visual acuity and auditory capability of the casts along with the sexually dimorphic results on the basis of the weight (Prof. Dunbar's analysis of the brain to body size ratio) and age (the maximum age where the brain stops its enlargement is 15 yrs of age). A study has revealed that the body height generally depends on the brain weight. Tall individuals have low relative weight of the brain (Chrzanowska et al., 1975). If this study can be imposed on the fossil remains, then it seems theoretically that we can find a correlation of the brain weight to that of the body weight ratio which depicts that weighty brain has more convolutions enhancing more complexity.

Pruvic (2009) observed that the left hemisphere is lighter than the right and these are connected with four neurotransmitters: dopamine, norepinephrine, serotonin and acetylcholine. The first and fourth are present in the left hemisphere and advance intellect.

Observation of tool making (Tab. 15.1)

Acheulian vs. Oldowan stone tool making: frontal cortex, right par orbitalis, left BA 45, bilateral BA 44, right anterior insula, left posterior middle frontal gyrus and left recentral gyrus, left middle temporal gyrus and right inferior temporal gyrus, anterior part of the left intraparietal sulcus, left prefrontal sulcus with inferior frontal sulcus.

Naïve: left frontal cortex, dorsal superior frontal gyrus and ventral BA 44 of inferior frontal gyrus.

Experts: right medial frontal and parietal cortices and inferior parietal lobule, anterior intraparietal sulcus.

Tool making with control condition

Cortical regions: including inferior frontal gyrus, dorsal premotor cortex, intraparietal sulcus and inferior parietal lobule, anterior intraparietal sulcus and inferior frontal sulcus on the left hemisphere.

Naïve: motor resonance activation with ventral prefrontal cortex.

Trained: strong response of motivational set with training.

Expert: medial prefrontal cortex inferring high level of intentional reading motor planning and execution, working memory, cognitive flexibility and creativity. Amongst the four, the dopamine is useful in advancing the intelligence and noted to have increased during primate and hominid evolution, found exclusively concentrated in non-human species, and associated with our mathematical, reasoning and planning. Pruvic (op.cit) also considered that dopamine enhanced bipedalism from the *Australopithecus* to *Homo habilis*. He thinks that the choice of raw material played greater role during migration. In this regard, we believe that further research on the left handed people can be more meaningful.

We observe that Stout and others experiments are important and interesting to find differences in the activation levels of human brain in Oldowan and Acheulian hominins. But, the evolution of modern human brain has been a much more complicated process and cannot be explained by mere technological change. We briefly place below the important influencing factors briefly responsible for the same.

BEYOND TECHNOLOGY: INFLUENCES ON BRAIN DEVELOPMENT

The Increased Neocortex

Studies say that the neocortex increased in size during evolution of brain from simple to complex forms, but how and why neocortex became a main feature? Dunbar (1988) explained it through the social intelligence hypothesis that the increasing group size among Primates evolved the neocortex as they interacted and recognized other members of the groups born with nepotism or reciprocity. So, with the increasing group size, the complexity of social life also enhanced and has a positive correlation with the neocortex ratios (Boyd and Silk, 2006). Therefore, the ecological factors related to home range and diet are also related to the complexity of social life. Thus, the brain size is dependent on the behavioural flexibility and complexity – the number of neurons as indicators of brain complexity.

The Brain/Body Ratio

The brain/body ratio increases in a negative allometric way meaning that when the brain becomes absolutely larger the size becomes relatively smaller (Roth et al., 2005). The 18,000 years old Indonesian *Homo floresiensis* is exceptional. Its brain was just a little larger than that of the Chimpanzee (Brumm et al., 2006), but it made sophisticated stone tools like long blades, spears and points similar to that of modern man. While its virtual endocast was comparable to human pygmy/human microcephalic, its brain/body size ratio is similar to Australopithecus, and the endocast similar to Homo erectus having the frontal and temporal lobes possessed with high cognitive processing (Falk et al., 2005; Blater, 2005). This species had an unusual combination of ape-like and human-like foot for bipedal locomotion but the foot bone described a more primitive design. Its body and brain evolution is explained through heterochrony (van Heteren & Sankhyan 2009) due to island dwarfing of an early *Homo erectus* lost in isolation. Thus, here we can see that there can be various reasons for the evolution of a master mind.

The evolution of brain through dietary pattern

The big brains are evolved not only through an exceeded benefit of high cost climate but also through the ancestral development of good metabolic systems which continuously supported the generations (Parker et al., 2003). The richer the diet, the bigger becomes the brain volumenously reflecting that the metabolic demands enhance the body composition preceded with greater encephalization.

The brain size between *Australopithecus robustus* (~530cc) and *Homo erectus* (~1000cc) differed from each drastically, besides that the latter has a reduction in craniofacial and mandibular robusticity and a probability of maintaining a human like stature, body mass and body proportions. The animal bones in archaeological records of the latter revealed that *Homo erectus* consumed more animal foods than plant (Leonard et al., 2003). So, it can be said that the consumption of animal food made the brains much richer. Barham et al. (2008, p. 83) infers that "the systematic incorporation of animal protein into the diet of early hominin traits, including large brain size, increased intelligence, increased body size, and delayed maturation with an extended childhood supported by food provisioning."

The *Paranthropus* (Swartkrans) had the fossilized larger molars, pre-molars and thick enamel which aimed at crushing and grinding the food and revealed that 25% of their diet were derived from C4 plants (best known example is the Swartkrans tool with wear-marks dated about ~1.7-1.5 Mya); whereas the Australopithecus was merely human like, with reduced molars and thin enamel, justifiable between the omnivorous and carnivorous, generally, used to derive its nutrition from C4 plants as well C3 sources like grasses, sedges, insects etc. and reflecting it to be lived during the savanna habitat in Africa (Barham et al., 2008). Now when we try to correlate the food habit with the utilization of the stone tools, it is highly emphasized that the two are related which can be deciphered from the utilization marks and big game hunting strategy. In the Oldowan technology the core provides a hallmark with the cut marks on the bones which explains the sharpness of the tools used to take out the bone marrow from the cavity of the bone with full of protein.

Influence of Raw Material

Widely used definition of tool use is "the external employment of an unattached environmental object to alter more efficiently the form, position, or condition of another object, another organism, or the user itself when the user holds or carries the tool during or just prior to use and is responsible for the proper and effective orientation of the tool" (Beck, 1980:10).

The stone knapping generally includes the operational features like the detachment, conjunction and reduction which forms the main basis for the technological advancement in both spatial and temporal locations of an archaeological site which enhance the modelling of the stone tools. Best examples are seen from Lokalalei 2C (West Turkana, Kenya) hominids (dated back ~2.34 Mya) who performed stone tools from the cobble reduction sequences of the refitting groups knapped specifically from the lava cobbles (Barham et al., 2008). These hominids might have had a particular state of planning and foresight management of the raw materials where they had a controlled debitage flakes and specifies that it will not only deal with technical competencies but also with technical economic behaviour which when merged gives rise to a specific question of intersite diversity and intrasite complexity (Delagnes et al., 2005). The skilful knapping reveals well-controlled motor skills and a quick grasp in decision-making (chaine operatoire) with no probability of novice knappers (Barham et al., 2008). At the same point of time we had a quick view on the Gona assemblages. The raw materials were selective like those of the Lokalalei 2C with controlled bifacial flaking (Barham et al., 2008). These arrays of raw material utilization give a glimpse of the availability within the context as well as a bio-mechanical diversity among the hominids. But recently this view was rejected when Kanjera formation had the Oldowan stone tool making, transported raw materials from a distance source of about >10 km where the rock was different from their original habitation. The stone tools found were the associated fauna's mostly of conglomerates. This has been studied from the geochemical method Energy Dispersive X-Ray Fluorescence (ED-XRF) (Braun et al., 2008). This transport might have changed their lives with an advent of more energy utilization which finally led them to be more adaptive and with a change in their anatomy (Bramble & Lieberman, 2004; Lieberman et al., 2007; Pickering & Bunn, 2007). Therefore, the raw material is not a qualitative but a quantitative variable considered on the basis of mineralogical structure, size and shape of the material (Hounsell, 2005).

Thus, it reveals that individual raw materials used from the environment are selected for specific tool manufacture to provide a requisite shape to support them.

Influence of hand morphology on technology

"Making stone tools is deceptively simple act that involves a complex interplay between bone, muscle and brain. A skilled knapper understands the flaking properties of stone, the relationships among angle, force and placement of a blow, and the resulting length and thickness of a flake. These skills take time to learn, but also involve distinctive biomechanical features of human anatomy that are a likely consequence of the evolution of bipedalism." (Barham et al., 2008, pp. 74).

The human hand is the main organ which falls under the categories of the manipulative abilities and generally divides into two categories: prehensile movement where two grips are taken into account such as the precision and power grip and all these are possible for the opposable thumb and non-prehensile movement (Aiello et al., 1990). The Hadar fossil (site AL333/333w; ~3.2 Mya) discovery comprised of the carpal, metacarpal and phalange claimed to be a complete hand, but didn't have transverse ridges at the bases of the dorsal articular surfaces. It had a wide transverse diameter of metacarpal whose head is placed anteriorly while the radius lacks the dorsal ridge where the dorsoflexion muscle of the wrist is not identified and therefore revealed that Hadar fossil had a few similarities with the modern humans with a generalized hominoid hand. Therefore the tool making technique can be applied as an ad-hoc tool user (Aiello, 1990). The trapezoid bone of the fossil indicates a distinctive human grip responsible for the pad-to-side holding which holds the Oldowan culture pre-dated by 1 Mya (Tocheri et al., 2003).

The hand of *Paranthropus robustus* reveals that they were more human like than that of the A. *afarensis*. The species had a modern human-like thumb which would perform both the precision as well as power grip. The palmer surface had a clear insertion of the tendon which indicates the presence of flexor pollicis longus muscle, a help for the flexion of the thumb and therefore is adequate to analyse that they were the first proper tool user according to the evolutionary tree. Whereas the hand of *Homo habilis* at the Olduvai had three carpal bones, four proximal phalanges, four middle phalanges, three distal phalanges and a single second metacarpal but among them, two proximal phalanges seemed to belong to the juvenile individual so it is not yet considered (Aiello, 1990). The length of the thumb plays a vital role for the variation of the grips. The larger the thumb in length, the greater manual dexterity for making the stone tools (Barham et al., 2008).

Faisal et al. (2010) studied the Oldowan flaked tools and late Acheulian stone tools for analysing the manipulation skills of the hominids. They claimed Acheulian stage as an intentional knapping form which can be correlated to the language evolution inferring neural overlap in the inferior frontal cortex. Joint angles of hand helped them to analyse the brain activated areas. They found Oldowan stone tool making is associated with the activation of the left premotor cortex which is the point for judging the manual grip coordination and phonological processing whereas the Acheulian involves in the right hemisphere activation includes the supramarginal gyrus of the inferior parietal lobule, right ventral premotor cortex and right hemisphere homolog of the anterior Broca's area which is involved in higher order of hierarchical cognition and processing of linguistic context. They confessed that the skilled tool making comes into existence only when the stiffness of control between hand-arm systems is well grasped.

The authors also viewed that the thumb plays a vital role in the manipulative skills of both the hominins (where power grip and precision grip are well played) as well as non-human primates (where only precision grip is maintained). According to them, the makers of the Oldowan had the thumb muscles (Flexor pollicis brevis and Adductor pollicis transverse) and little finger (Flexor digiti minimi and Abductor digiti minimi) (pp. 8).

CONCLUSIONS

The broad overview presented here provides some important clues for future studied.

1. Recent advancement in scanning techniques, CT, SEM, TMS, PET and fMRI have indicated useful for the determination of brain evolution since different brain parts are activated during stone tool making in experiments conducted on the Oldowan vis-a-vis Acheulean.

2. They revealed that the Oldowan tool technique is associated with left premotor cortex for manual gripping, whereas the Acheulian is associated with inferior parietal lobule and right ventral cortex. But, both are correlated with the linguistic processing which obviously had greater control in brain development in modern humans.

3. However, brain evolution has been a long process influenced by numerous factors. The early hominids were very selective in choosing the raw materials from different zones based on their needs. The variation in diet ranging from the frugivorous, herbivorous, gramnivorous, omnivorous to flesh eating, etc., played great role, besides the gender in food collection, hunting, etc. The technological development was much influenced by the

availability of soft or hard raw materials in their habitats, besides that the hominid manual dexterity (power and precision grip) also constrained stone tool making.

4. Therefore, the present experimental studies on modern humans may not replicate the situation of the archaic humans of the Oldowan or Acheulian times over a million years ago. These experiment would become meaningful if extended across primate taxa, notably the extant hominoid apes, and primitive hunter-gatherer populations living different in ecological settings.

References

AIELLO, L. & DEAN, C. 1990. An introduction to human evolutionary anatomy. Academic Press Ltd. London.

BARHAM, L. & MITCHELL, P. 2008. The first Africans: African archaeology from the earliest tool makers to most recent foragers. Cambridge University Press.

BECK, B. 1980. Animal tool behaviour: the use and manufacture of tools by animals. Garland press, New York.

BLATER, M. 2005. Small but smart? Flores hominid shows signs of advanced brain. *Science* 307: 1386-1388.

BOYD, R. & SILK, J.B. 2006. How humans evolved. W.W. Norton and Co., London.

BRAUN, D.R., JOANNE, C.T., JOSEPH, V.F., SHIRA, L.A. et al. 2008. Oldowan reduction sequences: method-logical considerations. Journal of Archaeological *Science* 35: 2153-2163.

BRAMBLE, D.M., LIEBERMAN, D. 2004. Endurance running and the evolution of Homo. *Nature* 432, 345-352.

BRUMM, A., AZIZ, F, VAN DEN BERGH, G.D., MORWOOD, M.J. et al. 2006. Early stone tool technology on Flores and its implications for Homo floresiensis. *Nature* 441: 624-628.

CHRZANOWSKA, G. & KRECHOWIECKI, A. 1975. Deos the weight of the brain depend on the body height? *Gegenbaurs Morphol Jahrb* 121 (2): 192-208.

DELAGNES, A. & ROCHE, H. 2005. Late Pliocene hominid knapping skills: the case of Lokalalei 2C, West Turkana, Kenya. *J. Hum. Evol.* 48(5): 435-72.

DUNBAR, R.I.M. 1988. The Social Brain Hypothesis. *Evolutionary Anthropology* 6: 178-90.

FAISAL, A., STOUT, D., APPEL, J. & BRADLEY, B. 2010. The Manipulative Complexity of Lower Paleolithic Stone Toolmaking. *PLoSONE* 5(11): e13718. doi:10.1371/journal.pone.0013718.

FALK, D., HILDEBOLT, C., SMITH, K., MORWOOD, M.J. et al. 2005. The Brain of LB1, Homo floresiensis. *Science* 308: 242-245.

FREY, S. 2007. What puts the how in where? Tool use and the divided visual stream hypothesis. Cortex 43: 368-375.

GROVE, M. & COWARD, F. 2008. From individual neurons to social brains. *Cambridge archaeological journal* 18(3): 373-386.

HOUNSELL, S.P. 2005. Quantifying stone tool performance: with especial reference to the Oldowan and Acheulean lithic traditions. PhD dissertation, SACE, University of Liverpool, UK.

LEONARD, W.R., ROBERTSON, M.L., SNODGRASS, J.J. & KUZAWA, C.W. 2003. Metabolic correlates of hominid brain evolution. *Comparative biochemistry and Physiology part* A 136: 5-15.

LIEBERMAN, D., BRAMBLE, D.M., RAICHLEN, D.A. & SHEA, J.J. 2007. The evolution of endurance running and the tyranny of ethnography: a reply to Pickering and Bunn. *J. Hum. Evol.* 53: 439-442.

PARKER, S.T. 2003. Why big brains are so rare: energy costs of intelligence and brain size in anthropoid primates. In Parker, S.T., Gibson, K.R. (ed.) Language and intelligence in monkeys and apes. Cambridge University Press.

PICKERING, T. & BUNN, H. 2007. The endurance running hypothesis and hunting and scavenging in savannah-woodlands. *J. Hum. Evol.* 53, 434-438.

ROTH, G. & DICKE, U. 2005. Evolution of the brain and intelligence. *Trends in cognitive sciences* 9(5): 251-257.

STOUT, D. 2005. Neural Foundations of Perception and Action in Stone Knapping. In (Valentine Roux & Blandine Bril, Eds) Stone Knapping: the necessary conditions for uniquely hominin behaviour. McDonald Institute Monographs. Oxford: Oxbow books, pp. 273-286.

STOUT, D. & CHAMINADE, T. 2007. The evolutionary neuroscience of tool making. *Neuropsychologia* 45:1091-1100.

STOUT, D,, TOTH, N. & CHAMINADE, T. 2008. Neural correlates of early stone age tool making: technology, language and cognition in human evolution. *Philosophical Transactions of Royal Society London* 363:1939-1949.

STOUT, D. 2011. Stone tool making and the evolution of human culture and cognition. *Philosophical Transactions of Royal Society London* B, 366:1050-1059.

STOUT, D., PASSINGHAM, R.R., CHRISTOPHER FRITH, C. JAN APEL, J. & CHAMINADE, T. 2011. Technology, expertise and social cognition in human evolution. *European Journal of Neuroscience*, pp. 1-11, doi:10.1111/j.1460-9568.2011.07619.x

TOCHERI, M.W., MARZKE, M.W., LIE, D., BAE, M. et al. 2003. Functional capabilities of modern and fossil hominid hands: three-dimensional analysis of trapezia, *American Journal of Physical Anthropology* 122:101-112.

WYNN, T. 1993. Layers of thinking in tool behaviour. In Tools, Language and Cognition in Human Evolution. K.R. Gibson & T. Ingold, (Eds) Cambridge: Cambridge University Press.

16.
PLEISTOCENE BEADS AND COGNITIVE EVOLUTION

Robert G. BEDNARIK
International Federation of Rock Art Organizations (IFRAO),
P.O. Box 216, Caulfield South, Melbourne, VIC 3162, Australia
robertbednarik@hotmail.com

Abstract: *Among all the evidence archaeology is capable of providing about the cognitive evolution of hominins, beads and pendants are perhaps the most instructive. Other reportedly symbolic remains can in many cases be challenged, but perforated objects that are too small or too fragile to have served as pulling handles or similar can be safely described as beads. Several hundred such specimens have been excavated from Lower Palaeolithic strata in Libya, Israel, Austria, France and England. In many cases specific microwear has been detected on them that indicates that they were worn threaded onto string for prolonged periods of time. Their study and replicative experimentation have also provided empirical information about their technology, how they were made. But their most important scientific testimony is what they can tell us about the cognition of their makers and wearers. Beads and pendants demand self-awareness and a theory of mind in their users, as well as complex meanings of individual status, and research has shown that concepts of perfection were clearly involved in their production or use. The same applies to the oldest currently known rock art, located in India, which suggests that Lower Palaeolithic humans had developed relatively sophisticated cultural practices and advanced cognition several hundred millennia ago. This finding, confirmed by several others, shows that the hitherto dominant model of cultural evolution during the second half of the Pleistocene must be false, and that essentially modern human behaviour did not appear in the last third of the Late Pleistocene, but much earlier, during the Middle Pleistocene period.*

Keywords: *Pleistocene, beads, pendants, cognitive evolution*

INTRODUCTION

The most comprehensive hard evidence we are likely to glean from gestures in the Pleistocene human past is the information we can secure from production traces (chaînes opératoires) of artefacts, such as those on stone tools and other artefacts. However, there can be little doubt that the most detailed such traces are those related to, or found on, palaeoart objects. For instance, it is possible to secure considerable forensic detail about the production of rock art, be it petroglyphs or pictograms. We can analyse the chemical composition of paint residues of paintings or drawings; we can microscopically study the striations and other tell-tale markings in engravings; or examine the traces the artisan left on portable art objects of any type. These markings tell the specialist many details, such as the nature of the tools or materials used in the production of the palaeoart, the direction and sequence in which these tools were applied, even the pressure applied to tools, or whether modifications were made over a prolonged period of time. This type of information is more comprehensive than what we are likely to secure from practically any other class of archaeological objects of the Pleistocene.

I have conducted studies of this kind for several decades, including for the determination of the long taphonomic histories of manuports, the analysis of production traces of engravings in caves and at open-air sites, the authentication of very early engravings on such materials as bone, ostrich eggshell and ivory, and even for discriminating between anthropic and natural markings on numerous types of materials.

In this work I have learned to appreciate that the study of beads and pendants is particularly productive, in terms of the information it is likely to yield about the way the artefact was produced, how it was used, and what happened to it after it was deposited in what we now consider to be its archaeological context (taphonomy). But beads convey a great deal more information about their makers and users than their history. Technologically alone they illustrate not only the ability to drill through brittle or often very hard materials, but also they imply the use of cordage. The very essence of a bead or pendant is to be threaded onto a string; it would simply be pointless to perforate a small object for another purpose but to pass a string though it. However, the use of cordage also suggests the use of knots, because a string needs to be closed to form a loop to be effective. Although the ends of a string may be joined by means other than a knot, e.g. by the use of adhesive or by plaiting, these alternative means are either impracticable or they are technologically even more complex than the use of knotting (Warner and Bednarik 1996). It is relevant to note that seafaring, too, is practically impossible without the use of ropes and knotting. The diachronic availability of Pleistocene remains of cordage (Leroi-Gourhan 1982; Nadel et al. 1994; Pringle 1997) is of no relevance to the question, because that class of material evidence obviously possesses an exceptionally long taphonomic lag time (Bednarik 1994a). In short, what beads tell us about the technology of the people who used them is well in excess of deductions concerning their manufacture.

Without doubt the technological deductions beads permit us are of great interest, but of perhaps more importance are the cultural and cognitive assumptions they make possible. Beads can be used in a number of ways or for several purposes: they may be emblemic, for instance, and provide various forms of information about the wearer and his or her status in society.

Availability for marriage, political or social status, or state of mourning might be such possible symbolic meanings. At one level one might believe that beads indicate simply body adornment, but this is almost certainly an oversimplification. Even if vanity were the motivation for wearing such items, stating this explains not why such items are perceived as 'decorative'. The concept itself is anthropocentric; we do not assume that other animals perceive the information imparted by the beads as meaningful. In human culture, however, various forms or levels of meaning may be encoded in such objects, as well as in other kinds of body adornment (tattoos, body painting, cicatrices, infibulation, anklets, armbands etc.). In ethnography, beads sewn onto apparel or worn on necklaces may signify complex social, economic, ethnic, ideological, religious or emblemic meanings, all of which are only accessible to a participant of the culture in question. To name just one example: beads or pendants may function as charms; they may be a means of protection against evil spells or spirits. Clearly, no archaeological access exists to such complex meanings and practices. But there is another generic inference to be made from the use of beads: it is impossible to escape the deduction that the people using them must have a clear concept of the self. Without self-awareness, beads are entirely useless pieces of material.

Beads have been the subject of a great deal of anthropological and archaeological attention (e.g. Beck 1928, 1941; Biggs 1969; Chen 1968; Cheng 1959: 31; Indraningsih 1985; Karklins 1987; Nieuwenhuis 1904). Some of the perhaps most extensive research of pre-Historic beads might be that of Peter Francis and Randolph White. Mistakenly believing that the ivory beads of the French Aurignacian and contemporary Russian traditions are the earliest beads known to us, White (1989, 1992, 1993a, 1993b, 1995) is the principal protagonist of the view that the appearance of beads and pendants coincides with and marks the advent of the Upper Palaeolithic. The model of an explosion-like appearance of the Upper Palaeolithic derives a great deal of support from this fallacy, which I have tried to correct on various occasions (e.g. Bednarik 1992a, 1995a, 1995b). White describes in admirable detail the manufacturing processes of Aurignacian beads of just one material, without having seen or even considered Pleistocene beads outside of France and Russia, or outside the early Upper Palaeolithic period (Bednarik 1995a: 628). Moreover, his understanding of the early Upper Palaeolithic is probably severely mistaken: recent evidence from Germany and other parts of Europe render it much more likely that the Aurignacian was a tool tradition of either Neanderthals, or their direct descendents, and not as White and literally every other Pleistocene archaeologists has believed and claimed until very recently, of fully "modern" humans. His pronouncements concerning the beginnings of bead use, and what it means archaeologically, therefore need to be ignored.

Francis has examined aspects of both archaeological and ethnographic beads in various regions of Asia (1978, 1981, 1982a, 1982b, 1982c, 1985, 1986, 1989a, 1989b, 1989c, 1989d, 1990). In the present context, his experiments with shells (Francis 1982d) are of particular interest. They are the only replicative work with beads that I am aware of, other than what is mentioned in the present article. Francis considers five techniques of perforating shell beads that he found in the literature: gauging, scratching, sawing, grinding and hammering. He has applied each of these methods to some shell species, using in all nine species, but he has not applied the most obvious method of perforation, drilling or boring. He does not elaborate on this omission. In beads or pendants other than those made of shell, which are widespread, the perforations are made almost exclusively by rotating action, except for a number of specimens that exhibit some gauging around the perforation (especially teeth). It is to be noted, incidentally, that some pendants lack a perforation altogether, having instead been attached to the supporting string with the help of an incised groove.

Stone implements used for drilling are well known from Lower Palaeolithic cultures onwards (Keeley 1977) and Francis himself reports that in replicating scratching of perforations he found himself "applying rotary motion" (Francis 1982d: 714). Francis' five methods of perforation are generally unsuitable for all potential bead materials other than shells, including stone, amber, ivory, teeth and ostrich eggshell, therefore they are of no relevance to the manufacture of most surviving pre-Historic beads and pendants. Shell beads are among the earliest 'ornaments' found in many regions, including India (Francis 1981: 140), China (Cheng 1959: 31), Australia (Morse 1993) and South Africa (Henshilwood et al. 2004), and one of the earliest pendants of Europe, from the Châtelperronian of the Neanderthals, is even made of a fossil cast of a shell (Bednarik 1995b: Fig. 6).

Irrespective of their cultural purpose, beads convey complex information about the wearer, which it would be impossible to create a context for without the use of a communication system such as language. This needs to be emphasized because it leads to the postulate that the use of beads assumes the availability of a complex communication system. We have many other indicators of possible language use during the Lower and Middle Palaeolithic (e.g. other forms of symbolism, or successful ocean navigation), and the very early use of beads and pendants provides similarly crucial evidence that, collectively, renders the hitherto dominant model of cognitive evolution redundant. We can no longer afford to ignore this kind of evidence (Bednarik 1995a). I will briefly describe the evidence I refer to and will then focus on the production processes involved, to illuminate their role in exploring the 'gestures' their makers would have employed, among other things.

EARLY PENDANTS AND BEADS

Small, perforated objects of the Pleistocene may have been beads or pendants (Biggs 1969), or they could have been quangings, pulling handles or buckles as reported ethnographically (e.g. Boas 1888: Figs 15, 17, 121d; Nelson 1899: Pl. 17; Kroeber 1900: Fig. 8). However, most of such utilitarian objects are not only of a quite typical shape or design, they exhibit particular wear traces and material properties. To be more specific, small circular objects with central perforation are considered to be beads, especially where they occur repeatedly. Similarly, objects such as animal teeth, perforated near one end (near the root) are not thought to be pulling handles, nor are objects that are too fragile to function as such utilitarian equipment.

Middle and Lower Palaeolithic finds with both artificial and natural perforations are quite common, and have been found since the 19th century. Thousands of such objects are reported in the literature, although there is often no reliable evidence that the perforation is anthropic (cf. Klíma 1991). Some materials can be perforated by natural processes. For instance, bones can be chewed through by animal canines or partially digested by stomach acids, while mollusc shells are commonly perforated by parasitic organisms. To acquire experience in recognizing such natural perforations I have microscopically examined hundreds of specimens of the latter type. But before hastily omitting objects with natural perforations from all consideration in this context we would do well to remember that the cultural status of such an object is not contingent on whether the hole in it was made by human agency.

While it is preferable to rely on specimens bearing clear evidence of human work when dealing with a period from which bead use has not as yet been conclusively demonstrated, it is to be emphasized that the perforation of a bead or pendant certainly does not need to be man-made, as d'Errico and Villa (1997) erroneously assume.

On the contrary, naturally perforated objects are commonly used as ethnographic beads (as are perishable materials) and it seems highly likely that such natural objects were also used in the distant past. Indeed, the earliest beads ever used could quite reasonably be expected to have had natural perforations.

Thus the determining factor in recognizing pre-Historic beads is not an artificial perforation, but microscopic evidence of wear use. Such evidence consists of two types: the wear occasioned by the string on which the bead is threaded, and the facet-type wear around the hole that results from the rubbing of the adjacent bead on a string, in very prolonged use. I have studied both these forms (Bednarik 1997a, 2005).

The earliest presumed beads of the Lower Palaeolithic were mentioned by Boucher de Perthes (1847-64) one and a half centuries ago, occurring together with the first Palaeolithic tools ever reported, and from the very type site of the Acheulian. In the famous paper by Prestwich (1859), in which he recognized the authenticity of the St. Acheul stone tools Jacques Boucher de Perthes had been collecting for many years, the occurrence of possible beads is also mentioned. These were reported to be fossilized remains of a sponge, *Coscinopora globularis*, and Prestwich noted that '[S]ome specimens do certainly appear as though the hole had been enlarged and completed' (Prestwich 1859: 52). Numerous more apparent beads of the same species were found at Acheulian sites, in France and several decades later also in Britain (Smith 1894: 272-6). Intrigued by these reports, I examined microscopically 325 spherical specimens from Acheulian sites in both countries (Fig. 16.1a), and found that the fossils are not, as assumed until then, of *Coscinopora globularis*, but that they are of the species *Porosphaera globularis* Phillips 1829, a Cretaceous sponge (Bednarik 2005). To my surprise, many of them not only showed considerable human modification of the natural tunnel (the tunnel is thought to have been caused by a parasite), but also numerous specimens possessed clear evidence that they were worn on a string. Around both tunnel apertures there were more or less extensive wear facets, in the most pronounced cases amounting to conical depressions approaching the full size of the bead in question. This, amazingly, had not been noted before, and together with other forms of evidence it demonstrates beyond reasonable doubts the use of the fossils as beads.

Besides these spherical fossils, circular, disc-like fossil casts have also been found at another Acheulian site, the crinoid columnar segments (*Millericrinus* sp.) from Gesher Benot Ya'aqov, Israel (Goren-Inbar et al. 1991). Here, however, no evidence of wear has been reported. Some of the earliest objects with indisputably human-made perforations we know of are the two perforated pendants from the Repolusthöhle in Styria, Austria. If their age estimate is correct, they are in the order of 300,000 years old. One is a wolf incisor, very expertly drilled near its root (Fig. 16.1b). The second is a flaked bone point, roughly triangular and perforated near one corner. Both objects were first mentioned by Mottl (1951) but have received little attention since then. They were excavated with a lithic industry variously described as Levalloisian, Tayacian and Clactonian, which is in fact an undifferentiated Lower or Middle Palaeolithic assemblage, but clearly free of Mousterian elements.The occupation deposit was found well below an Aurignacian level, separated from it by substantial clastic deposits of stadial periods. There is no reliable dating evidence available, the age estimate is based on the faunal remains, especially the phylogeny of the bear remains. However, it is broadly supported by the typology of the accompanying lithics, which is easiest to reconcile with a late Lower Palaeolithic industry.

Apart from the Acheulian ostrich eggshell beads reported below, there are no further reports of bead-like finds from the Lower Palaeolithic. It has long been known, however, that some stone tools of that earliest period of human tool use were applied as borers or reamers, especially from micro-wear traces (Keeley 1977). The paucity of drilled

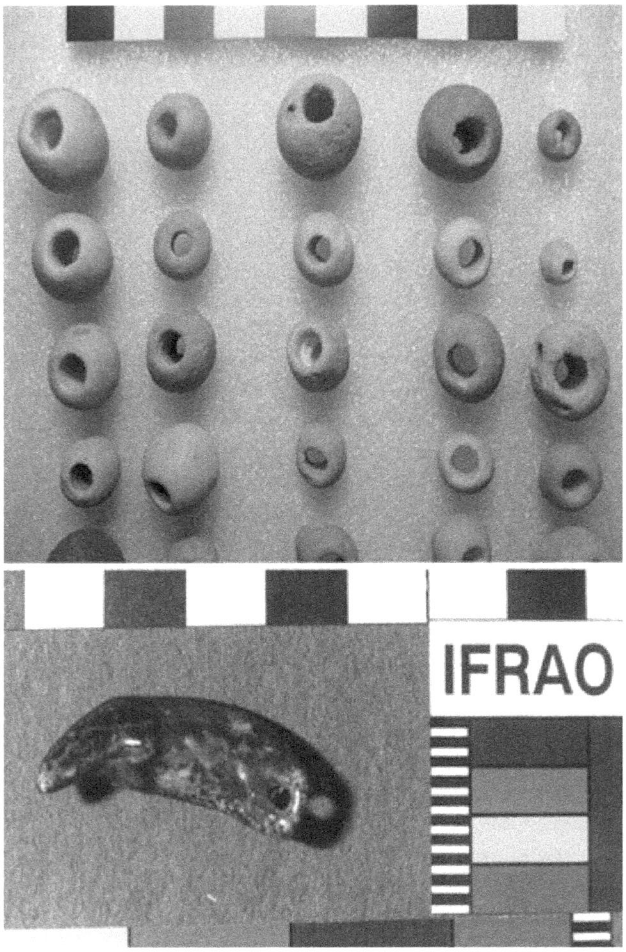

Fig. 16.1. (a) Some of the 325 Acheulian stone beads examined from France and Britain; (b) Perforated wolf incisor from the Lower Palaeolithic of Repolusthöhle in Austria

objects is therefore probably attributable to a preference for softer materials to work with, especially wood. There can be no doubt that the Acheulian was a tool industry concerned primarily with woodworking. Among the wooden implements or fragments of the Lower Palaeolithic are the seven objects from Schöningen, Germany (Bednarik 1996); the spears and spear fragments from Bad Cannstatt (Wagner 1990), Bilzingsleben, Lehringen, Torralba (Howell 1966: 139) and Clacton-on-Sea; and the polished willow plank from Gesher Benot Ya'aqov, Israel (Belitzky et al. 1991).

There are numerous perforated objects also from the early Middle Palaeolithic/Middle Stone Age, and many of them may have served as beads or pendants. Forty-one perforated snail shells have been excavated from the Howieson's Poort levels of Blombos Cave, South Africa, and are about 75,000 years old (Henshilwood et al. 2004). The Micoquian has yielded an artificially perforated wolf metapodium as well as a wolf vertebra from the Bocksteinschmiede, Germany (Marshack 1991). The Micoquian of Prolom 2, Crimea, produced no less than 111 perforated animal phalanges, besides four engraved palaeoart objects (Stepanchuk 1993). Although there is no proof that the phalanges were perforated by human hand, the fact that they are all of one species, *Saiga tatarica*, and that no perforated bones of other species were found in the cave, suggests that these may also be anthropic perforations.

The Mousterian of France has yielded a partly-perforated fox canine and a perforated reindeer phalange from La Quina (Martin 1907-10), and another perforated bone fragment from Pech de l'Azé (Bordes 1969). The two pierced canines from Bacho Kiro, Bulgaria (Marshack 1991), too, are of the Middle Palaeolithic. As we approach the end of this technological phase, beads and pendants become increasingly common, and materials of stone are now drilled, first appearing in Russia and China. Thirteen such specimens from the lower occupation layer of Kostenki 17, found below a volcanic horizon that is about 40,000 years old, include not only polar fox canines and gastropod shells with perforations, but also stone and fossil cast objects (Bednarik 1995b: Fig. 4). From an intermediate Middle to Upper Palaeolithic site in China, wenhua Shiyu, comes a broken stone pendant (Bednarik and You 1991), and the oldest beads found in Australia, from Mandu Mandu Creek rockshelter, are about 32,000 years old (Morse 1993). That country's earliest known stone pendant is from the final Pleistocene of Devil's Lair, still belonging to a Middle Palaeolithic technology (Bednarik 1997a).

With the advent of the Upper Palaeolithic in Eurasia, beads become more numerous and are increasingly manufactured from unwieldy materials, especially ivory. Just three human inhumations at the Russian site Sungir', related to a stone tool technology that is transitional between Middle and Upper Palaeolithic implement types, the Streletsian, contained more beads than have been found in the entire Pleistocene sites of the rest of the world. The three graves yielded 13,113 small ivory beads and over 250 perforated canine teeth of the polar fox. By this time, perhaps 28,000 years ago, the art of bead making had reached an extraordinary level, in which the results of thousands of hours of labour were lavished on three burials.

This synopsis of Pleistocene bead remains might convey the impression that beads were produced infrequently for 200,000 or 300,000 years, and then became much more numerous with the advent of the Upper Palaeolithic. While this is remotely possible it must be cautioned that this pattern of distribution in time provides a typical parabolic curve as demanded by taphonomic logic (Bednarik 1994a: Fig. 2). Accordingly the advent of the Upper Palaeolithic should NOT indicate the advent of frequent bead manufacture, but merely the taphonomic threshold of this phenomenon category. This is almost certainly the correct explanation of the evidence available to us, in which case that record must be tempered by taphonomic logic before it can be interpreted.

Ostrich eggshell beads of pre-History

So far we have ignored one particular type of pre-Historic bead, the ostrich eggshell bead. To understand the significance of flat disc beads manufactured from this material, and their role in interpreting the cognitive evolution of humans, we need first to consider two factors: the distributions, in both time and space, of such finds, and the taphonomic explanation of both these distributions.

Disc beads such as those made from ostrich eggshell are a form of artefact that is not likely to have been made singly or in very small numbers. To provide such symbolic objects with a social meaning it would have been essential that they were made in quite large numbers, because it is repeated and 'structured' use which confers meaning on symbolic artefacts. The role of beads, as well as pendants, would have always been non-utilitarian, ideological, emblemic or symbolic. Moreover, very small beads such as those made from ivory or ostrich eggshell were probably not worn singly, because to achieve a decorative effect they are generally worn as sets in ethnographic specimens.

This renders it necessary to explain why — wherever ostrich eggshell beads have been found in Pleistocene contexts — only extremely small numbers were recovered. Moreover, why are the few known occurrences so extremely isolated in both time and space? Major intervening time spans have yielded no such artefacts, nor have vast geographic regions in which the ostrich is known to, or can be assumed to, have occurred. Taphonomic logic offers the most realistic explanation for this pattern (Bednarik 1986, 1992b, 1994a). Accordingly we are almost certainly dealing with a phenomenon of a very long taphonomic lag time. The extreme paucity of Pleistocene finds can readily be explained by postulating that they survived from beyond the taphonomic threshold of the phenomenon category in question (Bednarik 1994a: Fig. 2).

In India we have only a few specimens from the entire Palaeolithic (Bednarik 1993a, 1993b). Two are from Bhimbetka, south of Bhopal, and three from Patne, Maharashtra. Two of the latter are not perforated, although one is centrally scored. The Bhimbetka specimens were found in the neck region of an Upper Palaeolithic human burial (in shelter No. III A-28), so it has been suggested that they formed part of a necklace made up of beads of perishable materials. While the Patne specimens range from 7 mm to about 10 mm diameter and are rather angular, those from Bhimbetka measure about 6 and 7 mm respectively and are well rounded. In all, some forty-one Indian sites have yielded fragments of Pleistocene ostrich eggshell (Kumar et al. 1988). Radiocarbon dates ranging from about 39,000 to 25,000 years BP have been cited as relating to these finds. Of the 46 marked fragments I have examined, which are all those that have been found in India so far, 45 bear no anthropic decoration. A natural process I have described in detail, involving mycorrhyzal organisms, marked them and also affects other mineralized calcium carbonate-dominated substances of animal origin (ivory, limestone, bone; Bednarik 1992c, 1993b).

Other Asian regions producing ostrich eggshell beads are Siberia (Krasnyi Yar, Trans-Baykal), Inner Mongolia (Hutouliang) and the Gobi desert in northern China and Mongolia. In particular, an Epipalaeolithic or perhaps Mesolithic stone tool industry of the Gobi, usually named after the site of Shabarak-usu, has produced many disc beads, made of freshwater shells as well as ostrich eggshell (Narr 1966: 366). This tradition, typically of non-geometric microliths, is not dated but seems to precede the local Neolithic (Bednarik and You 1991). The ostrich (*Struthio camelus* ssp.), now extinct in Asia (Andrews 1911), seems to have been widely distributed to the end of the Pleistocene and even well into the Holocene (in Arabia; Bednarik and Khan 2005). Depictions of it have been reported from the rock art of Inner Mongolia but their identification has been questioned (Bednarik and Li 1991; Tang 1993).

Both southern and northern Africa have produced finds of worked ostrich eggshell. The southern African sites yielding such finds date from the Middle Stone Age right up to the proto-Historic period. Decorated specimens from the Howieson's Poort phase in Apollo 11 Cave, Namibia (Wendt 1974), may well be 70,000-80,000 years old, even older. This site has also yielded beads made of eggshell from a layer thought to be 22,000 years old. Diepkloof Cave in the south-western Cape, South Africa, contained about a dozen supposedly decorated ostrich eggshell fragments of the Middle Stone Age (Beaumont 1992; Bednarik 1993c). Ostrich eggshell beads from Bushman Shelter near Ohrigstad, Transvaal, have been suggested to date from somewhere between 12,000 and 47,000 years ago (Kumar et al. 1990). Such beads still occur in much more recent periods in southern Africa. For instance they are found in the Smithfield B, a tool complex of the subcontinent's interior regions of the 14th to 17th centuries (Hirschberg 1966). The use of ostrich eggshell for a variety of purposes, including the production of disc beads and as water vessels, continued to be practised by the Bushmen of southern Africa until recent times, and has been described ethnographically (e.g. Forde 1934).

In the far north of Africa, where the ostrich has been extinct for millennia, two pre-Historic periods have provided evidence of the past use of ostrich eggshell: the Capsian and the Acheulian. The Capsian is an Epipalaeolithic blade and burin industry in northern Algeria and Tunisia, dating from the first half of the Holocene. It includes not only numerous figurative and non-figurative engravings on ostrich eggshell fragments (Camps-Fabrer 1966), but also beads of snail shells, teeth and small stones (Camps-Fabrer 1975: 280-2). Almost any excavation of major Capsian deposits produces ostrich eggshell beads, usually well rounded with central perforation. Containers of wholly preserved ostrich eggshells, too, have been recovered from the Capsian. The decoration they bear suggests that the engraved fragments found in the Capsian deposits may well be from such containers. Saharan rock art depictions convincingly resembling the ostrich are known and may well be of the mid-Holocene. Examples are from Wadi

Tilizahren (Jelínek 1985a: Figs 4, 6, 31, 34, 55, 56; 1985b: Figs 5, 28) and Wadi Mathendous, Fezzan (Striedter 1984: Fig. 7); Tzeretegem, Niger (Striedter 1984: Fig. 187); Iheren, Tassili-n-Ajjer (Striedter 1984: Fig. 125); and North Thyout, Atlas (Muzzolini 1995: Fig. 200).

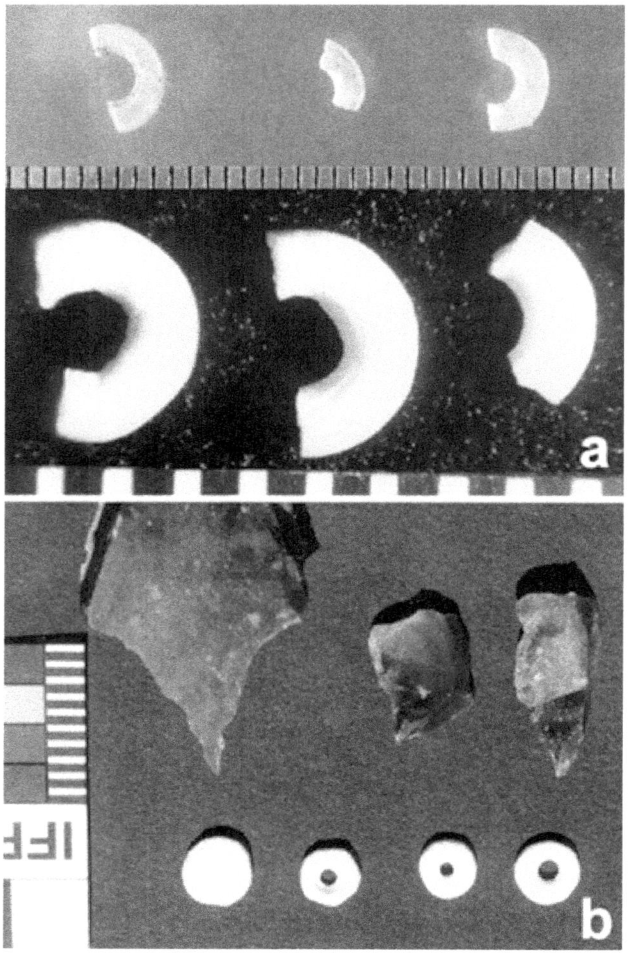

Fig. 16.2. (a) The first three Acheulian beads found at El Greifa in Libya (top) and the replicas made of them (bottom); (b) Progressive stages in making experimental ostrich eggshell beads, with some of the chert borers or reamers used in drilling them

Of very considerable greater age than the Capsian are the three fragments of disc beads from a major Libyan occupation site of the Acheulian (Ziegert 1995). Also made from ostrich eggshell, they closely resemble those from other regions and later periods (Figure 16.2a). These first Acheulian ostrich eggshell beads ever reported come from the El Greifa site complex (Wadi el Adjal, near Ubari). More recently another forty ostrich eggshell beads have been recovered from the site's Acheulian. The site is located on what was a peninsula of the huge Fezzan Lake of the Pleistocene, which then occupied a large part of south-western Libya, measuring about 200,000 km^2. The alkaline and calcareous sediments have provided excellent preservation conditions for insect remains, seeds, bone and ostrich eggshell fragments. The remains of what appears to be a round semi-permanent dwelling structure, about 180,000 years old, have been found on the former lakeshore. There is ample evidence of quarrying of quartzite, and substantial ash beds indicate that the reed belt was annually burnt for a period of many millennia. The sites' lithic inventory includes generally 'handaxes', scrapers, borers and burins, but is dominated by large Acheulian types.

The favourable conditions also led to the preservation of over forty ostrich eggshell beads from the Late Acheulian of El Greifa site E. Dated by the U/Th isotopes of the calcareous sediments they are from, they appear to be in the order of 200,000 years old. The near-perfect rounded circumference and perforation of the El Greifa ostrich eggshell beads demonstrate that even hominins of the Acheulian possessed a well-developed technology of working this fragile medium with the greatest possible confidence and skill. These perfectly made artefacts also imply the existence of the social structures necessary to provide an ideological context for the production and use of complex body decoration. The beads share a similar perforation diameter of about 1.7 mm, and even their external diameter is very consistent (5.8-6.2 mm). This consistency in size and the near-perfect rounding of all preserved edges, internal and external, suggests the use of a standardized manufacturing process, a characteristic these beads seem to share with the much later beads of the Upper Palaeolithic as well as those of various cultural traditions of the Holocene.

The technology of ostrich eggshell beads

The immediate purpose of my experimental replication work between 1990 and 1996 was to determine the technological processes involved in the production of beads of, and engravings on, ostrich eggshell. The results relating to engravings have been reported (e.g. Bednarik 1992c); here I will summarize my findings relating to beads, and their implications in terms of the cultural context of their production.

Kumar has conducted experimental replication work with heavily weathered ostrich eggshell fragments collected from Chandresal, which are in the order of 36,000-39,000 years old (Kumar et al. 1990: 36). He used Mesolithic stone tools to produce the perforations of two experimental beads, which each took him 10 to 12 minutes to drill through, working from both sides. In my own replication work I have always used fresh ostrich eggshell, because that is what was presumably used in the distant past, and I applied freshly made stone tools of different types and materials to establish relative suitability (Bednarik 1991, 1992c, 1993b). I found it difficult to economically drill through the unweathered shell using thin pointed tools of cryptocrystalline sedimentary silica. The most effective tools for this purpose were found to be rather coarse-grained quartzites and quartz. With them I initially reported drilling through the shell of a complete ostrich egg in times ranging from 70 to 90 seconds, i.e. working from just one side (Bednarik, 1991).

I have subsequently found it easy to reconstruct the production processes for these beads. The raw material is of unusually consistent properties: the shell thickness is uniform, as is the three-layered morphology of the shell (described in admirable detail by Sahni et al. 1990). The only significant material variable is attributable to the shell's curvature, which is of a smaller radius at the ends of the egg than it is along the sides. My replication work soon established that the manufacture procedure used followed a specific pattern, as demanded by the morphology and dimensions of the end product, work traces and the nature of the available stone implements (Figure 16.2b). For instance I found that it was difficult and uneconomical to first shape the bead and then drill it, and that it was marginally easier to drill from the concave side than from the convex. Thus experimentation succeeded in reconstructing the work process quite convincingly, which it seems was as follows.

Once drained of its contents, an ostrich egg was dried and broken into fragments. These were then reduced further, into polygonal pieces of about 1-2 cm^2 area. This was done by carefully breaking the shell between fingers, probing for already existing fracture lines. The small fragments were then drilled individually, which is a little more difficult than drilling into the complete egg. An experienced operator takes between 70 and 145 seconds (average 121 secs, n = 11) to perforate the dry shell from one side. (I consider that I became an 'experienced operator' after attempting to produce 25 or 30 beads, and quantitative production details reported here refer only to subsequent work.) No significant differences in drilling time were noted according to direction (from outside or inside), but the outer veneer (<0.1 mm; Sahni et al. 1990) is somewhat harder to start from, and is of course of convex surface, so I came to prefer the concave mammilary innermost layer (Sahni et al. 1990: Fig. 2) to start drilling from. Contrary to various opinions stated, I do not believe that ostrich eggshell beads were usually drilled from both directions, as it is very difficult to meet up with the centre of the first opposite indentation. It is much easier to ream out the opening once the boring tool breaks through, using the point of a thin prismatic sliver of chert. I propose that this is the way ostrich eggshell beads were customarily perforated.

I also drilled shell fragments soaked in water for 24 hours, taking from 80 to 140 seconds (average 118 secs, n = 11), which suggests that this does not affect workability of the shell. The principal variable in drilling time is clearly the quality of the stone tool point, and this can vary considerably. In my replicative work I used a variety of stone tool materials, including cryptocrystalline flint, microcrystalline cherts of various types, chalcedony, coarse and fine quartzites, and quartz crystal. I also tried out a variety of tool morphologies, finding that thin points became blunt very quickly, as did finely-grained materials. Nevertheless, all materials I used necessitated the application of two or more points to produce a single perforation economically, so the time of making or resharpening borers has to be added to production time. Stout angular points on flakes or blades of 1-2 mm thickness at their end were found to be the most effective, and excessive pressure is counterproductive as it accelerates the wear of the tool point exponentially.

Once the perforation is complete it is reamed out from the other (convex or outer) side, using slender bladelets or prismatic points, which may be more fragile. The duration of this process depends on the desired hole diameter, but in about one minute an even diameter of around 2 mm, eliminating much of the drilling cone, can be attained. It is clear from my work that the three perforated beads of the Indian Upper Palaeolithic were reamed out by alternating rotation of the borer: this usually results in a slightly oblong perforation, as already noted by Semenov (1964: 78) in drilling through other materials with stone tools.

Before commencing the abrading of the still angular fragment, the excess area is trimmed off by gripping the piece firmly between two fingers in the area that is to form the final bead, and pressing its convex side against a stone surface. This process of snapping off small angular fragments until the actual bead blank is obtained requires skill and judgment: if the bead is incorrectly held or handled, it can easily crack through the perforation. The average time of the trimming process is 34 seconds.

Grinding the excess material from the fragment's edge is easy, although very demanding on the operator's finger tips. I found it convenient to divide this process into two steps, first grinding the bead blank into a roughly circular shape of under 10 mm, resembling the Patne specimen. This requires between 65 and 270 seconds (mean 217 secs, n = 12), the duration being related directly to the amount of excess material to be removed. Siliceous sandstone, silcrete and quartzite provide excellent grinding surfaces, and an experienced craftsman should not break any pieces in this process.

Ethnographic specimens of disc beads are sometimes manufactured by a method called the *heishi* technique, named after the Santo Domingo Pueblo Indian word for 'shell bead' (New Mexico, U.S.A.). The *heishi* technique was a widespread method of mass-producing beads from ostrich eggshell and other thin materials, in which the perforated blanks are threaded onto a rod or stiff fibre, the entire set is ground together, resulting in very consistent sizes and shapes (Francis 1990: 47). I emphasize, however, that I have observed no evidence that this method was used in the Palaeolithic period, anywhere in the world. Most particularly, the few Indian specimens we have were made singly (contra Francis 1982a, 1990).

In attempting to replicate the Acheulian specimens from El Greifa, I found that I had to further refine the product of the last step. It takes between 580 and 645 seconds to reduce the <10 mm beads to almost perfectly round specimens of about 6 mm diameter (mean 618 secs, n = 12). On this basis we can estimate that the time it took to produce one of the El Greifa ostrich eggshell beads, assuming that the maker was a skilled craftsman, was in the order of 17 minutes, or about 25 minutes if we include the time of preparing and resharpening stone points.

Both the beads and the stone tools used in their manufacture were examined under a stereoscopic optical microscope at low to medium magnifications. The information so gained is not only useful in the microscopic study of pre-Historic bead specimens and stone borers, it also explained the surprisingly rapid blunting I experienced with the stone tools. Expecting to find significant microscopic spalling on working edges, I was surprised to see that the 'blunting' of borers was not so much due to wear, but due to clogging up of recesses with compacted calcium carbonate. Nevertheless, a characteristic type of wear sheen was also noted on the edges at the point of many tools.

The ground and powdered eggshell material was also examined carefully, and was found to contain surprisingly large chips of eggshell layer, commonly measuring 0.1-0.5 mm, but in rare cases of up to 1.8 mm length. However, over half the volume of the white powder is of much smaller grainsize, most of it 2-20 µm. Differences in its composition were noted according to the rock type used: a gritty siliceous sandstone and a silcrete produced slightly different cumulative grain size distribution curves than a dense central Indian quartzite.

DISCUSSION

The replication of archaeological specimens is part of experimental archaeology, without which interpretation in this discipline is of very limited use. It is through the experimentation with technologies that we gain credible insights into how materials must have been utilized to produce the kind of record the archaeologist encounters. In this sense experimental archaeology is related to the study of the taphonomy of archaeological remains, and together these two areas of research can bring archaeological interpretation to life. I will try to illustrate this with the presently considered evidence.

The most important deductions we can draw from the present study concern Acheulian beads from Libya, and what we can learn about the circumstances of their manufacture, in terms of illuminating the conceptual world of their makers. The first observation we can make concerns the considerably finer workmanship of these Acheulian specimens in comparison to those we have of the Upper Palaeolithic. This may be unexpected, but it mirrors an experience we had recently with European rock art: the most sophisticated we have found so far, that of Chauvet Cave in France (Chauvet et al. 1995) turned out to be also the earliest we know of in the European Upper Palaeolithic (Clottes et al. 1995). Hence the idea of evolution towards increased sophistication is a Eurocentric myth in rock art development, and may well be so in other areas of archaeology.

The near-perfect roundness of the Acheulian beads can be obtained only by constant checking of the shape during the final abrading process, using not just a developed sense of symmetry, but possessing a very clear concept of a perfect geometric form. This roundness cannot be the result of chance or some 'instinct' driven by a mere desire to reduce the size of the beads. It is the outcome of a very clear abstract construct of form — a concept-mediated, geometrically perfect form. Moreover, it is the result of a determined effort to produce high-quality work. To extract the full potential information offered by these few beads, I find the following point particularly important, and it also demonstrates vividly the enormous benefits of replication studies.

During my experiments I found that as the beads are ground to a diameter of 8 or 7 mm it becomes increasingly difficult to hold them while grinding them, and after a time it becomes a rather painful task. The fingertips not only have to maintain a tight grip, they are also subjected to abrasion from the siliceous stone. About 6 mm is the diameter at which it becomes uneconomical to continue reduction further, and this is precisely the size of Acheulian beads we have. This, too, is not a coincidence, but the result of a deliberate decision to reduce the beads to the smallest realistically possible size. It must be considered also that at sizes of less than 6 mm, the beads become increasingly fragile: with a perforation of almost 2 mm, their rim width falls to under 2 mm. Moreover, because of what remains of the bi-conical perforation profile, the innermost part of the rim is never of full eggshell thickness. I found that if the beads were ground to a smaller size, they would become susceptible to fracture, either during manufacture or during subsequent use.

So we have two limits on minimum size imposed by practical considerations, and we need to ask: why did the makers of these beads push their technology to its practical limits? After all, a larger bead is much easier to see, yet a smaller bead represents a significantly greater investment of effort. This observation coincides with the already mentioned geometric perfection of the form, which is certainly deliberate. The most parsimonious explanation for both the size and the form of these objects is that these characteristics reflect a highly developed abstract value system and a considerable social complexity in the society that made and used these beads. Without a cultural impetus placing value and meaning on such perfect forms, and on a standard of craftsmanship that pushes the available technology to the utmost limit, it seems simply impossible to account for the empirical characteristics of the evidence. There is certainly no utilitarian explanation to account for them, so the motivation of these artisans is to be found in ideology.

The strong hypothesis that humans of the Late Acheulian period, about 200,000 years ago, possessed such a cultural system is at massive odds with the currently dominant paradigm. Not only does it postulate a value system concerning purely abstract criteria, there must have been a socially shared and communicated meaning regarding the significance of the characteristics of these symbolic products. There can be no purpose in producing technological perfection if there is no comprehension and appreciation of its ideals.

Another insight provided by the replication of Acheulian ostrich eggshell beads concerns their technological perfection. It suggests that their makers drew from the experience of a long tradition of manufacturing such products of which we know almost nothing. We do know that perforation of hard objects (e.g. teeth) was probably already practised earlier, and very competently. Bearing in mind that most ethnographically known beads are of perishable materials, we may reasonably assume that this also applied in the distant past. Naturally perforated small objects may have been used as beads, such as crinoid columnar segments (Goren-Inbar et al., 1991), fish vertebra or the ear-bone of the cave bear (Marshack 1991: Fig. 6), and were certainly used in the form of *Porosphaera globularis* fossils. Finally, but perhaps most importantly, taphonomic logic simply demands a much earlier commencement of the use of beads than can be detected on the fossil record (Bednarik 1994a).

The excellent rounding of the circumferential edge of the Acheulian beads and the even width of the ring indicate a conscious appreciation of an essentially abstract, geometric form by 200,000 BP at the latest, an appreciation which is amply evident from the later Middle Palaeolithic technological traditions. The latter period has provided such evidence from Hungary (the Tata nummulite; Bednarik 1992a: Fig. 4) to Australia (the extensive geometric rock art of that country's Pleistocene tradition, which is the world's most recent Middle Palaeolithic).

Mainstream archaeologists may find such evidence of early sophistication extraordinary, but seen in the context of other finds of the general period in question it should be neither unexpected nor controversial (Bednarik 1995a, 1997b). The question to ask is: why, for instance, are orthodox archaeologists still speculating whether language was possible prior to 35,000 BP (Davidson and Noble 1989) or 60,000 BP (Noble and Davidson 1996)? They are unaware that even *Homo erectus* must have had language to navigate the sea and colonize new islands (Maringer and Verhoeven 1970, 1977; Bednarik 1995c, 1997b, 2003). They may be unaware that petroglyphs, too, were produced in the Acheulian, that haematite or other iron compounds were used as pigment up to 900 millennia ago (Bednarik 1994b), that hafted tools with wooden handles, stone-walled dwellings and portable engravings date from the Lower Palaeolithic (Bednarik 1992a, 1995a, 1996). It is unfortunate that the dominant models in archaeology, since the time of the rejection of the Altamira art well over a century ago, remain largely determined by scholars who are unfamiliar with the relevant evidence. The most urgent task in archaeology is to introduce a systematic study of the limitations of knowledge of its practitioners 'concerning existing data – how language barriers and other biases limited the flow of information in this field, or how false constructs – flourished in archaeology' (Bednarik 1995d: 120). This should be done as one of the several strategies of introducing metamorphology (op.cit.), the scientific version of archaeology. The example illustrated in the present paper confirms this need for major reappraisal.

References

ANDREWS, C.W. 1911. Notes on some fragments of fossil egg shell of a large struthious bird from southern Algeria, with some remarks on some pieces of egg shell of an ostrich from northern India. In H. Schalow (ed.), *Verhandlungen des 5. Internationalen Ornithologen-Kongresses*, Berlin 1910, p. 173. Berlin.

BEAUMONT, P. 1992. The time depth of aesthetic and symbolic behaviour in southern Africa. Paper presented to the Biennial Conference of the Southern African Association of Archaeologists, 1992, Johannesburg.

BECK, H.C. 1928. Classification and nomenclature of beads and pendants, *Archaeologia* 77: 1-76.

BECK, H.C. 1941. The beads from Taxila. *Memoirs of the Archaeological Survey of India* 65, Delhi.

BEDNARIK, R.G. 1986. Parietal finger markings in Europe and Australia. *Rock Art Research* 3: 30-61.

BEDNARIK, R.G. 1991. Comment on F. d'Errico, 'Microscopic and statistical criteria for the identification of prehistoric systems of notation', *Rock Art Research* 8(2): 89-91.

BEDNARIK, R.G. 1992a. Palaeoart and archaeological myths. *Cambridge Archaeological Journal* 2 (1): 27-43.

BEDNARIK, R.G. 1992b. The stuff legends in archaeology are made of: a reply to critics, *Cambridge Archaeological Journal* 2(2): 262-265.

BEDNARIK, R.G. 1992c. Natural line markings on Palaeolithic objects. *Anthropologie* 30(3): 233-240.

BEDNARIK, R.G. 1993a. Palaeolithic art in India, *Man and Environment* 18(2): 33-40.

BEDNARIK, R.G. 1993b. About Palaeolithic ostrich eggshell in India. *Indo-Pacific Prehistory Association Bulletin* 13: 34-43.

BEDNARIK, R.G. 1993c. Early cognitive evidence in Africa. *The Artefact* 16: 61.

BEDNARIK, R.G. 1994a. A taphonomy of palaeoart. *Antiquity* 68: 68-74.

BEDNARIK, R.G. 1994b. Art origins. *Anthropos* 89: 169-180.

BEDNARIK, R.G. 1995a. Concept-mediated marking in the Lower Palaeolithic. *Current Anthropology* 36(4): 605-634.

BEDNARIK, R.G. 1995b. Towards a better understanding of the origins of body decoration, *Anthropologie* 33(3): 201-211.

BEDNARIK, R.G. 1995c. Seafaring *Homo erectus*, *The Artefact* 18: 91-91.

BEDNARIK, R.G. 1995d. Metamorphology: in lieu of uniformitarianism, *Oxford Journal of Archaeology* 14(2): 117-122.

BEDNARIK, R.G. 1996. Lower Palaeolithic hafting evidence, *The Artefact* 19: 99.

BEDNARIK, R.G. 1997a. Pleistocene stone pendant from Western Australia. *Australian Archaeology* 45: 32-34.

BEDNARIK, R.G. 1997b. The initial peopling of Wallacea and Sahul, *Anthropos* 92: 355-367.

BEDNARIK, R.G. 2003. Seafaring in the Pleistocene. *Cambridge Archaeological Journal* 13(1): 41-66.

BEDNARIK, R.G. 2005. Middle Pleistocene beads and symbolism. *Anthropos* 100 (2): 537-552.

BEDNARIK, R.G. & FUSHUN, LI 1991. Rock art dating in China: past and future. *The Artefact* 14: 25-33.

BEDNARIK, R.G. & KHAN, M. 2005. Scientific studies of Saudi Arabian rock art. *Rock Art Research* 22: 49-81.

BEDNARIK, R.G. & YUZHU, YOU 1991. Palaeolithic art from China. *Rock Art Research* 8: 119-123.

BELITZKY, S., GOREN-INBAR, N. & WERKER, E. 1991. A Middle Pleistocene wooden plank with man-made polish. *J. Hum. Evol.* 20: 349-353.

BIGGS, H.E.J. 1969. Mollusca from human habitation sites and the problem of ethnological interpretation. In D. Brothwell and E. Higgs (eds), *Science in archaeology*. Thames and Hudson, London.

BOAS, F. 1888. *The Central Eskimo*, reprinted 1964. University of Nebraska Press, Lincoln.

BORDES, F. 1969. Os percé mousterién et os gravé acheuléen du Pech de l'Azé II, *Quaternaria* 11: 1-5.

BOUCHER DE PERTHES, J. 1847-64. *Antiquités celtiques et antédiluviennes*, 3 Vols. Paris.

CAMPS-FABRER, H. 1966. *Matière et art mobilier dans la préhistoire nord-Africaine et Saharienne*. Mémoires du CRAPE, Paris.

CAMPS-FABRER, H. 1975. Un gisement Capsien de faciès Sétifien, Mendjez II, El-Eulma (Algérie). Éditions du Centre National de la Recherche Scientifique, Paris.

CHAUVET, J.-M., BRUNEL-DESCHAMPS, E. & HILLAIRE, C. 1995. *La grotte Chauvet à Vallon-Pont d'Arc*. Le Seuil, Paris.

CHEN, CHI-LU 1968. *Material culture of the Formosan aborigines*. Taiwan Museum, Taipei.

CHENG, TE K'UN 1959. Archaeology in China, Volume 1, *Prehistory China*. Heffer, Cambridge.

CLOTTES, J., CHAUVET, J.-M., BRUNEL-DESCHAMPS, E. et al. 1995. Les peintures paléolithiques de la Grotte Chauvet-Pont d'Arc, à Vallon-Pont-d'Arc (Ardèche, France): datations directes et indirectes par la méthode du radiocarbone, *Comptes Rendus de l'Académie des Sciences de Paris* 50: 1133-1140.

DAVIDSON, I. & NOBLE, W. 1989. The archaeology of perception: traces of depiction and language. *Current Anthropology* 30: 125-155.

D'ERRICO, F., & VILLA, P. 1997. Holes and grooves: the contribution of microscopy and taphonomy to the problem of art origins. *Journal of Human Evolution* 33: 1-31.

FORDE, C.D. 1934. *Habitat, economy and society*. Methuen, London.

FRANCIS, P. 1978. The historical import of beads. *Bead Journal* 3(3-4): 28-29.

FRANCIS, P. 1981. Early human adornment in India, Part 1. The Upper Palaeolithic. *Bulletin of the Deccan College Postgraduate and Research Institute* 40: 137-144.

FRANCIS, P. 1982a. The ostrich, ostrich eggshells, and ostrich eggshell beads. *Man and Environment* 7: 142-146.

FRANCIS, P. 1982b. *Indian agate beads*. World of Beads Monograph Series 6, Lapis Route Books, Lake Placid.

FRANCIS, P. 1982c. *The glass beads of India*. World of Beads Monograph Series 7, Lapis Route Books, Lake Placid.

FRANCIS, P. 1982d. Experiments with early techniques for making whole shells into beads *Current Anthropology* 23: 713-714.

FRANCIS, P. 1985. A survey of beads in Korea. *Occasional Papers of the Center for Bead Research* 1, Lake Placid, N.Y.

FRANCIS, P. 1986. Bead report XVII: The Asian bead study tour, part III: A tale of two bead making cities. *Ornament* 9 (3): 53-58.

FRANCIS, P. 1989a. *Beads and bead trade in Southeast Asia*. Contributions of the Center for Bead Research 4, Lake Placid, N.Y.

FRANCIS, P. 1989b. *The type collection of beads in the Phillipine National Museum*. Contributions of the Center for Bead Research 5, Lake Placid, N.Y.

FRANCIS, P. 1989c. *Heirloom and ethnographically collected beads in Southeast Asia*. Contributions of the Center for Bead Research 6, Lake Placid, N.Y.

FRANCIS, P. 1989d. The manufacture of beads from shell. In C.F. Hayes III (ed.), *Proceedings of the 1986 Shell Bead Conference: Selected papers*, pp. 25-36. Research Records 20, Rochester Museum and Science Center, Rochester.

FRANCIS, P. 1990. Two bead strands from Andhra Pradesh, India. *Asian Perspectives* 29(1): 45-50.

GOREN-INBAR, N., LEWY, Z. & KISLEV, M.E. 1991. Bead-like fossils from an Acheulian occupation site, Israel. *Rock Art Research* 8: 133-136.

HENSHILWOOD, C., D'ERRICO, F., VANHAEREN, M., VAN NIEKERK, K. & JACOBS, Z. 2004. Middle Stone Age shell beads from South Africa. *Science* 304(5669): 404.

HIRSCHBERG, W. 1966. Späte Jäger-und Sammler-kulturen Süd- und Ostafrikas. In K.J. Narr (ed.), *Handbuch der Urgeschichte*, Band 1: Ältere und Mittlere Steinzeit. Jäger- und Sammlerkulturen, pp. 368-381. Francke Verlag, Bern and München.

HOWELL, F.C. 1966. Observations on the earlier phases of the European Lower Paleolithic, *American Anthropologist* 68: 88-201.

INDRANINGSIH, J.R. 1985. Research on prehistoric beads in Indonesia. Paper presented to the 12th Congress of the Indo-Pacific Prehistory Association Symposium, Manila.

JELÍNEK, J. 1985a. Tilizahren, the key site of the Fezzanese rock art, part I. *Anthropologie* 23(2): 125-165.

JELÍNEK, J. 1985b. Tilizahren, the key site of the Fezzanese rock art, part II. *Anthropologie* 23(3): 223-275.

KARKLINS, K. 1987. Some comments on mulberry and twisted square beads. *Bead Forum* 11: 12-14.

KEELEY, L.H. 1977. The function of Paleolithic flint tools. *Scientific American* 237 (5): 108-126.

KLÍMA, B. 1991. Zur Frage von gelochten Knochen im mährischen Paläolithikum, *Anthropologie* 29(1-2): 73-78.

KROEBER, A.L. 1900. The Eskimo of Smith Sound. *Bulletin of the American Museum of Natural History* 12: 265-327.

KUMAR, G., NARVARE, G. & PANCHOLI, R.K. 1988. Engraved ostrich eggshell objects: new evidence of Upper Palaeolithic art in India. *Rock Art Research* 5: 43-53.

KUMAR, G., SAHNI, A., PANCHOLI, R.K. & NARVARE, G. 1990. Archaeological discoveries and a study of Late Pleistocene ostrich eggshells and eggshell objects in India, *Man and Environment* 15(1): 29-40.

LEROI-GOURHAN, A. 1982. The archaeology of Lascaux Cave. *Scientific American* 246(6): 80-88.

MARINGER, J. & VERHOEVEN, T. 1970. Die Steinartefakte aus der Stegodon-Fossilschicht von Mengeruda auf Flores, Indonesien. *Anthropos* 65: 229-247.

MARINGER, J. & VERHOEVEN, T. 1977. Ein paläolithischer Höhlenfundplatz auf der Insel Flores, Indonesien. *Anthropos* 72: 256-273.

MARSHACK, A. 1991. A reply to Davidson on Mania and Mania. *Rock Art Research* 8: 47-58.

MARTIN, H. 1907-10. Récherches sur l'évolution du Moustérien dans le gisement de la Quina (Charente). *Industrie osseuse*, Vol. 1. Schleicher Frères, Paris.

MORSE, K. 1993. Shell beads from Mandu Mandu Creek rock-shelter, Cape Range Peninsula, Western Australia, dated before 30,000 B.P., *Antiquity* 67: 877-883.

MOTTL, M. 1951. Die Repolust-Höhle bei Peggau (Steiermark) und ihre eiszeitlichen Bewohner, *Archaeologica Austriaca* 8: 1-78.

MUZZOLINI, A. 1995. *Les images rupestres du Sahara.* Préhistoire du Sahara 1, Toulouse.

NADEL, D., DANIN, A., WERKER, E., SCHICK, T., KISLEV, M.E. & STEWART, K. 1994. 19,000-year-old twisted fibers from Ohalo II. *Current Anthropology* 35: 451-457.

NARR, K.J. 1966. Miolithische Kulturen von Westafrika bis Ostasien. In K.J. Narr (ed.), *Handbuch der Urgeschichte, Band 1: Ältere und Mittlere Steinzeit. Jäger- und Sammlerkulturen*, pp. 349-367. Francke Verlag, Bern and München.

NIEUWENHUIS, A.W. 1904. Kunstperlen und ihre kulturelle Bedeutung, *Internationales Archiv für Ethnographie* 14: 136-154.

NOBLE, W. & DAVIDSON, I. 1996. *Human evolution, language and mind. A psychological and archaeological inquiry.* Cambridge University Press, Cambridge.

NELSON, E.W. 1899. The Eskimo about Bering Strait. In *Eighteenth Annual Report of the Bureau of American Ethnology to the Secretary of the Smithsonian Institution*, 1896-97, Part I, 1899, pp. 19-526.

PRESTWICH, J. 1859. On the occurrence of flint-implements, associated with the remains of extinct mammalia, in undisturbed beds of a late geological period, *Proceedings of the Royal Society of London* 10: 50-59.

PRINGLE, H. 1997. Ice Age communities may be earliest known net hunters. *Science* 277: 1203-1204.

SAHNI, A., KUMAR, G., BAJPAI, S. & SRINIVASAN, S. 1990. A review of Late Pleistocene ostriches (Struthio sp.) in India. *Man and Environment* 15(1): 41-47.

SEMENOV, S.A. 1964. *Prehistoric technology*, Londres, Cory, Adams and Mackay, London.

SMITH, W.G. 1894. *Man the primeval savage.* London.

STEPANCHUK, V.N. 1993. Prolom II, a Middle Palaeolithic cave site in the eastern Crimea with non-utilitarian bone artefacts, *Proceedings of the Prehistoric Society* 59: 17-37.

STRIEDTER, K.H. 1984. *Felsbilder der Sahara.* Prestel-Verlag, München.

TANG HUISHENG 1993. Theory and methods in Chinese rock art studies. *Rock Art Research* 10: 83-90.

WAGNER, E. 1990. Ökonomie und Ökologie in den altpaläolithischen Travertinfundstellen von Bad Cannstatt, *Fundberichte aus Baden-Württemberg* 15: 1-15.

WARNER, C. & BEDNARIK, R.G. 1996. Pleistocene knotting. In J.C. Turner and P. van de Griend (eds), *History and science of knots*, pp. 3-18. World Scientific, Singapore.

WENDT, W.E. 1974. Art mobilier aus der Apollo 11 Grotte in Südwest-Afrika, *Acta Praehistorica et Archaeologica* 5: 1-42.

WHITE, R. 1989. Production complexity and standardization in Early Aurignacian bead and pendant manufacture: evolutionary implications. In Mellars, P. and Stringer, C. (eds), *The human revolution: behavioural and biological perspectives on the origins of modern humans*, pp. 366-390. Edinburgh University Press, Edinburgh.

WHITE, R. 1992. Beyond art: toward an understanding of the origins of material representation in Europe. *Annual Review of Anthropology* 21: 537-564.

WHITE, R. 1993a. A social and technological view of Aurignacian and Castelperronian personal ornaments in SW Europe. In V. Cabrera Valdés (ed.), *El origen del hombre moderno en el suroeste de Europa*, pp. 327-357. Ministerio de Educacióny Ciencia, Madrid.

WHITE, R. 1993b. Technological and social dimensions of Aurignacian-age body ornaments across Europe. In H. Knecht, A. Pike-Tay and R. White (eds), *Before Lascaux: the complex record of the early Upper Paleolithic*, pp. 277-299. CRC Press, Boca Raton.

WHITE, R. 1995. Comment on R.G. Bednarik 1995, *Current Anthropology* 36(4): 623-25.

ZIEGERT, H. 1995. Das neue Bild des Urmenschen. *Uni hh forschung* 30: 9-15.

17.
THE ANDAMAN PYGMY: ORIGINS AND NEW ADAPTATIONS

Anek R. SANKHYAN
Palaeo Research Society, Ghumarwin, Dstrict Bilaspur, H.P.-174021, India
arsankhyan@gmail.com

Ramesh SAHANI
Department of Anthropology, Panjab University, Chandigarh-160014, India
rameshrksahani1966@gmail.com

Abstract: *Ever since known the Andaman Pygmy is a centre of attraction and curiosity for their tiny body, prehistoric life style, affinities with other people and mystery of their entry into the Andaman archipelago. Present study takes stock of the historical, archaeological, palaeoanthropological and genomic evidences, and present-day bio-cultural adaptations of the four major population of the Negrito pygmy – the Onge, Jarawa (Ang), Greater Andamanese and the Sentinelese. A re-synthesis of various evidences sheds new remarkable light on the hitherto confusing picture of their origin. New fossil discoveries from Narmada valley document a 'short and stocky' lineage of the late archaic or early modern Homo sapiens that colonised Indian heartland since ~150 to ~70 Kya and formed the ancestral substratum to the subsequent similar populations, like the Rajbanshi, the Pauri Bhuiya, the Munda, the Lodha, the Chenchu as well as the Andaman Pygmy. This is also supported by the archaeological and recent genomic evidences, which tell that Andaman Islanders were not differentiated from the mainland Indians until between 25-35, and that their entry to the Andaman Island was quite late during Holocene time. Among the four Andaman Pygmies, the Great Andamanese are highly acculturated followed by the Onge, and they have shed off their traditional life style and consequently shown morbid maladaptation to obesity and decreased PAL. The Jarawa, though now in touch with modern people, have still retained their traditional hunter-gatherer life style and maintained good physique, but, the recent contacts have brought in urban diseases to them like common cold or an ordinary flu, let alone pneumonia, measles or venereal diseases, including the malaria for which they have no immunity and can be life threatening.*

Keywords: *Andaman, pygmy, origins, Indian ancestry, genome, fossils, archaeology, life styles, health changes*

INTRODUCTION

The Andaman and Nicobar archipelago is spread in a north-south direction in the Bay of Bengal consisting of a total of 204 islands falling between 10°30'-13°30'N: and 92°15'-95°10'E. The Andaman constitutes North, Middle and South divisions. 64 km south of the South Andaman, is the Little Andaman Island, the abode of the Onge pygmy. The Jarawa inhabit the South and Middle Andaman, the Great Andamanese occupy the North and Middle Andaman, whereas the Sentinelese inhabit the Sentinel Island in complete isolation on coral-fringes where the stormy waters prevent outsiders to access them. They drive off the outside intruders with swarms of arrows.

The Negrito pygmies of the Andaman Islands fascinate the modern world for their unique tiny body, nakedness, the hunter-gatherer way of life in the deep island forests and unique language, now accepted as Andamanese. The term 'Negrito' is used generally for the pygmies of Southeast Asia, living in Philippines, peninsular Malaysia, Thailand, the Papuan, and the New Guinean/Melanesian groups. Some scholars also contend that the Negrito type is also found in Congo region of Africa and designates all pygmies with hunting and gathering economy. In general the term 'pygmy' means 'small or tiny people' as they have low average height just less than 150 cm in men and 137 cm in women. They have a stocky build and infantile face and pepper comb hair like the Africans with whom their women also share a genetic trait of steatopygia ("fat bottom"). They are also comparable to the aboriginal Australians. Nei and Ray Choudhury (1982) and Omoto (1984) suggested a close affinity between the Philippine Negrito and their neighbouring Negrito population of South-East Asia than the African pygmies. Biological affinities between the Jarawa, Semang of Malaysia and Aeta of Philippines have been suggested with regard to anthropometric (Sarkar 1989) and dermatoglyphic (Sarkar 1987) characteristics. Tylor (1865, 1871) believed that in the world the pygmy are found in the pure form only in the Andaman Islands where the entire indigenous population is Negrito. In the Andaman Islands they have four major tribes, namely, the Greater Andamanese, the Onge, the Jarawa and the Sentinelese. The Onge inhabit the Little Andaman Island, and before the colonization, they were cited an example of the classic foragers as they were exploiting both terrestrial and aquatic resources of the Islands (Portman,1888; Radcliffe-Brown, 1922, 1948; Cipriani, 1951, 1966; Chatterjee, 1955; Bose, 1964; Basu, 1990).

The Jarawa live in South and Middle Andaman region and lead a semi-nomadic way of life. They lived almost in total isolation even though their islands lie in the

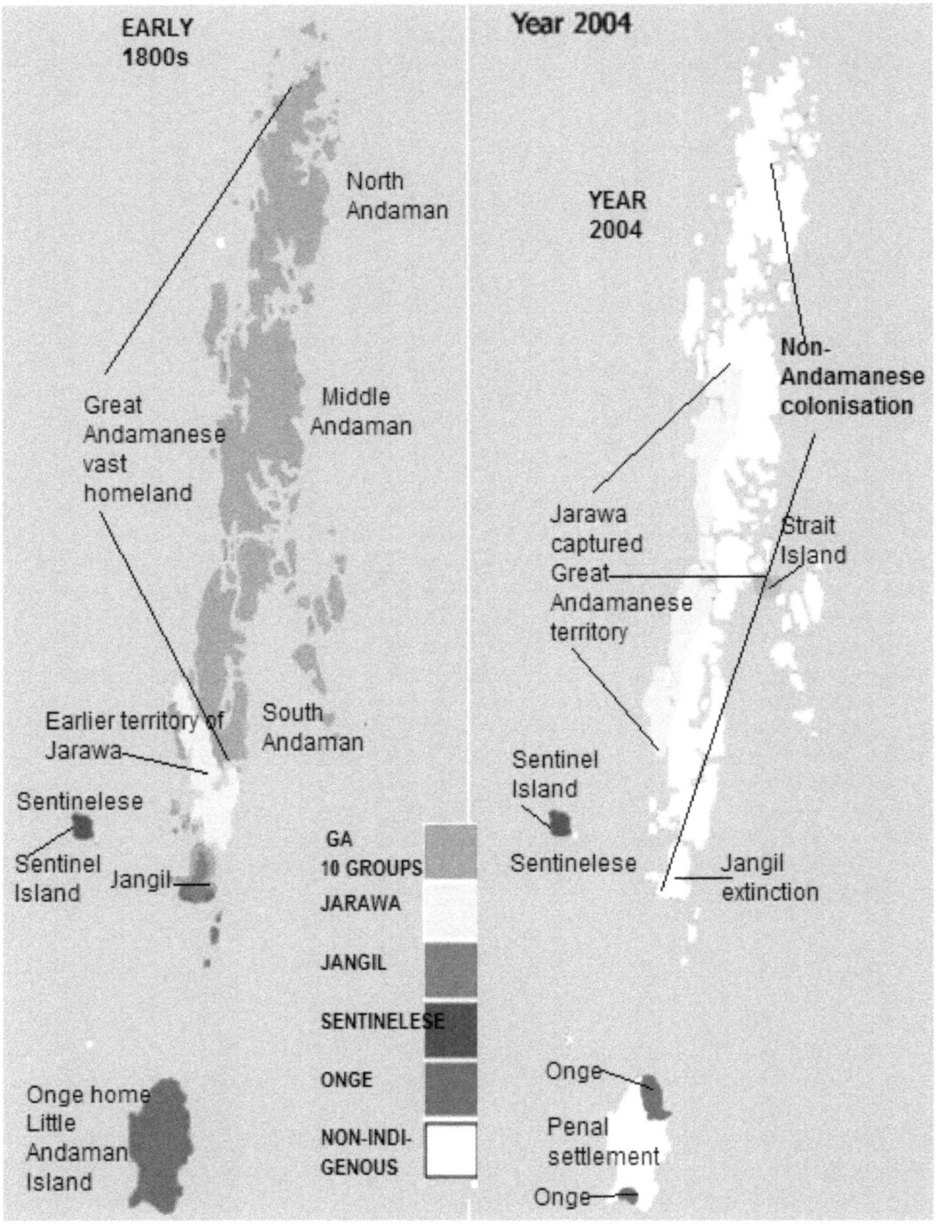

Fig. 17.1. Distribution of four Andaman Pygmies during early 1800s and in 2004 (modified from Jarawa: Wikipedia: Andaman Islands); GA=Great Andamanese

ancient sea trading routes, since no outsider could dare to enter into their habitats as they would shoot with the arrows.

It was not until 1858, that the British established a penal colony of the convicts, jailers, and the contact could be established with the mainlanders. Recently, the Andaman Trunk Road has also enabled a handshake with the Jarawa.

PREHISTORY

The rest of the world knew very little about when the pygmies peopled the Andaman islands. It was in the 2nd century AD that Claudius Ptolemy mentioned about the islands known to the sailors and travellers, also became evident from the accounts of the Chinese in 17th century AD. Until 19th century all writers described the islanders as some kind of anthropomorphic beasts practicing cannibalism; the pirates operating in those parts used to capture the islanders and sell them as slaves (Portman 1899:51-52, 177). He further opined (Portman 1899:722) that the Jarawa resemble very closely to the Little Andaman tribe but differ markedly from those of the Great Andamanese; but the linguistic affinity between the Andaman Negritos and the collective group of South-East Asia Negritos is yet to be established.

The hunting tool-kits of the pygmy are quite simple but perfectly adapted to hunting with bow and arrows. They also use canoes, and of late, they have learnt simple pottery too. But, in general the technology is of a sort that was in use in tropical areas around the world during Upper Palaeolithic period 40,000 years ago.

Are They of African origins?

For long scholars considered the Andaman pygmies of African origins. However, recent genomic evidence indicates their distant relationship with the Mbuti pygmies of Central Africa and the KhoeSan – the Bushmen/Hottentots of Southern Africa. The Khoe-San possess the oldest DNA signatures such that Veeramah et al. (2011) concluded that the "autosomal resequencing data from multiple intergenic regions.....support a demographic model that incorporates an early diver-gence of the lineage leading to the KhoeSan from a population that gave rise to both the ancestors of Pygmy and non-Pygmy NKs groups". They contended a very early divergence of the Khoe-San from a "proto-Pygmy non-Pygmy" common ancestor at 110 Kya and argued that it was very wide range (187-103 Kya) event. But, they contended the Pygmy/non-Pygmy divergence quite late around 48 Kya (ranges 10-105 Kya) and a 32 Kya date (range 5-62 Kya) for Eastern/Western Pygmy split. Such wide-range divergence dates are, however, practically meaningless unless supported by fossil and archaeological evidences. Africa has yielded no such fossil which could attest an early presence of the KhoeSan over 100 Kya nor there is any fossil evidence for the existence of the Pygmy lineage beyond 20 Kya on African land. But, the 'Out of Africa' view has been followed sacrosanct. Under this shadow the general contention has been that the Pygmy first expanded to the equatorial Africa, and later migrated to Southeast Asia where they colonised New Guinea and Philippines (Aeta), Malaya (Semang), Thailand (Mani). But, their entry to the Andaman archipelago (Jarawa/Ang, Onge, Greater Andamanese and Sentinel) as well as to the Flores Island (Rampasasa) and Vanuatu archipelago, is considered quite late when these islands were not separated or were fissured by narrow seas.

Omoto (1984) suggested that the African pygmies moved to South East Asia 35 Kya inferred from this antiquity of their collateral groups living in the Philippines. He, however considered that their movement to the Andaman Islands was considerably later.

Of course, the earliest evidences of human habitation on Andaman Islands are observed in the form of small stone tools (=microliths), bone and shell artefacts, and potsherds deposited in mounds that are known as 'kitchen middens' (Dutta, 1963, 1978, 1983). The excavations carried out in the 'kitchen middens' in the habitats of the Great Andamanese at Beehive Island yielded a radiocarbon date of 1400 ± 100 years B.P., but those at Chauldary were dated to 2,200 years before present (Cooper, 1989, 2002). However, there are no human skeletal remains of that time. This date may enhance if we carry out systematic explorations and excavations since the author (ARS) on a brief leave visit witnessed considerable thickness of the Quaternary deposition in Andaman Islands which may yield fossil/skeletal remains of much more older dates.

Bellwood (1983) objected to the African theory of Andaman Pygmy origins. He contended that the Andamanese are the early Austro-Melanesian settlers of Southeast Asia and Oceania, and not closely related to the African pygmy populations. This means they had no connection with the African pygmy. The D-loop and protein-coding data have also revealed that Andaman pygmy phenotypic similarities with the African Pygmies are likely convergent (Endicott et al., 2003, 2006; Thangaraj et al., 2002).

Thangaraj et al. (2005) studying the genomic signatures revealed that the Andaman Islanders possess mtDNA haplogroup M31 and M32 shared with the South-East Asians as well as on Indian mainland Rajbanshi and Pauri Bhuiya rather than with the Africans (Fig. 17.1). This view is furthered by (Chandrasekar et al., 2009) that the earliest mtDNA M31 signatures of 45 ± 12 Kya are found in the Pauri Bhuiya/Munda inhabiting the easternmost fringe of the Narmada Valley. The same group of Indian scholars (Barik et al., 2008) also confirmed Thangaraj et al. (2005) that the Andaman pygmy lineage branched off from Indian mainlanders around 25-35 Kya. But, they could not interpret their important findings considering that:

> "like Africa, South East Asia has not yielded any fossil evidence to support the existence of the Pygmy in Southeast Asia", and concluded that, "whether M31/32 ultimately originated in South or South-East Asia is still unresolved, but both possibilities are potentially of great importance because of the need to evoke inter regional movements of people during a period of human prehistory about which little is known" (Barik et al., 2008:25).

But, Chaubey & Endicott (2013) also ruled out any near ancestry of the Andaman pygmy in African continent and favours the splitting of the Pygmy lineage at least by 26 Kya or earlier, and their early arrival in the Andaman ~50 Kya thought by Thangaraj et al. (2006) if M31a1 was a founding lineage provided that the separation of the two M31a clades occurred on the mainland India. Chaubey & Endicott (2013) also support Wang et al. (2011) who ruled out Myanmar origins of the Andaman Pygmy and suggested that Andaman-specific M31a1, could in fact be traced to northeast India and the Andaman archipelago was likely settled by modern humans from northeast India via the land-bridge which connected the Andaman archipelago and Myanmar around the Last Glacial Maximum (LGM), a scenario in well agreement with the evidence from linguistic (Abbi, 2009) and palaeoclimate studies. But, northeast ancestor is not supported by fossil/skeletal and archaeological evidence. Of course, northeast India was on the prehistoric corridor to South East Asia, which is why the Rajbanshi also share the genomic signatures with the Pygmy.

Where to find the pygmy fossils?

A solitary and the startling discovery from the Flores Island of Indonesia is of a tiny "Hobbit" or micro-pygmy designated as a debated new species, *Homo floresiensis* (Brown et al., 2004). Several interrelations have been given to its tiny skull of the little above that of a chimpanzee and body not tall beyond 100 cm. The skull and body is however, modern morphologically, though there are primitive features in the cranium, mandible, brain, shoulder, wrist, pelvis, limb bones, and foot, and in its general body proportions; the foot is more similar to *Australopithecus* rather to genus *Homo.* van Heteren & Sankhyan (2009) and van Heteren (2012) attributed it to heterochrony arguing that reversal to a more primitive condition should be viewed as supportive of the insular dwarfing in island situation. So, she considered that 'Hobbits' evolved from a branch of *Homo erectus* that got dwarfed in Flores Island.

But, another fact is that the date of the Hobbit skeleton is about 18,000 year only when modern humans had already colonised the entire globe. So, on the basis of its tiny body size and date the 'Hobbits' cannot be the ancestors of the modern pygmies. Rather, there is an alternate interpretation offered that the dwarfing of the nearby Rampasassa pygmy might have led to the tiny size of the 'Hobbits'. However, Hobbits skeleton was associated with very sophisticated Upper Palaeolithic tool-kit, which none other than an archaic or modern *Homo sapiens* (AMHS) could manufacture. Hence, the Hobbits have evolved neither from the *Homo erectus* nor from the modern Ramapsassa pygmy, but from a "pygmoid" or 'short and stocky' archaic *Homo sapiens.* Their earliest appearance is not known in the fossil record of South East Asia. But, their lineage could have split from a 'short and stocky' archaic *Homo sapiens* sometimes around the Youngest Toba eruption ~75 Kya to escape the climatic adversity by taking shelter in the Liang Bua cave in Flores Island.

We cannot go to the *Homo erectus* level and have to restrict our search for the Pygmy ancestry within the archaic or anatomically modern *Homo sapiens.* The only clue if the Pygmy had shared genomic signatures with the central-eastern mainland Indians as early as >60 Kya, we may look at the fossil evidences from Indian heartland. They have come so far only from the Central Narmada valley. Sankhyan & Rao, (2007) had argued even for Hobbit's ancestors origins in Indian heartland during the international seminar on SEA Palaeoanthropology held at Yogyakarta in July 2007. Since then the Narmada hominin fossil evidences have increased, and the associated faunal and archaeological assemblages recovered in the excavated contexts by Sankhyan et al. (2009) got piled up in the Anthropological Survey of India at Kolkata. A systematic study on the hominin fossils and select associated fauna and Palaeolithic collection was carried out by Sankhyan (2010) and briefly reported by Sankhyan et al. (2012a, b). Subsequently, Sankhyan (2013) distinguished the Narmada hominin fossils into two main categories, also supported by Palaeolithic typo-technology, faunal and stratigraphic contexts as follows:

(1) The 'large-bodied' robust Hominin

It was represented by the Hathnora hominin skullcap and the Netankheri femur recovered from the lower part of the boulder conglomerate horizon in association with mega-terrestrial mammals – *Equus namadicus, Hexaprotodon namadicus, Stegodon insignis ganesa, Elephas namadicus, Elephas hysudricus, Bos namadicus, Bos planifrons, Bubalus palaeindicus, Cervus duvauceli,* and 'large flake Acheulian' Mode 2 industries recovered in excavated contexts. Parsimoniously it is more akin to *Homo heidelbergensis,* and colonised Narmada valley between 250-150 Kya, and gradually disappeared due to sedimentological and ecological changes, which also impacted the faunal richness and diversity.

(2) The 'short-stocky' Hominin

It was represented by two clavicles and a 9th rib recovered from the U2 (middle layer) of Hathnora boulder conglomerate, associated with non-Acheulian

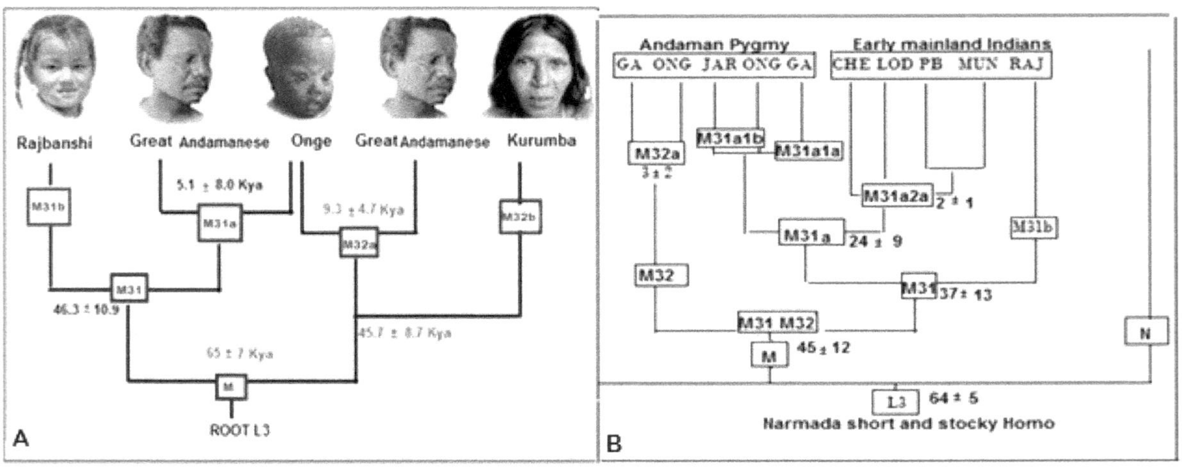

Fig. 17.2. mtDNA trees of the Andaman Pygmy and some ancient Indian populations sharing M31/M32 signatures. A. adapted from Laji Singh; ***B****. simplified by ARS from Barik* et al. *(2008). GA=Great Andamanese, JA=Jarawa, ONG=Onge, CHE=Chenchu, LOD=Lodha, PB=Pauri Bhuiya, MUN=Munda, RAJ=Rajbanshi*

Tab. 17.1. Three decade anthropometric data for the Andaman Islanders 1999 (males)

Population	Year	Sample size	Stature (cm)	Body weight (kg)	MUAC (cm)	Triceps (mm)	BMI (kg/m^2)
Onge	1969	29	149.80	43.2	24	5.9	19.25**
Onge	1989	36	151.80	47.7	26.1	7.7	20.5
Onge	2001*	27	150.70	49.07	26.85	9.39	21.59
GA	2001*	11	154.76	54.68	27.11	11.63	22.78
Jarawa	1999	25	153.68	44.1	23.39	8.68	18.9

* Present study ** Onge 1969 BMI computed from the mean values of weight and stature.

Tab. 17.2. Three decade anthropometric data for the Andaman Islanders 1999 (females)

Population	Year	Sample size	Stature (cm)	Body weight (kg)	MUAC (cm)	Triceps (mm)	BMI (kg/m^2)
Onge	1969	13	140.4	43	24.9	11.7	21.81
Onge	1989	31	140.4	43.3	25.8	17.8	22.00
Onge	2001*	18	139.71	43.97	25.31	16.16	22.56
GA	2001*	9	143.00	45.44	24.21	16.02	22.22
Jarawa	1999	16	141.62	39.78	24.7	11.6	19.79

* Present study ** Onge 1969 BMI computed from the mean values of weight and stature.

Tab. 17.3. Distribution of body mass index according to WHO, 1995 classification

BMI Category	Male			Female		
	Jarawa	Onge	GA	Jarawa	Onge	GA
>16.0 Severe thinness	02 (8.00)	0 (0.00)	0 (0.00)	0 (0.00)	0 (0.00)	01 (11.11)
(16.0-16.9) Moderate thinness	03 (12.00)	02 (7.41)	0 (0.00)	03 (18.75)	02 (10.00)	0 (0.00)
(17.0-18.50) Mild thinness	07 (28.00)	0 (0.00)	0 (0.00)	02 (12.50)	03 (15.00)	0 (0.00)
(18.51-25.00) Normal	13 (52.00)	22 (81.48)	9 (81.82)	11 (68.75)	10 (50.00)	6 (66.67)
25.0-30.0 Overweight	0 (0.00)	2 (7.41)	2 (18.18)		01 (5.00)	2 (22.22)
31.0-35.0 Overweight-I	0 (0.00)	1 (3.70)		0 (0.00)	0 (0.00)	0 (0.00)
35.1-40 Overweight-II	0 (0.00)	0 (0.00)	0 (0.00)	0 (0.00)	4 (20.00)	0 (0.00)

Figure in parentheses are percentages.

Mode 3 Middle Palaeolithic tools, and Upper Group fauna – *Equus hemionus khur, Axix axix, Hystrix crassidens, Cervus sp., Crocodylus sp., Tryonix sp.* and *Lessomus sp.* It likely emerged in Narmada valley at around 150 Kya. Sankhyan (2013) provisionally distinguished it by a new nomen, *Homo "narmadensis"* – considering it a sub-species or distinct population of early *Homo sapiens,* which was the earliest modern human coloniser of Indian heartland (and perhaps the entire South Asia as well) throughout the Middle to Upper Palaeolithic time at least until ~70 Kya as documented by a humerus fossil from the same horizon at Netankheri. Besides the late Mode 3 or early Mode 4 stone tools, the humerus was also associated with unique bone artefacts. It is, therefore argued that *Homo "narmadensis"* was quite innovative and culturally adaptive and capable of withstanding the "Volcanic Winter" (Chesner et al., 1991; Ambrose, 1998; Oppenheimer, 2002, 2003). Its 'short-stocky' body was also of added advantage due to lesser metabolic requirements to withstand the adversity of the "Volcanic Winter". So, it is not unlikely that it attained early anatomical modernity and had colonized South Asia.

The 'short bodied' *Homo "narmadensis"* had a stature of about 135 cm and a shoulder width (=acromial diameter of 30 cm) that was remarkably scaled to that of the shortest female Onge and Greater Andamanese. Sankhyan (2013), therefore postulated that the 'short-stocky' lineage continued throughout Upper Pleistocene and early Mesolithic time, and later expanded to various short-bodied mainland Indian populations, like the Munda and Pauri Bhuiya, including the pygmies.

Probably, the early Pygmies on the mainland India were outcompeted culturally by other dominant groups and therefore pushed off from the mainland to Southeast Asia, and later to the Andaman Islands.

Tab. 17.4. Indices showing distribution of normal and obese (overweight and obesity) category

Ratio & Obesity limit	Male		Female	
	Onge	GA	Onge	GA
Waist/Hip ratio: O >0.85	8 (29.63)	2 (18.18)	18 (100.00)	6 (66.67)
	19 (70.37)	9 (81.82)	0 (0.00)	3 (33.33)
Abdomen/Hip ratio: O >0.85	6 (22.22)	2 (18.18)	13 (72.22)	4 (41.44)
	21 (77.77)	9 (81.82)	5 (27.78)	5 (55.56)
Waist/Height ratio: O >0.50	21 (77.78)	6 (54.55)	14 (77.78)	6 (66.67)
	6 (22.22)	5 (45.45)	4 (22.22)	3 (33.33)
Sum of skin fold (T+S): O >35 mm	24 (88.88)	7 (63.64)	12 (66.67)	5 (55.56)
	3 (11.11)	4 (36.36)	6 (33.33)	4 (44.44)

Figure in parentheses are percentages.

In the light of both fossil and genomic evidences, it is unlikely that the Asian pygmy has a late African ancestry. Rather, it could be the other way round. The African KhoeSan and Pygmies might have had their ancestry in *Homo narmadensis* in Indian heartland about 150 Kya as briefed above. Here, it may not be out of place to mention Bednarik (2013), who has questioned the "Out of Africa" theory on modern human origins (termed it a hoax) through a criticism of various archaeological, palaeontological and genomic evidences. Similarly, Klyosov (2014) has recently come out heavily on the "Out of Africa" theory on genomic evidences. She advocated that the split between the "African" and 'non-African' haplogroups has occurred about 160 Kya.

But, the last common ancestor of anatomically modern humans (AMHS) was non-African and split to various lineages around 64 ± 6 Kya. From this Europeoids (Caucasoids) diverged ~58 Kya, the Mongoloid and Austronesians ~36 Kya, the Middle Easterns ~42 Kya). She questioned all announcements on discoveries of skeletal remains of AMH older than 50 Kya, and more so, older than 150 Kya, in the sub-Saharan area, are distorted or incorrect in the first place. She did not rule out the possibility of "reciprocal migrations". She referred to Rightmire (2009) who noted that "neither the Herto hominins, nor others from Late Pleistocene sites such as Klasies River in southern Africa and Skhul/Qafzeh in Israel, can be matched in living populations. Skulls are quite robust, and it is only after ~35,000 years ago that people with more gracile, fully modern morphology make their appearance". She also cited Hammer (2013) who noted that: "the roots of modern humans trace back to not just a single ancestral population in Africa but to populations throughout the Old World".

THE RE-SYNTHESIS ON THE PYGMY ORIGINS

The above discussion indicates that, if at all there are African connections, they must have been very archaic and could be traceable to the first appearance of *Homo "narmadensis"* (if that was an early African migrant) probably during 150-70 Kya. Thus, Indian heartland had common origins of the Andamanese lineage and other 'short and stocky' populations of South Asia, is better validated by fossils, archaeological, morphological and genomic evidences. Indirectly, the great bio-cultural diversity of India in itself is also an indicator of her deep prehistory and hybridization. The totality of evidences—social, cultural, historical, archaeological, linguistic, phenotypic, epigenetic and genetic—support a conclusion that the Andaman Islanders have only archaic relatedness with Africans which is true for all humans over 100 Kya. It is also clear that South Asia was first inhabited for a long time by diverse 'short-bodied' populations.

Some of these populations ranged from the pygmy to the micropygmies, like the tiny Indonesian "Hobbits" (*Homo floresiensis*), who occupied the Liang Bua cave in Flores Island, the present habitat close to the Rampasasa Pygmy. The movement continued down to the Near Oceania / Sohul (PNG and Australia). This scenario is unlike the previous view that they bypassed South Asia via southern route to arrive Australia (Lahr and Foley, 1994).

RECENT CHANGES

The Andaman Islanders represent the most authentic, purest form of the remnant foragers retaining their traditional life styles, food items, like sea fish, wild boar, crab, dugong, shellfish, turtle and its eggs, different types of wild tubers and fruits. But the contact with the outsiders has caused changes in their food habits, cooking practices, and biological fitness.

Among the four foragers of the Andaman Islands, the Great Andamanese (GA), the Onge and the Jarawa are acculturated to different levels. The Great Andamanese are the most acculturated and interact with the main-stream populations. In the year 2009 the senior author (ARS) met a Great Andamanese girl married to a Bengali boy. The next comes the Onge (less than 100 in number) in the level of acculturation, followed by the Jarawa who are the least in cultural contact with the mainstream people. The Sentinelese is rarely in contact with the outside world.

THE JARAWA

The Jarawa inhabit South and Middle Andaman in good harmony with the nature and has a deep sense of

Tab. 17.5. Comparative dietary studies of the Onge over four decades

	Year of Survey & Methods				
Food Intake	***1964 (30 day wmt)***	***1969 (24 hour recall)***	***1989 (3 day wmt)***	***1997 (3 day wmt)***	***2001* (5 day wmt)***
Protein (gm)	140	127	69	124.43	88.91
Fat (gm)	(61.08)	(55.78)	(60.80)	215.25	88.33
Calories (kcal)	1571	1862	2263	4864.01	3604
Calorie from protein	588	533	290	523	373
% Calorie from protein	37.43	28.65	12.81	10.74	10.36
% Calorie from fat & carbohydrate	63	71	87	89	90

Figure in parentheses are the computed value from the dietary records, * present study; wmt = weightment.

Tab. 17.6. Dietary comparison of the Onge, Great Andamanese (GA) and the Jarawa

	Population and year of study					
Major nutrients	**Onge 1964**	**Onge 1969**	**Onge 1989**	**Onge 2001****	**GA 1997**	**Jarawa 2001-02**
Protein (g)	140	127	69	88.91	124.3	56
Fat (g)	61.4*	55.58*	60.78*	88.33	62.8	140
Calories (kcal)	1571	1862	2263	3604	2901	2502
Calories: protein	588	533.4	289.8	373.42	522.06	235.2
Calories: carbohydrate & fat	983	1328.6	1973.2	3230.58	2378.9	2266.8
% Calorie: protein	37.43	28.65	12.81	10.36	18.00	9.40
% Calorie: carbohydrate & fat	62.57	71.35	87.19	89.64	82.00	90.60

* computed value from the dietary records, ** present study.

preserving nature as is evident from their method of collection without destroying the sources of collection whether a plant or the animal resource. For instance, while collecting the bark of the tree for preparing their chest guard, they peel off only the required amount and allow the tree to rejuvenate. While collecting tubers, they pick up only the mature plants and leave the young plants undisturbed for future use. In the same way they cut the required part of the Sageraea *elliptica* tree for making their bow.

The Jarawa behaved in a friendly manner with the colonisers during the first phase of British settlement in 1789-1796, and since 1858 onwards the Jarawa were described as a "peacefully disposed" people. But, the hostilities increased gradually to 75 acts between 1950s and1970s. However, there was an unprecedented rise since mid 1980s and 1990s with a record of 189 attacks against the Public Works Department, Forest officials, Bush Police personnel, etc. despite of friendly contacts since April 1974. Nevertheless, a section of the Jarawa could befriend since October 1997. However, construction of the Andaman Trunk Road (ATR) allowed outside encroachments and poaching, and therefore increase in violence by the Jarawa (*Report of the Expert Committee on the Jarawa of Andaman islands, July 2003*). Even then, it enabled the Jarawa to have easy access and they themselves come forward to meet the travellers on the ATR, of course expecting gifts. There are no exact population figures available so far, but our assessment is of 350-400 individuals, though at the time of present study they were estimated to be 250 individuals spread to 638 sq. km area.

Nutritional Status

A recent detailed work is by Sahani (2003, 2004, and 2013) on the nutrition, health and anthropometric parameters of the Jarawa. Their diet is well balanced and wide ranging from the terrestrial to aquatic resources, and they never experienced food crisis. Their traditional food sources include the wild boar, big monitor lizard, various kinds of fruits, roots and tubers in the forest (Sarkar 1990). For marine resources they trap the fish with nets, and also hunt fish with bow and arrows as they do the wild boar. They also catch crabs, dugong, shell-fish, and turtle and eat their eggs too. They get most of the essential amino acids from the animal sources, such that the protein content of their diet is very rich both in quality and quantity. For the same reason, the intake of vitamin C, folic acid, vitamin B12, iron and zinc contents in their diet are satisfactory.

Their carbohydrate sources include two kinds of tubers, viz. *Dioscoria bulbifera (*Nad-Hada*)* and *Dioscoria Pentaphylla* (Pug/Puk) consumed frequently which are rich in starch content as the cereals. The tubers are also of medicinal value and occupy a vital place in their daily diet. Food processing is by roasting, baking as well as by boiling; sometimes food items are also consumed raw. Honey is another important food item of the Jarawa. Since 1998, the outside contact has lured them to incorporate external items, such as banana and coconut which have induced some changes in their lifestyle.

Health Status

We knew very little about Jarawa's biological and medical aspects earlier. Gates (1940) had reported about their ABO blood groups, but it was the researchers of the Anthropological Survey of India who contacted the Jarawa and conducted studies on their body measurements, dermatoglyphics, nutrition and health (Sarkar, 1985, 1987; Awaradi, 1992; Prasad, 1997 unp.) as well as on ethnography (Sarkar, 1990, Mukhopadhyay). The Jarawa diet contains plenty of fibre, less fat, more protein and they perform very high physical activity and also spend more leisure time. Therefore, in general, they are nutritionally well balanced and healthy population. Following Amador et al. (1975), the means Energy/Protein Index in the male Jarawa is 119.78 (range 91.62-149.95) and in the female 121.76 (108.35-134.22). The energy intake and energy expenditure of the Cameroonian Baka hunter-gatherers however is relatively low which enables them to adapt to low food availability in the village camps during the rainy season (Yamauchi et al., 2000).

The Jarawa males maintain a general body linearity, i.e. narrow shoulders, chest and hip and overall leanness. Tobias (1964) and Lee (1979) interpreted the small stature and general thinness of the KhoeSan as an efficient adaptation to the demands of hunting and gathering subsistence. The same holds good for the Jarawa males doing hunting which involves running, long distance walk, swimming and climbing in daily routine. However, due to their hunter-gatherer life style, most of the Jarawa males possess marks of accidental injury.

Diseases among the Jarawa

Nutritional deficiency signs are almost absent among the Jarawa, so is the obesity. But, some of their children show a bulging belly due to ringworm infestation, which was epidemical in November-December 1998 onward and engulfed 70% population.

The contact also brought to them a number of diseases hardly experienced by them earlier. A health survey on the Jarawa estimated that 71% haemoglobin content is normal, except the exception of a pregnant woman who suffered from severe anaemia. Incidence of measles and malaria due to outside contact was also reported. They also showed a high prevalence of Hepatitis B carriers, which increased from the initial record of 30% to 50%. However, its carriers were healthy. But, due to their long isolation they could not develop immunity against outside diseases, even the common cold or ordinary flu, let alone pneumonia, measles or venereal diseases, which can be deadly to them, including the malaria.

THE ONGE

The Onge inhabits the Little Andaman Island, and until recently, they were cited as an example of the classic foragers, exploiting both terrestrial and aquatic resources of the Islands (Portman, 1888; Radcliffe-Brown, 1922, Cipriani, 1966 & 1951; Basu, 1990; Bose, 1964; Rao et al., 1989). Their traditional food items included fish, wild boar, crab, dugong, shell-fish, turtle and its egg, different types of wild tubers and fruits. But the contact with outsiders has caused changes in their food habits and cooking practices. In the year 1976, the Onges were forced to settle in two settlements viz. Dugong Creek and South Bay of Little Andaman Island.

Comparative nutritional and health status

Compared to the Jarawa and the Kung! San, the Onge has changed remarkably from their traditional foraging lifestyle since 1976 onward, and became considerably sedentary. They have started coconut plantation, and remain partial foragers. The Onge as well as the Great Andamanese now receive free ration and other necessary assistance from Indian Government agency which made them sedentary. In addition, they have also taken to smoking and liquor. All this contributed to obesity among the Onge and the Great Andamanese. While the Jarawa cook their food by boiling or roasting and do not use oil and spices, the Onge and the Great Andamanese, on the other hand have taken to oils and spices received from Govt. ration shops.

Due to higher carbohydrate consumption, the energy (calorie) intake is becoming higher among the acculturated Onge compared to the least acculturated Jarawa. Their protein intake has declined compared to the Jarawa who consume more fat. It is apparent that the declining protein is the major causes of overweight and obesity in the Onge. In addition, the estimated average physical activity level (PAL) of the Onge as well as of the Great Andamanese falls in medium level compared to the Jarawa showing very active PAL, >2.2 (Sahani, 2004). The Great Andamanese have comparatively higher protein intake but less fat and less carbohydrate (82%) compared to the Onges (89.64%) and the Jarawa (90.60%).

Therefore, the observed changes in anthropometric variables, the prevalent overweight and obesity among the Onge and the Great Andamanese are the results of a change from semi-nomadic to settled life and due to incorporation of the alien foods, which include grains (carbohydrate) oils, coconut (more fat), etc. and decreased PAL. The external population pressure on their natural resources due to their forced re-settlements after the penal settlement during the colonial regime is responsible for the mal-adaptations in the marginalized Onge and the Great Andamanese, which will advance their extinction.

According to Boydon (1972, 1973, 1988, 1994, 2002) the percentage of energy obtained from fat in hunter-gatherer society is lesser (15-20%) compared to a higher (40%) value in modern affluent society. This transition to the high-energy phase began about 200 years ago and is associated with spectacular technological developments, a massive growth of the human population; human population has increased tenfold since 1650 and doubled in the past 30 years. About 1/4 of the world's population lives in established high-energy societies, which has far-reaching changes in the ecosystems of the biosphere.

References

ABBI, A. 2009. Is Great Andamanese genealogically and typologically distinct from Onge and Jarawa? *Language Sciences* 31:791-812. www.elsevier.com/locate/langsci.

AMADOR, et al. 1975. Cited in: *Nutritional Anthropology*, ed. Francis F. Johnson, 1987; Alan R. Liss, Inc.

AMBROSE S.N. 1998. Late Pleistocene human population bottlenecks, volcanic winter, and differentiation of modern humans. *J. Hum. Evol.* 34: 623-651.

AWARADI, S.A. 1992. Anthropometric Re-study of the Onge of Little Andaman. *J. Anthropological Survey of India*, special issue 41:54.

BARIK, S.S., SAHANI, R., PRASAD, B.V.R., ENDICOTT, P. et al. 2008. Detailed mtDNA Genotypes permit a reassessment of the settlement and population structure of the Andaman Islands. *American Journal of Physical Anthropology* 136:19-27.

BASU, B.K. 1990. The Onges. The An.S.I. Andaman and Nicobar Island Tribe Series. Calcutta: Seagull Books.

BEDNARIK, R.G. 2013. African Eve: Hoax or Hypothesis? *Advances in Anthropology* 2013. Vol. 3, No. 4, 216-228 Published Online November 2013 in SciRes. (http://www.scirp.org/journal/aa); http://dx.doi.org/10.4236/aa.2013.34031.

BELLWOOD, P. 1983. Prehistory of the Indo-Malaysian Archipelago, Crossroads: An Interdisciplinary Journal of Southeast Asian Studies. 15, Northern Illinois University. Center for Southeast Asian Studies, p. 117.

BOSE, S. 1964. Economy of the Onge of Little Andaman. *Man in India*. 44: 4, 298-310.

BOYDEN, S.V. 1972. Biological view of problems of urban health. *Human Biology in Oceania* 1(3): 159-169.

BOYDEN, S.V. 1973. Evolution and Health. The Ecologist 3 (8): 304-309.

BOYDEN, S. 1988. Western civilization in biological perspectives: pattern in biohistory. Oxford, Oxford University Press.

BOYDON, S. 1994. The human situation in the biosphere: an overview. Nature & Society Forum February 1994.

BOYDON, S.V. 2004. The Biology of Civilisation: Understanding Human Culture as a Force in Nature A UNSW Press book, UNSW Press, 189 pages.

BROWN, P., SUTIKNA, T., MORWOOD, et al. 2004. A new small-bodied hominin from the Late Pleistocene of Flores, Indonesia. *Nature* 431: 1055-1061.

CHANDRASEKAR, A., KUMAR, S., SREENATH, J. et al. 2009. Updating phylogeny of Mitochondrial DNA Macrohaplogroup M in India: Dispersal of modern human in South Asian Corridor. *PLoS ONE*, e7447 4:1-13. www.plosone.org.

CHAUBEY, G. & ENDICOTT, P. 2013. The Andaman Islanders in a Regional Genetic Context: reexamining the evidence for an early peopling of the Archipelago from South Asia. Special issue on revisiting the "Negrito" hypothesis. *Human Biology* 85 (1) 7:11-27.

CHATTERJEE, B.K. 1955. A comparative study of the different body proportions of the Onges of Little Andamans. *The Anthropologist 2*(2): 12-21.

CHESNER, C.A., ROSE, W.I., DRAKE, A.D.R. & WESTGATE, J.A. 1991. Eruptive history of earth's largest Quaternary Caldera (Toba, Indonesia) clarified. *Geology* 19: 200-203.

CIPRIANI, L. 1955. Census of India, 1951; On the origin of the Andamanese.

CIPRIANI, L. 1966. The Andaman Islanders, edited by D. Taylor Cox and Linda Cole. London: Weidenfed and Nicolson.

COOPER, Z. 1989. Analysis of the nature of contracts with the Andaman Islands during the last two millennia. *South Asian Studies* 5:133-147.

COOPER, Z. 1989. Analysis of the nature of contracts with the Andaman Islands during the last two millennia. *South Asian Studies* 5:133-147.

COOPER, Z. 2002. Archaeology and history: early settlements in the Andaman Islands. London, New Delhi, Oxford: Oxford University press.

DUTTA, P.C. 1963. Recent research on kitchen-middens in the Andamans. *Current Anthropology* 4:356.

DUTTA, P.C. 1978. The Great Andamanese: Past and Present. Anthropological Survey of India, Kolkata.

DUTTA, P.C. 1983. Research in biological anthropology of the Andaman Negritos: A critical review. *Anthropologie* 21: 259-268.

ENDICOTT, P., GILBERT, M., STRINGER, C., LALUEZA-FOX, C. et al. 2003. *American Journal of Human Genetics* 72: 178-184.

ENDICOTT, P., METSPALU, M., STRINGER, C., MACAULAY, V. et al. 2006. Multiplexed SNP typing of ancient DNA clarifies the origin of Andaman mtDNA haplogroups amongst South Asian tribal populations. *PLoS ONE* 1:e81.

GATES, R.R. 1940. "Blood Groups from the Andamans." *Man* 40:55-57.

HAMMER, M. 2013. Human hybrids. *Sci. American* 308: 66-71. http://dx.doi.org/10.1038/scientificamerican0513-66.

KLYOSOV, A.A. 2014. Reconsideration of the "Out of Africa" Concept as not having Enough Proof.

Advances in Anthropology 4 (1):18-37. Published Online February 2014: http://www.scirp.org/journal/aa) http://dx.doi.org/10.4236/aa.2014.41004.

LEE, R.B. 1979. The Kung! San. Cambridge: Cambridge University Press.

LAHR, M. & FOLEY, R. 1994. Multiple dispersals and modern origins. *Evolutionary Anthropology* 3:48-60.

NEI, M. & ROY CHOUDHURY, A.K. 1982. Genetic relationship and evolution of human races. *Evolution Biology* 14:1-59.

OMOTO, K. 1984. The Negrito: Genetic origins and mcroevolution. Acta Anthropogenetica 8:137-147.

PORTMAN, M. 1884. The Andaman Islanders. Calcutta, India: Office of the Superintendent of Government Printing.

PORTMAN, M.V. 1899. A History of our Relation with the Andamanese. Govt. of India, Kolkata.

OPPENHEIMER, C. 2002. Limited global change due to largest known Quaternary eruption, Toba ~74 Kyr BP. Quat. *Sci. Rev.* 21: 1593-1609.

OPPENHEIMER, C. 2003. Ice core and palaeoclimatic evidence for the great volcanic eruption of 1257. *Internat. J. Climat.* 23: 417-426.

RADCLIFFE BROWN, A.R. 1922. The Andaman Islanders. A study in Social Anthropology. Cambridge: Cambridge University Press.

RADCLIFFE BROWN, A.R. 1948. The Andaman Islanders. Free Press, Illinois.

RAO, D.H., BRAHMAN, G.N.V. & RAO, N.P. 1989. Health and nutritional status of the Onges of Little Andaman Island. *Journal of Indian Anthropological society*: 24 (1): 69-78.

REPORT OF THE EXPERT COMMITTEE ON JARAWAS OF ANDAMAN ISLANDS. July 2003. http://www.and.nic.in/C_charter/Dir_tw/ecr/contents.htm.

RIGHTMIRE, G.P. 2009. Middle and later Pleistocene hominins in Africa and Southwest Asia. *PNAS* 106:16046-16050. http://dx.doi.org/10.1073/pnas.0903930106.

SAHANI, R. 2003. Nutritional and Health Status of the Jarawas: A preliminary Report: *Journal of Anthropological Survey of India 52*: 47-65.

SAHANI, R. 2004. Seasonal Variation in the Nutritional Status of the Jarawas. *Journal of Anthropological Survey of India.* 53: 99-116.

SAHANI, R. 2013. Foraging to settled life: A comparative study of anthropometry and nutrition of Onges of Little Andaman Island. HOMO – *Journal of Comparative Human Biology* 64: 391-397.

SANKHYAN, A.R. 1997a. Fossil clavicle of a middle Pleistocene hominid from the central Narmada Valley, India; *J. Human Evolution* 32: 3-16.

SANKHYAN, A.R. 1999. The place of Narmada hominin in the Jigsaw puzzle of human origins. Gondvana Geol. Magz. Spl. Publ. 4: 335-345.

SANKHYAN, A.R. 1997b. A new human fossil from the Central Narmada Basin and its chronology; *Curr. Sci.* 73:1110-1111.

SANKHYAN, A.R. 2005. New fossils of Early Stone Age man from Central Narmada Valley. *Curr. Sci.* 88: 704-707.

SANKHYAN, A.R. 2007. Significance of Human Post-cranial Fossils from Narmada with Remarks on the Skullcap, In A.R. Sankhyan & V.R. Rao (eds) *Human Origins, Genome & People of India: Genomic, Palaeontological & Archaeological Perspectives*, pp. 193-217. Allied Publishers, Pvt. Ltd, New Delhi.

SANKHYAN, A.R. 2009a. On the status of Indian hominoid and hominid fossils, in R. Ray & V. Jayaswal (eds), *Status of Prehistoric Studies in the Twenty-first Century in India*, pp.13-23. Proceedings of the 15th UISPP Congress, Lisbon, Portugal, 2006. BAR International Series 1924, BAR Publishing, England.

SANKHYAN, A.R. 2009b. Human Bio-Cultural Diversity in Prehistoric-to-Protohistoric India, in R. Ray & V. Jayaswal (eds) *Status of Prehistoric Studies in the Twenty-first Century in India*, pp. 67-74. Proceedings of the 15th UISPP Congress, Lisbon, Portugal, 2006. BAR International Series 1924, BAR Publishing, England.

SANKHYAN, A.R. 2010. *Pleistocene Hominins & Associated Findings from Central Narmada Valley bearing on the Evolution of Man in South Asia.* Unpublished Ph.D. dissertation. Panjab University, Chandigarh.

SANKHYAN, A.R. & RAO, V.R. 2007. Did ancestors of the Pygmy of Hobbit ever live in Indian heartland? In Indriati, E. (ed.) *Recent advances on Southeast Asian Paleoanthropology and archaeology*, pp. 76-89. Laboratory of Bioanthropology and Paleoanthropology Faculty of Medicine Gadjah Mada University, Yogyakarta.

SANKHYAN, A.R. & TALWAR, I. 2009. Indian Palaeolithic Transitions, In A.R. Sankhyan (ed), *Asian Perspectives on Human Evolution*, pp. 163-171. Serials Publications, New Delhi.

SANKHYAN, A.R., CHAKRABORTY, S., DEWANGAN, L.N., PRABHA, S. et al. 2009. Cultural, Faunal and Stratigraphic Contexts of Narmada Hominins. In A.R. Sankhyan (ed), *Asian Perspectives on Human Evolution*, pp. 73-91. Serials Publications, New Delhi.

SANKHYAN, A.R., BADAM, G.L., DEWANGAN, L.N., CHAKRABORTY, S. et al. 2012a. New Postcranial Hominin fossils from the Central Narmada Valley, India. *Advances in Anthropology* 2 (3): 125-131, Published Online August 2012 in SciRes DOI:10.4236/aa.2012.23015.

SANKHYAN, A.R., DEWANGAN, L.N., CHAKRABORTY, S. et al. 2012b. New human fossils and associated findings from the Central Narmada. *Current Science* 103 (12): 1461-1469.

SARKAR, B.N. 1987. Dermatoglyphic study of the Jarawas: A Negrito tribe of Andaman Islands. *Human Science* 36:346-358.

SARKAR, B.N. 1989 Anthropometric study of the Jarawa of Andaman Islands. *Journal of Indian Anthropological Society* 24:79-83.

SARKAR, N. 1985. The Jarawa of the Andaman Islands: An anthropometric study. *Human Science* 34:180-194.

SARKAR, J.K. 1990. *The Jarawa.* The Anthropological Survey of India, Andaman and Nicobar Island Tribe Series, Series ed. Singh, K.S., Seagull Books, Calcutta, on behalf of An.S.I.

SONAKIA, A. 1984. The skull cap of early man and associated mammalian fauna from Narmada Valley alluvium, Hoshangabad area, Madhya Pradesh, India; *Rec. Geol. Surv. India* 113: 159-172.

THANGARAJ, K., SINGH, L., REDDY, A.G., RAO, V.R. et al. 2002. Genetic affinities of the Andaman Islanders, a vanishing human population. *Current Biology* 13: 86-93.

THANGARAJ, K., CHAUBEY, G., KIVISILD, T., REDDY, A.G. et al. 2005. Reconstructing the origin of Andaman Islanders. *Science* 308:996.

THANGARAJ, K, CHAUBEY, G., SINGH, V., VANNIARAJAN, A. et al. 2006. In situ origin of deep rooting lineages of mitochondrial Macrohaplogroup 'M' in India. BMC *Genomics* 7:151.

TOBIAS, P. 1964. "Bushmen hunter-gatherer: a study in human ecology." In: *Ecological studies in Southern Africa,* ed. D.H.S. Davia, The Hague, pp. 69-86.

TYLOR, E.B. 1865. *Researches into the early history of mankind the development of civilization.* London: John Murray.

TYLOR, E.B. 1871. *Primitive culture: researches into the development of mythology, philosophy, religion, art, and custom.* 2 volumes. London: John Murray.

VEERAMAH, K.R., WEGMANN, D., WOERNER, A. et al. 2011. An early divergence of KhoeSan ancestors from those of other modern humans is supported by an ABC-based analysis of Autosomal Resequencing Data. Molecular Biological Evolution. doi:10.1093/molbev/msr212, Advance Access publication, September 1, 2011, pp. 1-14.

WANG, H.-W., MITRA, B., CHAUDHURI, T.K., PALANICHAMY, M.G. et al. 2011. Mitochondrial DNA evidence supports northeast Indian origin of the aboriginal Andamanese in the Late Paleolithic. *Journal of Genetics and Genomics* 38 (2011) 117e122. doi: 10.1016/j.jgg.2011.02.005.WHO,1995. Physical Status: The use and interpretation of Anthropometry. *WHO Technical Report Series 854: Geneva.*

YAMAUCHI, T., SATO, H. & KAWAMURA. K. 2000. Nutritional status, activity pattern, and dietary intake among the Baka hunter-gatherers in the village camps in Cameroon. *African Study Monographs 21*(2): 67-82.

18.
AMAZING SKILLS: PRACTICE OF TREPANATION AROUND THE WORLD

Alexandra COMŞA
Senior Researcher at the "Vasile Pârvan" Institute of Archaeology, Romanian Academy, Bucharest, Romania
alexcomsa63@yahoo.com

Anek R. SANKHYAN
Ex-Senior Anthropologist at Anthropological Survey of India, Kolkata,
President, Palaeo Research Society, Ghumarwin (H.P.)-174021, India
arsankhyan@gmail.com

Abstract: *Surgical intervention upon human brain case is very risky and scaring even today with modern technology, but the spectacular Trephination or Trepanation cranial surgical practice was followed globally as early as 5000-4000 years ago with awesome primitive stone and bronze instruments and archaic anesthetics and painkillers! As this amazing practice always stirred a great interest among anthropologists, archaeologists, the historians of medicine and religion, and the public at large, we present here its brief overview, noting its global history and continuity, the causes, the beliefs, the significance, the operational process and kinds of instruments and methods used, responses and complications of the operation, the symbolic and pseudo-trepanations, etc.*

Keywords: *Trepanation, distribution, cause, methods, instruments, complications, pseudo-trepanations*

INTRODUCTION

Trephination or trepanation is usually a spectacular practice as people are amazed to find that such a surgical operation could have been performed in archaic times. Such a topic, like the intervention upon the brain case, which is very risky, even today, when being done with modern equipment, did always stir a large interest, among anthropologists, historians of medicine, archaeologists and historians of religion, not to mention the public.

This archaic practice, may be one of the oldest surgical ones, is even more thrilling, considering that people in times immemorial performed it with "primitive" anesthetics, painkillers and less specialized implements. We should take into account that a few thousand years ago, just a flint knife, or a similar cutting object, was employed in order to carry out an intervention on the head, which could easily affect the brain. Yet, the abilities of the medicine people of those times made successful operations and their subjects lived for a shorter, or longer time afterwards, more often being killed by other ailments. It is also spectacular to find trephination cuts used for ritual purposes, namely to make bone discs that, afterwards, served as amulets and worn accordingly. It is interesting to see the variation in techniques and implements used for such a kind of surgery all around the world. As usual, in their everyday life, the archaic communities used to be very practical. For the mentioned instruments, they utilized the convenient raw materials they had around them, such as stone, shells and, later on, metal. Even if we hardly accept this idea, the healing of the wounds resulted from such interventions occurred even in immemorial times. Even then, illnesses affected each individual, depending upon certain influences of environmental factors, or weaknesses of the body. It is a fact that, people can get a disease in certain moments of their lives. Still, we should emphasize that a comparison between the archaic and recent communities is not very reliable, due to the fact that in older times, the immunity of the individual was better than today, when the body is permanently under the "pressure" of pollution and stress. Besides, the people of those times were gathered in small groups that could not be permanently in contact with each other, except within a longer time. At present, one can reach any part of the world, with the modern transportation means that exist. Yet, even today, a disease specific to a certain region can be taken over by a person from someone who lives far away and she can die from that, despite the advanced medical treatment. This is due to the fact that the immune system of her body is not adapted to the respective germs.

Distribution of the Trephination Practice

In Europe, almost every country has discovered its own cases of trepanned skulls. This practice seems to have been flourishing among certain groups. Anyway, we have good examples especially from the Neolithic sites of France, but elsewhere too. In Egypt, but also in other parts of Africa such skulls are also known. In recent times, in the Northern part of the mentioned continent, the Kabyles and Berbers, the Kisii from Kenya, the Tende from Tanganika and the Massai, still use this medical practice. In Asia, the use of trepanation was also wide-spread in Afghanistan, China, southern Siberia and Japan. In India Sankhyan & Weber (2001) reported multiple trepanation in a Neolithic skull from Burzahom of Kashmir valley (Fig. 18.1), and Sankhyan & Robbin Schug (2011) in a Bronze Age skull from Harappa (Fig. 18.2).

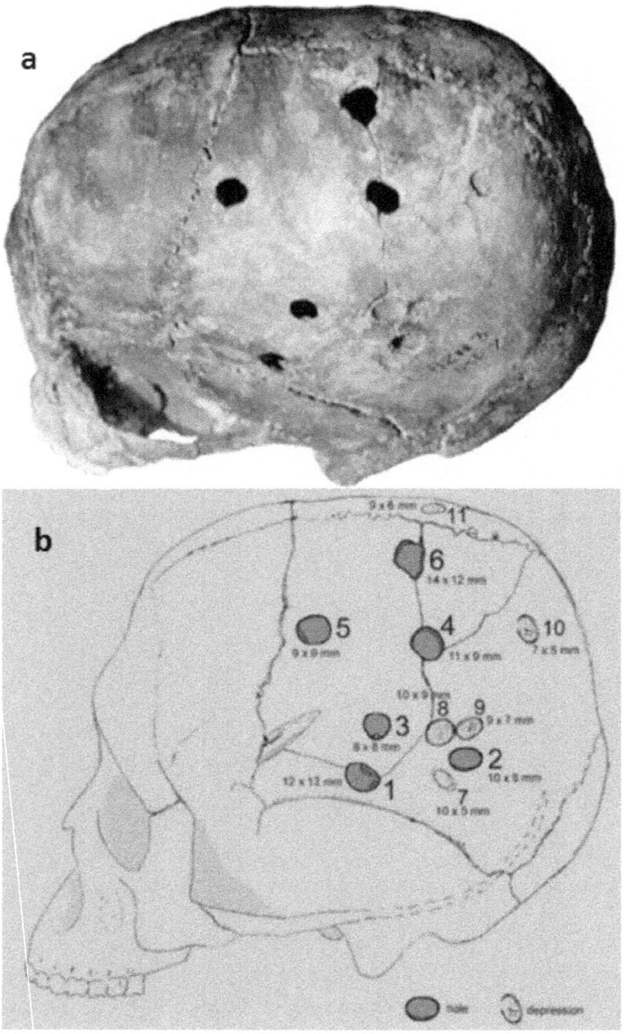

Fig. 18.1. a. Multiple trepanations on Burzahom female skull at AnSI, Kolkata; b. The sketch showing trepanned sites (after Sankhyan & Weber (2001))

In Oceania, the inhabitants of the Polynesian Islands, as well as others, from Bismark, New Britain, New Ireland, Tahiti and others, have also used trephination. In Melanesia, this kind of surgery was also very frequently used.

On the American continent, only a few cases of such interventions are found in the northern part, but a large number of cases are found in South America. They concentrated in Andean Mountains in Peru, mostly on its Southern coast at Paracas, central highlands, Huarochiri Jauyos provinces, as well as in the Southern highlands at Cuzco. In Bolivia most of such sites date between 5th and 6th A.D.

The expansion of this medical practice into the Middle East, India, Pakistan and Afghanistan is viewed by some scholars as a cultural diffusion, while others consider

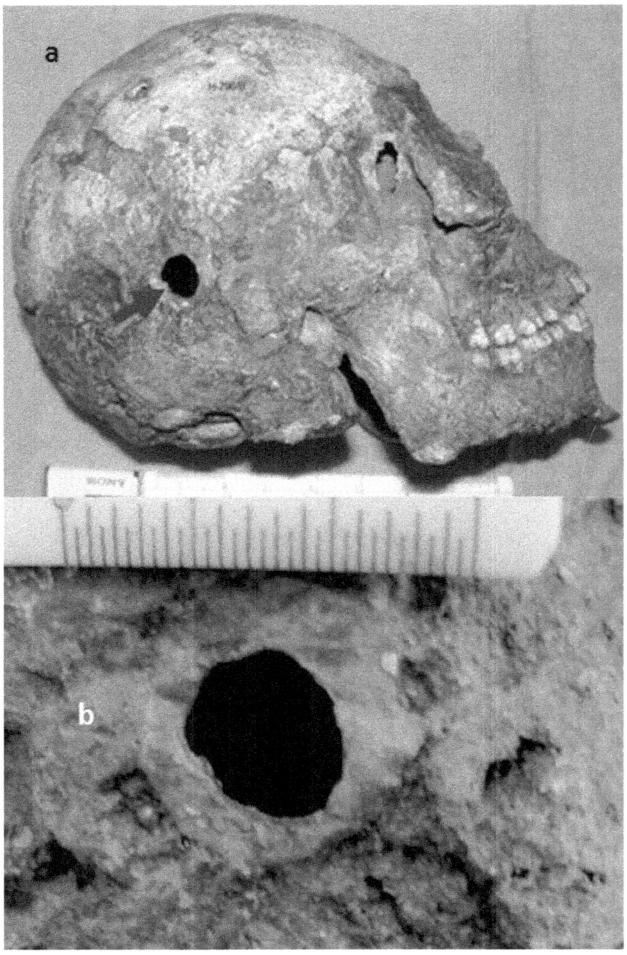

Fig. 18.2. a. The trepanned Harappa male Skull H-796/B at AnSI Kolkata; b. enlarged view of the trepanned site (after Sankhyan & Robbin Schug (2011))

it as a multilocal invention.[1] In our opinion, there were, indeed, several places where trephination could have occurred, but its distribution must have been by cultural means. As a surgery, in most of situations, it was meant to rescue lives, and therefore, the medical practitioners must have been very receptive to any improvement of the technique in any quarter of the world. And, this facilitated innovation and exchange among different populations, and was a significant path towards its fast acceptance and assimilation from one community to another.

TREPHINATION METHODS

Trephination consisted in cutting off the skin above the selected place if no lesion existed there. Afterwards, the scalp was flapped by incision and cut through, then taking off a fragment of bone. Finally, the membranes of the brain (*dura mater*) would have been exposed. Stewart had suggested that in certain cases an entire portion of the scalp could be totally removed.[2] The survival of the patient was highly influenced by the capability of the

[1] A.C. Aufderheide, C. Rodriguez-Martin, 1998, p. 31: Ž. Mikić, 1998, p. 146; A. Andrushko, J.W. Verano, 2008, p. 4.
[2] D.J. Ortner, W.G.J. Putschar, 1985, p. 97.

healer to avoid perforation of the *dura* or the blood vessels of the brain.

Various procedures were followed around the world, as mentioned below:

(i) *The Scraping method*. It was most frequent in Europe and in the southern coast of Peru in South America around 400 BC.[3] The Bronze Age Harappan skull (Fig. 18.2) also indicates use of the scraping method practiced over 4000 years ago (Sankhyan & Robbins Schug, 2011).

(ii) *The gouge method*

(iii) *The saw method*, more frequent in South America.

(iv) *The drilling method.*[4] For this, the Peruvian cases are well known instances where small holes were drilled in a circular pattern followed by cutting off the narrow connection zones between them. Burzahom skull from Kashmir (Sankhyan & Weber, 2001) was also trepanned by the drilling method (Fig. 18.1). We consider that a drilling method could have been taken over from the craftsmen who made various antler or bone implements, and used to pierce them.

(v) *The Canoe-shaped depressions* with a rectangular pattern are mostly found in Peru, and in the Neolithic French and Palestinian sites.[5] This kind of intervention also appears in symbolic trephinations.

(vi) *The "grit like" or reticulated trephinations*. These were mostly practiced by the Inca in Peru, done for healing cephalalgies, abscesses, tumors, or mental illnesses with symptoms like syncopa, convulsions, corresponding with impression *cranii.* This technique is also called cross-*cut sawing* and it is seldom found in the Old World, being first identified there in 1936.[6] Even if considered as being less precise for the surgery performed upon living subjects, we have such prehistoric examples at: Troy VI Ikiz Tepe (Turkey) with two cases (one mature man of 40-45 years, with a trephination located on the right parietal bone and another of 30-48 years with a trephination made by the cross-cut sawing technique). Both individuals had no traces of bone reaction, other pathological indicators, or trauma. At Lidar Höyük (Turkey), a mature individual had survived for several years after a similar intervention.[7] For later times, namely the 7th-8th centuries B.C. three cases of trepanned skulls had been found in Palestine, at Tell Duweir, two of them being done by the cross-cut sawing technique.

(vii) *The scraping and incision combined technique* – seen in some situations of some Bohemian skulls.

[3] J. Verano, 2003. p. 227.

[4] Charlotte Roberts, K. Manchester, 1995, p. 93.

[5] D.R. Brothwell, 1981, p. 123.

[6] H. Kiesewetter, 2002, p. 73, 78.

[7] H. Kiesewetter, op.cit. p. 78.

(viii) *The linear (straight) cutting* – often found in the central highlands of Peru.[8]

The frequent survival of the operated subjects, but also the risk of using the intervention in the area of a sanguine sinus (i.e. the skull fragment Sărata Monteoru P.71 dated in the Middle Bronze Age from Romania), point out the specific training and special abilities of those who carried out the surgery. It is also remarkable to note the resistance to pain and infections of the people in older times. We should also presume the strong solidarity of the people in those communities, as the post-surgery of the patient required a special "medical" care in order to assure his survival.

CAUSE OF TREPHINATION

Ullrich considered that not necessarily around or in connection with the trephination area, there had been detected pathological symptoms such as: (i) parietal-temporal processes that demanded space expansion; (ii) impression of the blood vessels upon the inner table of the trepanned area which indicate an increased inner pressure of the skull; (iii) massive damage of the teeth or maxillaries by caries or cysts located in the area of the molars, especially on the upper jaw, which show a diffusion of the pain towards the suture close to the trepanned area.

Ullrich (1971) reported that among the 40% of trephinations that occurred in the Neolithic and Bronze Age Germany had pathological hints on the skull or post-cranial skeleton. The trephination must have been initiated following the experience of the archaic people. After a trauma upon the cranial vault, the patient felt relieved and had a better general condition after the bone splinters were removed, smoothed, or the opening edge enlarged. This resulted in the desire of healing other illnesses with similar symptoms.[9] Finally, the indication for trephination was caused by various symptoms, such as: trauma; osteomyelitis; osteitis; periostitis;[10] convulsions; melancholia; epilepsy; paralysis,[11] tumors. Except for benign osteomata, tumours in prehistoric man are apparently rare'.[12]

Yet, some forms of cancer metastasis, myeloma, Ewing sarcoma, osteosarcoma, meningioma, echinococcosis, Hand-Schüller-Christianische illness, eosinofiles granuloma, lymphogranolomatose[13] were present.

Trephination could also be done in cerebral abscesses,[14] ailments with similar symptoms to the impressive lesions,[15] congenital scoliosis (on a skeleton from the Beaker period of Britain with the mentioned ailment a

[8] J. Verano, op.cit., p. 227.

[9] H. Ullrich, 1971, vol. II. p. 1282-1283.

[10] H. Ullrich, op.cit. p. 1281.

[11] T. Anda, 1951, p. 280.

[12] M. Goldstein, 1969, p. 484.

[13] P. Boev, 1968, p. 129; D.J. Ortner and W.G.J. Putschar, op.cit. p. 95.

[14] T. Anda, op.cit. p. 280.

[15] Ibidem.

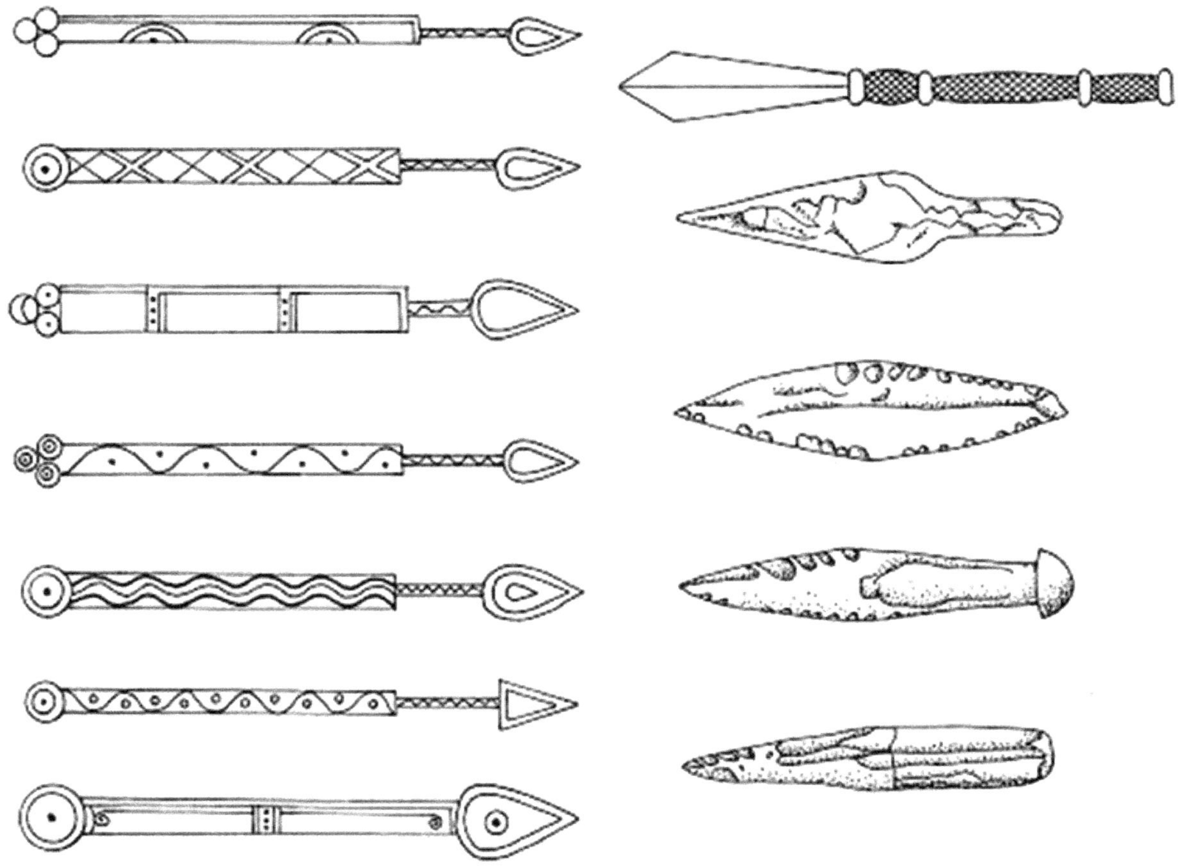

Fig. 18.3. Prehistoric stone instruments and relatively recent improved metal "scalpels" used in craniotomy – the bibliographic source is after Anda (1951)

trephination was done in order to ease the condition of the patient). Other troubles with the spine, or skeletal parts could be also treated by trephination,[16] haematoma, neurological disorders,[17] mental retardation,[18] infections.[19]

In the case of the large trepanning holes, the cause of the intervention, namely the fracture or trauma, had been removed by surgery itself.

INSTRUMENTS USED FOR TREPHINATION

We will not give too many details about such implements, which were very important for the successful accomplishment of the surgery. We emphasize that in prehistoric times the flint or obsidian blades had been used for such a purpose. Later, with technological evolution, the "scalpels" got improved and made of metal (Fig. 18.3). During the prehistoric times, trephination resulted in a low mortality, which was possibly determined by the fact that the freshly knapped flint has been shown to serve as a sterile surgical instrument.[20]

[16] M. Goldstein, op.cit., p. 485; H. Ullrich, op.cit., p. 1281.
[17] J. Verano, op.cit. p. 233-235.
[18] Zs. Bereczki, Antónia Marcsik, 2005, Vol. 12, no. 1, p. 65.
[19] D.J. Ortner, W.G.J. Putschar, op.cit. p. 95.
[20] B. López, L. Caro, A.F. Pardiñas, 119, 3, 2011, p. 249.

According to some studies, the metal scalpels used later performing trephinations were more dangerous, as they customary were washed between interventions but were still carriers of bacteria.[21]

The shape of the cutting edge was obviously determined by the capacity of the instrument to make a precise incision, but also depending upon the operator's training and skills.

STEPS OF THE TREPHINATION SURGERY

As described in the Hippocrates collection, namely in the work *"About the cerebral trauma"* the intervention was done by the following pattern:

At the beginning, the operator had to take care about the following aspects:

1. If the treatment of the patient was initiated by him, the bone had to be cut just until reaching the brain coverings because otherwise, there existed the risk of inducing a lesion at that level.

2. If, on the contrary, the patient was taken over from another, less experienced medic, the correct treatment

[21] B. López, L. Caro, A.F. Pardiñas, op.cit., p. 249.

being delayed, the intervention had to be done faster until reaching the brain coverings, often extracting the instrument from the wound so that some indications about the *dura mater* could be obtained.

In a situation when the bone had suppurations, the trepanation should have been carefully done as the bone was thinner. In the places where the bone was thicker, the "crowned trepan" had to be applied. If the treatment was done from the beginning, the bone piece had to be completely cut and removed.[22]

With the appearance of cutting marks upon the skull, the mode of surgical intervention could be reconstructed accordingly, especially as concerns the position of the "operator". The first step ought to be, the removal of hair on the affected part of the skull. Then it was cutting of the skin, turning aside its flaps to allow a good "access" to the skull. Initially, the cut could have been rather painful, but from this moment on the sensations were less perceived by the patient since bone and brain have no pain receptors. Besides, it is stated that the incision of the scalp is a very bloody procedure, "but the hemorrhage could be minimized by turning back the scalp flaps created; no doubt this was realized and carried out by early surgeons".[23] The bone of that area was cut with a scalpel and brain was exposed. The intervention either cleared the wound resulted from a violent blow upon the head, in order to take away the splinters broken from the skull, or in order to alleviate the suffering of a person with a mental illness.

After the skin flaps were put back in place, they were maybe even sewed with each other before applying the post-operating treatment. It is important to mention that in ancient times there were no aseptic conditions to perform the surgery. It is also relevant here to render the description made by the Romanian anthropologists who made the analysis with regard to the Early Bronze Age skull from Diniaş belonging to a woman aged at 20-25 years. The anterior "corner" in the circular plan showed the place where the manual labor had been started for perforating the bone. The "operator" placed himself behind the ill female, who lied on her right side, had thrust the point of the knife into the same spot and surrounded the outline of the bony area.

In this mode he was constrained to apply the knife on its upper side (his right), to lead it more vertical in its lower part (the left of the "operator") and to take it out again at the place where it started, fact that created the irregularity of the peripheral stripe. The cicatrisation of the wound had stressed this oblique plan and induced the closure of the diploë with a compact plaque. The central area represents the trephination wherein the interior table progressed faster; the trephination orifice had closed due to the bone produced by its proliferation.

[22] V. Crişan, 1957, p. 320.

[23] Charlotte Roberts, K. Manchester, op.cit. p. 93.

A evidence of this supposition is given, both by the aspect of the exocranial side and by the endocranial surface".[24] This is just an example, as the technique is different, according to the condition of the patient and to the place where the surgery took place. The trepanned area could have been protected by certain materials. The Inca, for instance, used to employ the tree bark, golden plaques, or bone fragments.

In Melanesia, the cut was covered with banana skin, coconut, or tree bark, as well. Yet, in our opinion, this was meant to be a treatment, carried out in order to prevent the trephination from getting infected,[25] even if in some cases, the situation was completely reverse, such substances applied upon the wound, facilitating the infection.

Four main types of such responses had been mentioned by Ortner and Putschar (1985) occurring after such an intervention:

- death, induced by the initial trauma;
- the deficient surgery, that can induce death;
- the intervention can cause infection which can possibly result in the death of the individual;
- the surgical procedures have good results and the individual survives for a shorter or longer time afterwards.

TYPES OF RESPONSES TO THE TREATMENT

Possible complications of the surgery

Such reactions could have been determined by the unskillful surgeon, the contaminated instruments, or the surgery performed in improper conditions. The most important ones could be:

- lesions of the *dura mater*, determined by the unappropriate method or unskillful operator;
- infection – trephine cut sequestrum;[26]
- lesions of the venous sinuses.

BONE VITAL REACTION

The cranial vault does not have a remarkable capacity of healing and this is also under the influence of some factors like metabolic condition, age and environment. Anda (1951) showed that the maximal opening in a skull should have a diameter of 25 mm which would allow the bone to heal properly. Therefore, the healing capacity of the cranium is reversely connected to the width of the hole. This means that a hardly perceptible layer is being produced in order to isolate the wound. It often happens that the above mentioned regeneration layer results from a conjunctive tissue, instead of a bony one, not elastic. In

[24] G.I. Russu, M. Şerban, N. Motioc, T. Farcaş, 1957, p. 316.

[25] E. Moldovan, 2003, p. 17.

[26] D.J. Ortner, W.G.J. Putschar, op.cit. p. 96.

time, the bone edges get rounded and eventually they fuse together, completely covering the wound.

Under such circumstances, we can imagine the vital force consuming and efforts made by the body of the female individual with a huge trephination hole discovered in Romania. In the Early Bronze Age site from Livezile (Alba County), in a tumulus, the skull, mandible and two teeth of a woman aged at 35-40 years have been discovered. The dimensions taken in the intact zone pointed to a larger antero-posterior diameter of about 10 cm and a smaller one, vertically placed, with dimensions that varied between 4.4 and 4.7 cm. The slope of the cut was rather uniform. Taking into account the aspect of the bones, it was considered that the intervention was carried out upon an individual in full health, by a skillful person. The edges of the trepanned hole indicated that the regeneration process was initiated and the individual had a healing process located in the paramedian postcoronal area which evidently "affected the diploic cells". This would indicate the survival of the individual longer than one year. There are no elements to imply the motivation of that intervention but, the presence of a severe cranio-cerebral trauma at that level is excluded.

The curing, or ritual purpose of this trephination is unknown.[27] Considering the aspect of the incision, it is possible that it was executed by a scraper. It is worth mentioning that the case was the largest trepanning hole found on the territory of Romania (Fig. 18.4).[28]

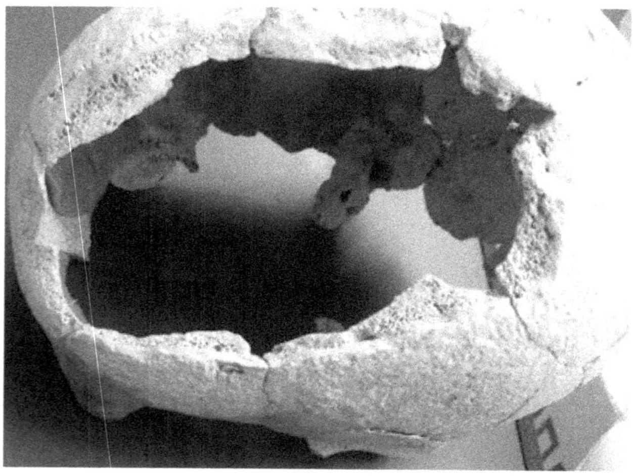

Fig. 18.4. Largest trepanned hole found in a Romanian skull; photo courtesy Horia Ciugudean

Roundel Removing

For the bony roundel taken from an individual, we could find two distinct situations:

i. A living person in order to take over her spiritual powers.

ii. A dead person, to use the roundel as a trophy or amulet.

Besides precious stones which were always considered to have both magic and healing properties, the amulets made of human skull (or other body parts) must have provided special spiritual powers, taken over from the dead and transferred to the living person who wore them. They sometimes accompanied another skull with the purpose of protection against the evil spirits, or, we should add, in order to prevent against a subsequent human intervention, which would profane the burial. Some interpretations believe that the trephination was performed upon the dead, in order to save them from being possessed by evil spirits.[29] A few scholars (Basu & Pal, 1980) also believed that some of the multiple trepanations on the Burzahom skull were performed for extracting roundels as amulets for magical purpose. In such a situation, both the community of the living and the one of the dead must have been protected by repelling the demons which, otherwise, could have come and kill the living or make the dead relentless. In our opinion, the roundels had played a significant role in ancient times.

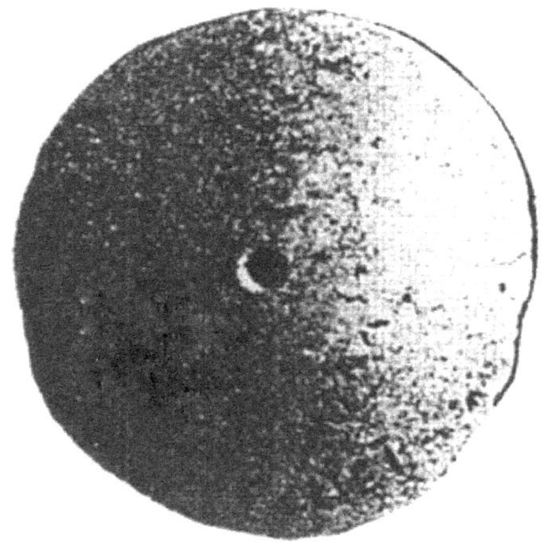

Fig. 18.5. Trepanation for a Roundel in Raklo (Bulgaria); bibliographic source is after Mikić (1998)

The interpretations of the archaic communities about the world were mostly connected with the unknown forces of nature and this is why the magical religious practices were very important for those people. Another argument in this sense is the fact that, although the cutting off of such a bone roundel was an act that did not demand a lot of care and precision, the amulet had to meet some certain criteria. Since the ancient people were mostly practical, we do not see any reason they would have made such amulets, or trophies of human skulls or pieces, unless they were really important. In the Eastern and Central Europe we have some good examples of such roundels, from Raklo (Fig. 18.5) in Bulgaria, or from sites like Nynice, near Plzeň (Hallstatt), Prosmyky, near Lovosice, Tuchoměřice, near Prague (Latene) in the

[27] E. Moldovan, op.cit. p. 8.

[28] D. Botezatu, H. Aldea, H. Ciugudean, 1987, p. 3-7.

[29] Al. Lenghel, 1926, Cluj-Napoca, p. 8.

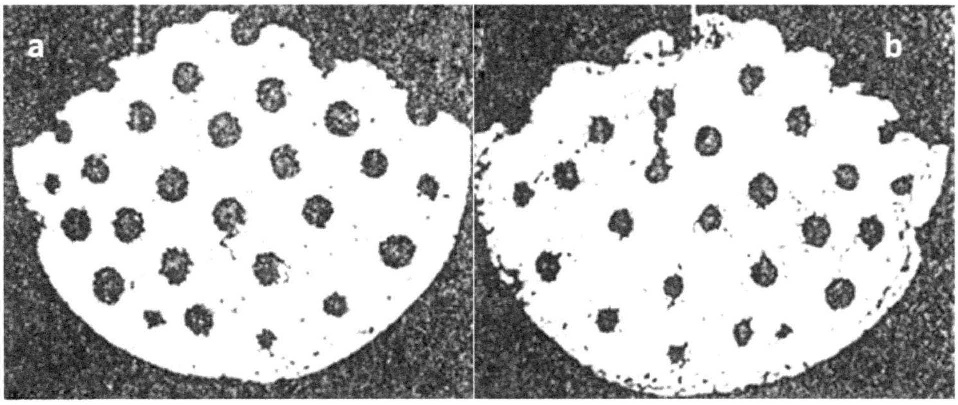

Fig. 18.6a & b. Trepanations done for extracting multiple Roundels in Kunzig-Ost (Germany); bibliographical source is after Moldovan (2003)

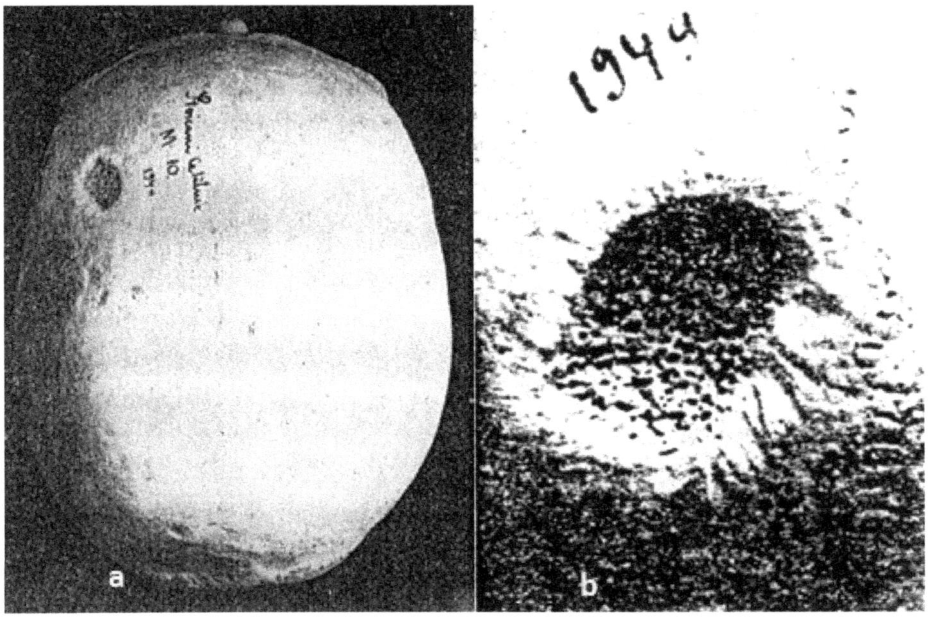

Fig. 18.7a & b. Pseudo-trephination on skull No. 10 from Stoicani Cetăţuie (Romania)

Čech Republic, or Kunzig-Ost in Germany (Fig. 18.6a, b).

PSEUDO-TREPHINATIONS

Sometimes, holes of different dimensions have been detected on skulls, sometimes being mistaken for trephinations. This is the reason why they had been also called pseudo-trephinations. Yet, they were the result of physiological[30] (acquired disorders, metabolic and endocrine malfunctions), pathological (tumors, infections, syphilis etc.), cultural (mutilations), or taphonomic conditions (destructions carried out by rodents as it appeared on the prehistoric skull from Stoicani Cetăţuie no. 10 – Romania – (Fig. 18.7a, b), stones pressing upon the cranial vault etc.). Among the pathological conditions, a congenital illness mentioned by T.D. Stewart in 1975 was the encephalocele or disraphysm, which is determined by a pulsating tumor which takes the form of a depression in the *bregma* point and can look like a partly healed trephination.[31] Another pathological condition with similar effect is the meningocele,[32] but others have been also identified by the specialists, e.g., tuberculosis, syphilis, infections, neopolasms, cysts, etc.

Of course, the human intervention could cause a lot of changes affecting the ancient skulls, either during the life of the respective individuals (by projectiles, or violent blows), or after their death (during the excavation of the skeletons).

[30] Pia Bennike, 2003, p. 112 and 113, fig. 20, 21; Josephina Palop Martinez,1970, p. 56; F. Fachini, E. Rastelli, L. Homo, 53/3, p. 248. Ferrero, E. Fulcheri.

[31] T.D. Stewart, 1975, p. 435.

[32] D.R. Brothwell, op.cit., p. 120.

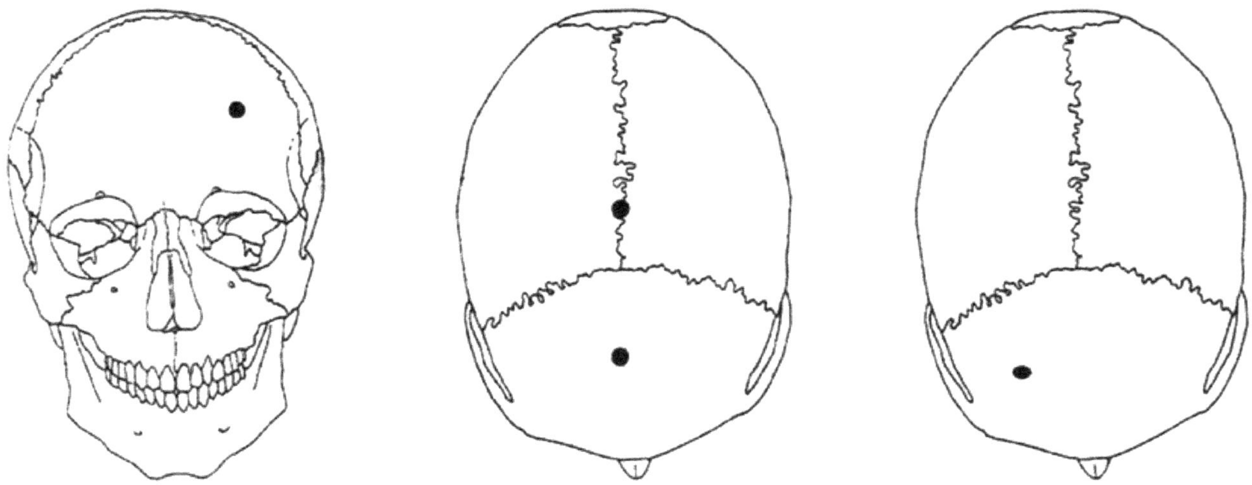

Fig. 18.8. Symbolic trephinations on different portions of the skull vary according to cultural period and gender (Mednikova, 2003) For the Eneolithis site near Vukovar, its location according to gender is the following one; a. typically male; b. typically female; c. atypically female

SYMBOLIC TREPHINATION

This kind of trephination had been documented even beginning with the Upper Paleolithic time in North Africa.[33] In Europe it is documented in the Gravettian period and goes further to the more recently dated time sequences. The origin of this kind of trephinations seems to be traumatic, having a cultural value and interpretation.

It consists in making usually rounded deepenings on the exocranial part of the skull, by using a hot iron or other procedures (incision of the lamina *externa*, deepening of the incision; removal of the bone piece from the surface delimited by incision; smoothing the remainder surface on the skull).[34] Such interventions were non-verbal messages to those who knew their meaning and this was the reason why they had to be placed in visible places. For the Eurasian territories Mednikova (1973) mentioned the following distribution of the trephination location, according to the historical time sequence:

During the Neolithic time, those lesions were usually found in the middle part of the frontal bone. In the Bronze Age, the usual place was the right parietal, while in the Urnfield Culture, the left parietal bone was most frequently chosen. In the Laténe times, the crania had symbolic trephinations on both parietals and sometimes, also, on the occipital bone.

During the period of the great migrations, those interventions were placed in the central part of the frontal bone and near the *bregma* point.[35]

In certain situations, the location of the intervention also varied according to the sex of the trepanned individual, as evident at the mentioned skulls from Croatia (Fig. 18.8).

[33] M.B. Mednikova, 2003, p. 150.

[34] P. Boev, 1968, p. 130.

[35] M.B. Mednikova, op.cit., p. 148.

Significance of Trephination

As already mentioned in this paper, in most of the cases, the intervention seems to have been done in order to cure an ailment or a wound. In others, it was done for ritual purposes, or for making a bone roundel.

On the other hand, we were wondering if such a procedure could have also not been applied to individuals who achieved a high rank in the society. We should not neglect the idea that such a situation could occur. In 1954, during the excavations carried out by Prof. Ion Nestor and Eugenia Zaharia at Sărata Monteoru (Buzău County) in Romania, a paved ritual platform had been uncovered. Upon it, among other human remains, a cranium fragment was found. This comprised the occipital squama and part of the parietal bones. The piece, marked as P.71, was cut away from a mature male skull (Fig. 18.9), whose age was established to have been around 50 years. The outline of the cutting was very regular and the *lambda point* was situated close to the center of the skull fragment. The trephination was done for some longer time before his death. Most probably, it was performed by using the sawing technique, as shown by the fine striations found upon the bone in the proximity of the cutting area. It seems that the respective man was very robust, if he had survived an intervention which affected his longitudinal venous sinus. There were found no traces to indicate the purpose of that trephination. No trauma, or pathological processes could be identified, either by using a magnifier, or a radiograph.

The regeneration traces, more intense on the inner table of the bone, show that the individual had lived a longer time after the surgery. By all means, the intervention had been done probably for treating a convulsive illness, or with a ritual purpose.[36] Of course, the fragment was taken away from the skull after the death of the individual. We must take into account the fact that the ritual platform

[36] Olga Necrasov, E. Floru, D. Nicolăescu-Plopşor, 1959, p. 14-16.

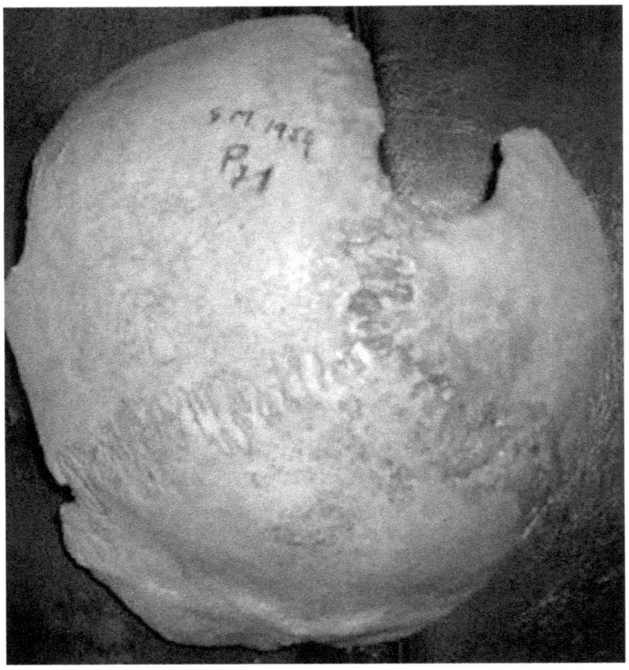

Fig. 18.9. ? Social cranial fragment with trepanation in Buzău County (Romania) – photo courtesy Georgeta Miu

was used for keeping some human remains which were probably mostly cherished by the community. The mentioned individual must have had important merits if his skull was preserved like that. It is possible that this very quality, of survivor to such a medical risk, but may be also the other social ones which, might have determined the community to preserve a part of his skull, mostly the one with a segment of the trephination, together with other valuable parts of the ancestors.

Then, we could ask ourselves: Was it possible to make trephinations on the skulls of important people and consider them after death as a source of power and spiritual fulfillment for the community? Otherwise, why should the trepanned piece of the skull need to be preserved? These are questions which need further explanations, probably offered by future research.

CONCLUSIONS

The trephination is a complex medical procedure and being practiced for a very long time made possible its permanent improvement, both concerning the methods and instruments employed, up to the recent times. In itself, it involves three main elements: (1) the health condition of the individual, (2) the skills of the operator, which, of course, is no longer necessary when used just the bone roundel prelevation, and (3) the conceptions and customs of the community to which the individual had belonged. All these "merged" with one another indicate that either the healing of the individual, or the prelevation of the bone roundel from the deceased was the case. We should image the complex surgery carried out by the medicine practitioners of the past, using herbs or different animal or mineral substances as painkillers, anesthetics or antibiotics. We should also emphasize the remarkable skills of such early surgeons, which were able to save the life of people under septic conditions. Anyway, in our opinion, the resistance of the human body, as well as its pain perception was different than it is today. We should not forget that the solidarity of the people in those times was also important in such cases, and here, we refer not only to the family but to the entire community (tribe), because the healing was a long and enduring process and the patient needed to be watched and cared for over a long period of time.

May be, given the importance of such a practice, it was adopted as a means of initiation in certain roles or with other social meanings. The pain endured when having a symbolic trephination done upon their skull, might have urged the people to have stronger feelings about their community, the way it happened in more recent time with the brothers in blood relations.

References

ANDA, T. 1951. Recherches archéologiques sur la pratique médicale des hongrois à l'époque de la conquéte du pays. Trépanation au ciseau et. au foret. *Acta Archaeologica Hungarica 1*: 251-316.

ANDRUSHKO, A. & VERANO, J.W. 2008. Prehistoric Trepanation in the Cuzco Region of Peru: A View into an Ancient Andean Practice. *American Journal of Physical Anthropology 137*: 4-13.

AUFDERHEIDE, A.C. & RODRIGUEZ-MARTIN, C. 1998. *The Cambridge Encyclopedia of Human Pathology.* Cambridge University Press, London.

BASU, A. & PAL, A. 1980. *Human Remains from Burzahom*, Memoir, No. 56. Anthropological Survey of India, Calcutta.

BENNIKE, P.I.A. 2003. Ancient trepanations and differential diagnoses: A re-evaluation of Skeletal Remains from Denmark. In R. Arnott, S. Finger, C.U.M. Smith (eds.) *Trepanation History-Discovery-Theory*, pp. 95-115.

BERECZKI, Z.S. & MARCSIK, A. 2005. Trephined skulls from ancient populations in Hungary. *Acta Medica Lituanica. 12* (1): 65-69.

BOEV, P. 1968. Die symbolischen Trepanationen. *Anthropologie und Humangenetik* Stuttgart.

BOTEZATU, D., ALDEA, H. & CIUGUDEAN, H. 1987. Consideraţii asupra unui craniu trepanat descoperit la Livezile (jud. Alba) aparţinând culturii Coţofeni (începutul mileniului II î.e.n.), Studii şi Cercetări de Antropologie *24:* 3-7.

BROTHWELL, D.R. 1981. *Digging up bones,* Cornell University press, Ithaca, New York.

BROTHWELL, D.R. & HIGGS, E. (eds.) 1969. *Science in Archaeology. A survey of progress and research,* Bristol.

CRIŞAN, V. 1957. Cîteva date şi informaţii cu privire la neurochirurgia antică, Istoria *medicinei. Studii şi cercetări,* Bucureşti, pp. 317-322.

FACHINI, E., RASTELLI, E., FERRERO, L. & FULCHERI, E. 2003. Cranial trepanation in two skulls of early medieval Italy. *Homo* 53/3: 247-254.

KIESEWETTER, H. 2002. Ein trepanierter Schädel aus Troia VI. *Studia Troica 12*: 73-80.

LENGHEL, AL. 1926. Trepanaţia pe craniu preistoric, pe craniu din sec. XI şi la popoarele naturale. *Clujul medical* 9-10: 1-11.

LÓPEZ, B., CARO, L. & PARDIÑAS, A.F. 2011. Evidence of trepanations in a medieval population (13th-14th centuries) of northern Spain (Gormaz, Soria). *Anthropological Science* 119 (3): 247-257.

MEDNIKOVA, M.B. 2003. Ritual Initiation in Prehistoric Eurasians Based on Cranial Data: Symbolic Trephination. *Archaeology, Ethnology & Anthropology of Eurasia* 1 (13): 147-156.

MIKIĆ, Ž. 1998. Über urgeschichtliche Trepanationen im Gebiet von der Adria bis zum Schwarzen Meer. *Ein anthropologischer Überblick*, PZ, Band 73 (Heft 2): 145-149.

MOLDOVAN, E. 2003. Trepanaţia craniană la culturi preistorice din Transilvania. *Buletinul cercurilor ştiinţifice studenţeşti, arheologie-istorie-muzeologie* 9 (Alba Iulia): 5-23.

NECRASOV, O., FLORU, E. & NICOLĂESCU-PLOPŞOR, D. 1959. Contribution à l'étude de la pratique de la trépanation chez les populations préhistoriques du territoire de la R.P. Roumanie. *Przeglad antropologiczny, tom* XXV (Wroclaw): 9-29.

ORTNER, D.J. & PUTSCHAR, W.G.J. 1985. *Identification of Pathological Conditions in Human Skeletal Remains*, Smithsonian Contributions to Anthropology. No. 28, Washington, London.

MARTINEZ, J. PALOP 1970. Distribucion mundial de la trepanacion prehistorica. *Revista Española de Antropologia Americana* 5 (Madrid): 51-66.

ROBERTS, C. & MANCHESTER, K. 1995. *Archaeology of disease*, Cornell Publishing Limited, Cornell University Press, Ithaca, New York.

RUSSU, I.G., ŞERBAN, M., MOTIOC, N. & FARCAŞ, T. 1957. Trepanaţie craniană vindecată datînd din epoca bronzului, Istoria medicinei. *Studii şi cercetări, Bucureşti*: 311-316.

SANKHYAN, A.R. 2008. Surgery in Ancient India. In *Encyclopaedia of the History of Science, Technology, and Medicine in Non-Western Cultures* (ed. Helaine Selin), Springer Link, part 19, pp. 2060-2063.

SANKHYAN, A.R. & WEBER, G.H.J. 2001. Evidence of Surgery in Ancient India: Trepanation at Burzahom (Kashmir) over 4000 years ago. *International Journal of Osteoarchaeology* 11: 375-380.

SANKHYAN, A.R. & ROBBINS SCHUG, G. 2011. First evidence of brain surgery in Bronze Age Harappa. *Current Science* 100 (11): 1621-22.

STEWART, T.D. 1975. Cranial Dyraphism Mistaken for Trephination. *American Journal of Physical Anthropology* 42/3: 435.

ULLRICH, H. 1971. Das Motivproblem der Trepanationsforschung im lichte neuer Funde. *Actes du VIIe Congrès International des Sciences Préhistoriques, Prague 21-27 août* 1966, Prague II: 1281-1283.

VERANO, J. 2003. Trepanation in Prehistoric South America: Geographic and Temporal Trends over 2.000 Years. In R. Arnott, S. Finger, and C.U.M. Smith (eds), Trepanation*: history, discovery, theory.* Swets & Zeitlinger B.V. Kisse, Netherlands, pp. 223-236.

19.
DECRYPTION OF ETHNIC IDENTITY OF THE WHITE MUMMIES IN TARIM BASIN, CHINA

Xinyan CHI
Nanjing Normal University, China
chixinyan@yahoo.com

Abstract: *Xiaohe (Little River) Cemetery was one of the major archaeological discovery of the world in the 21st century, whose archaeological era was about 2000 BC – 1500 BC. It was located in Tarim basin of China, and the dead were classified to white race. The landscape of the cemetery is extremely strange, mysterious and exaggerative. It is the unique discovery in the archaeological history, and is one of the hot spots highly concerned in the world. Focusing on its vegetation and digital phenomenon, this paper tries to decrypt the secret of Xiaohe people with the heavenly stems and earthly branches (HSEB), oracle-bone and bronze inscriptions, and historical documents, in addition of the comparative studies of the archeology and folklore and field information of Sichuan Basin.*

Keywords: *Xiaohe cemetery, White mummies, inscriptions, Tarim basin, China, legends*

INTRODUCTION

The central ideal of the Xiaohe Cemetery is recording the HSEB. Xiaohe Cemetery is located in an oasis of Lop Nur water system. With Taklimakan desert dust invasion by northwest wind and establishment of 5 layers of tombs by several generations, it was formed a oval sand hill of 7.7 meters high, 74 meters long, 35 meters wide, shown in Fig. 19.1. The sand hill clouds with fences more than 100 standing woods, seems incongruous in the slow flat sand terrain, like a mountain of woods with sands. A large amount of standing woods, branches, small twigs, seedlings, grass and burr bundles were placed around the coffins in a very regular way.

Fig. 19.1. Outlook of Xiaohe Cemetery

The paper tries to decrypt the secret of the white people unearthed in Xiaohe cemetery in Xinjiang, China, with the heavenly stems and earthly branches (HSEB), oracle-bone and bronze inscriptions, and historical documents, Xiaohe is believed to be the fief of the grandson of Emperor Huang, An, whose descendants were called "Ruo Mu", "white Qiang" or "Ruo Qiang".

1) Before each coffin stood a prismatic wood (Fig. 19.2b) or a paddle-shaped wood (Fig. 19.2a, c) with exaggerated proportion, indicating the symbol of the penis or the vulva, as the spouse of the opposite sex tomb. This was a typical HSEB recording to represent the paternal ethnic system as a foliage tree structure of branches and leaves for reproduction. The trunk was the father's spine evolved from his penis, and the branches referred to his ribs, indicating his offspring. The arms and leg indicated the full brothers, known as the "hands and feet". For example, National Speech • Zhou • recorded that family of Ji's was from Ximu (a name of a tree), referring that Fuxi was their ancestor. In Historical Book • Yu (Shangshu • Yugong) the records of Kunlun, Xizhi, Qusu and Xirong indicated the offspring of Fuxi. In Legends of Mountains and Seas (Shanhaijing) the words "3 Sang families with no branches" represented that they had no offspring.

The leaf was used to stand for vulva and then to female, comparative with branch for penis and then male. The Chinese idiom "open branches and leaves" means giving birth to sons and girls, and another idiom "golden branch and jade leaf" means prince and princess.

1) The long wooden sticks erected in the back of the male coffin, or placed by his leg, meaning his offspring in behalf of branches and sticks.

2) The Chinese oracle-bone inscriptions for branch (Zhi, 枝) and stick (Zhang, 杖) were the same, both representing offspring. In Legends of Mountains and

Fig. 19.2. Standing woods

Seas, the record of "Kuafu gave up his stick and the stick changed to forest of Deng" meant that the offspring of Kuafu established a country of Deng. The Chinese emperor scepter was called stick, meaning the emperor was the son of the supreme emperor.

3) The standing woods of male penises were surrounded with red willow branches, grass bundles, straw rings (Fig. 19.2a); the reeds, ephedra, grass thorn and seed grass-baskets were placed before the male coffins (Fig. 19.2b); and red willow branches were placed on leather packages in the coffins (Fig. 19.2c). All these were the pray for children. In HSEB records the male genetic was described as "seeding", and the gender descendants were called "fruit" and "seeds". Seedling grass and thorn represented the grandchildren and the generation descendant.

4) A whole tree was discovered in Xiaohe Cemetery, indicating the whole people belonged to a tribe of same surname, i.e., a same tree. The forest (Lin, 林) in HSEB recording refers to regional organization or emirate consisting of many tribes of different surnames. "Gui forest of eight trees" described in Legends of Mountains and Seas referred that there were 8 tribes in Gui Emirate.

HSEB's Reproductive idea and prayer for reproductive ability

1) Archaeological experts discovered that in Xiaohe Cemetery there was a lot of information of figure 7. There were 7 carved slots in the root of paddle-shaped woods in behalf of female vulva; there were 7 carved traces on the rods for hat; the wrist was surrounded with 7 rings of hand bracelets, shown in Fig. 19.4a, b, c. 7 layers of ox heads were placed in the cemetery. I also found that there were 7 red lines on the cloak, 7 red lines on the jade chain (Fig. 19.4d), 7 red lines on the carved wooden human face (Fig. 19.5a), and 7 slots on the carved bone human head (Fig. 19.5b), 7 red strings on the mud coffin (Fig. 19.5c).

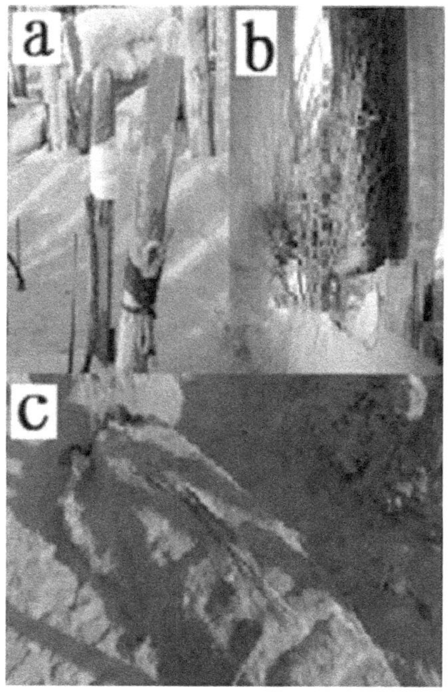

Fig. 19.3. Vegetation of Xiaohe

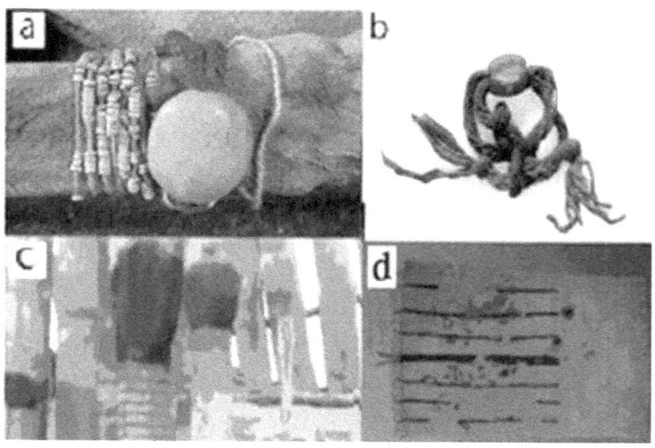

Fig. 19.4. "7" in Xiaohe (1)

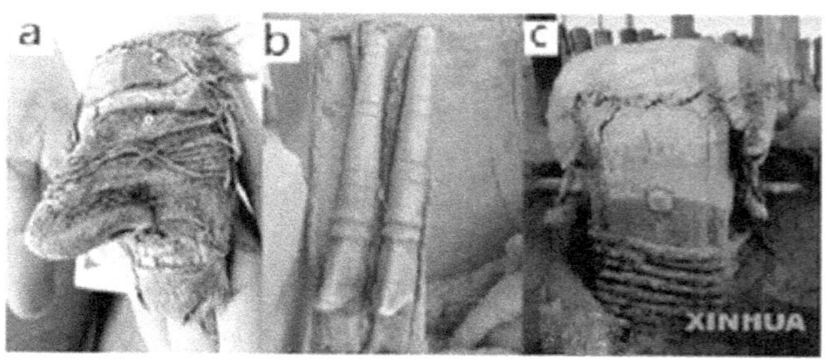

Fig. 19.5. "7" in Xiaohe (2)

Fig. 19.6. Sun Cemetery

Fig. 19.7. 8 slots on standing wood

2) **The universe reproductive number 7**. 7 is the number of 12 earthly branches, considered as a mysterious reproductive force due to the Big Dipper which consists of 7 bright stars, then 7 becomes a holy life figure. The ancients used the rotation of the mouth of the Big Dipper to observe the sky and measure the time, and with the Big Dipper they divided the whole sky into 12 divisions, known as the 12 earthly branches. In Han • Calendar (Hanshu • Lvlizhi, 汉书·律历志) it recorded that 7 was the original power of the time, life and the universe. The *Book of Changes* (Yijing, 易经) said that the rotation of the Big Dipper formed the rules of the heaven, and then produced the endless power to operate celestial bodies.

3) **The life figures 7 in Inner Canons of Emperor Huang**. Number 7 was regarded as the origin of the power of the female pregnant. Inner Canons of Emperor Huang (Huangdineijing) recorded "the woman has menstruation to in the age of 2 seven years, ready for pregnant and has menopause to terminate the pregnant ability in the age of 7 seven years. That is to say, the reproductive ability begins at 14 (2 x 7) years old and finishes at 49 (7 x 7) years old for a woman with the figure 7. Similarly, the number of male reproductive ability is 8. In Chinese medicine it is called "the blood of female and the sperm of male" or "seven of woman and eight of man".

4) **Two special case evidences**. a. There are 8 slots on a standing wood in behalf of male penis in cemetery, to indicate the eight of man in Inner Canons of Emperor Huang;

b. The design of the Sun Cemetery close to the Xiaohey is 7 geometric progressions. The graveyard center is a male, surrounding with short stakes like the light of the sun, where the inner and the outer both have 7 rings and there should be 49 rows in accordance of the figure 7.

12 +1 in HSEB indicator of prayer for children: The archeologists discovered that the red willow branches seems to be placed randomly on the coffin leather (shown in Fig. 19.8a), but every bundle has 12 red willow branches plus a reed and the archeologists were puzzled with 12+1. I also found in a cloak there are 12 longitude red lines and 1 latitude red line (shown in Fig. 19.8b). It indicates the first, seed or child (Zi, 子), of 12 earthly branches in HSEB. Zi refers the offspring of the sexual activity, the symbol of endless life.

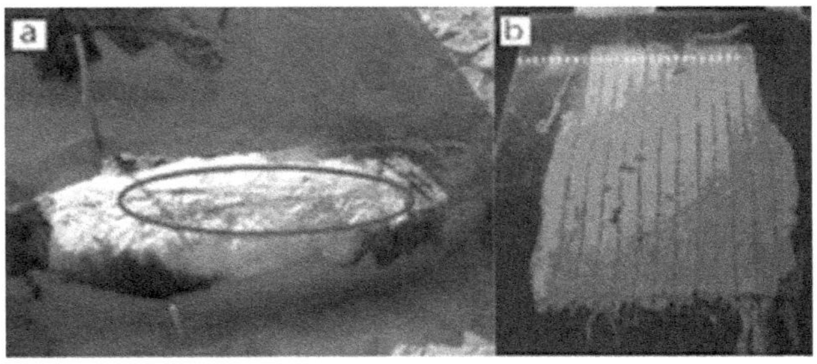

Fig. 19.8. 12+1 in Xiaohe

7 and 12 +1: the characters of + and ♀: In ancient Chinese characters, 7 is + to the reproductive process; 12 +1 is ♀ to indicate the reproductive result.

1) The oracle-bone inscription of 7 is +, shown in the Fig. 19.9 with its evolution.

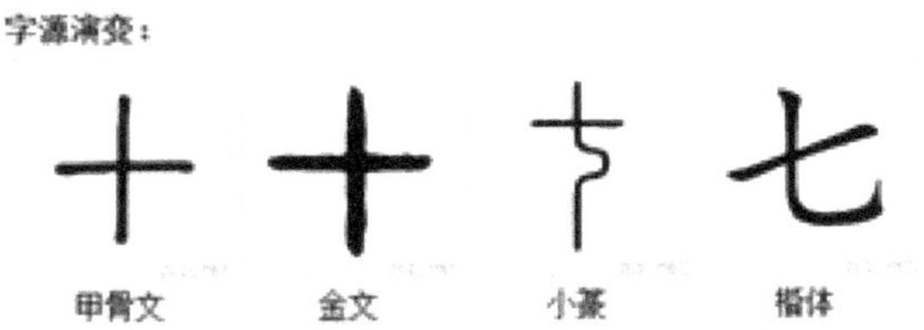

Fig. 19.9. Evolution of 七

a. The 7 was considered as the number of life due to the Big Dipper of 7 bright stars. In ancient time, the Big Dipper was observed as the center of heaven because it was close to the Polaris. The center of heaven indicated the center of 4 extremes of the heaven, and it was symbolized as +, shown in Fig. 19.10 with the connecting lines of extremes of east-west and south-north in the sundial;

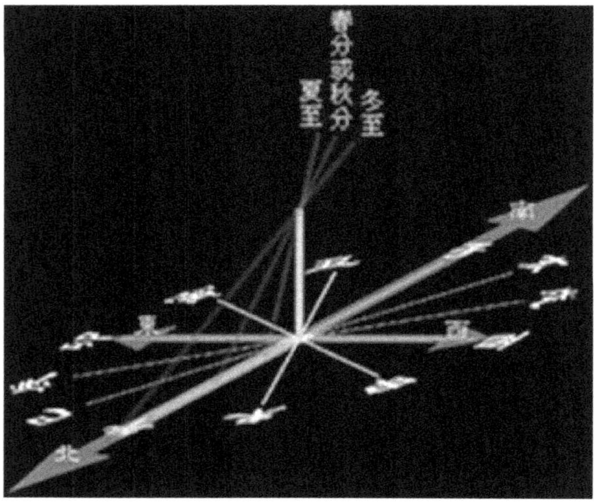

Fig. 19.10. The sundial

b. The 7 is the seventh branch, noon (Wu, 午), in 12 earthly branches, indicating the sexual activity. Its hieroglyph is +. The picture of Riverside Scene at Qingming Festival (Qingmingshanghetu, QMSHT, Fig. 19.11) showed 800 years ago the ancient Chinese to worship with the wooden 午.

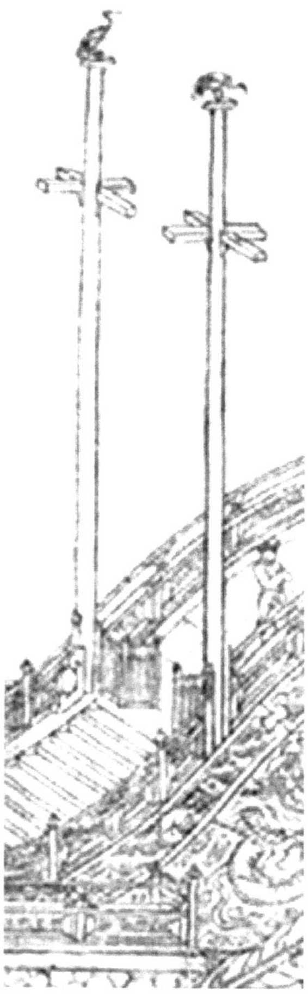

Fig. 19.11. QMSHT

2) The meaning of 12 +1 is seed (Zi, 子), whose oracle-bone inscription is ♀, indicating the result of sexual activity, with its evolution shown in Fig. 19.12. It showed there was a fruit on a crucifix to indicate the result of sexual activity, the child or the descendant.

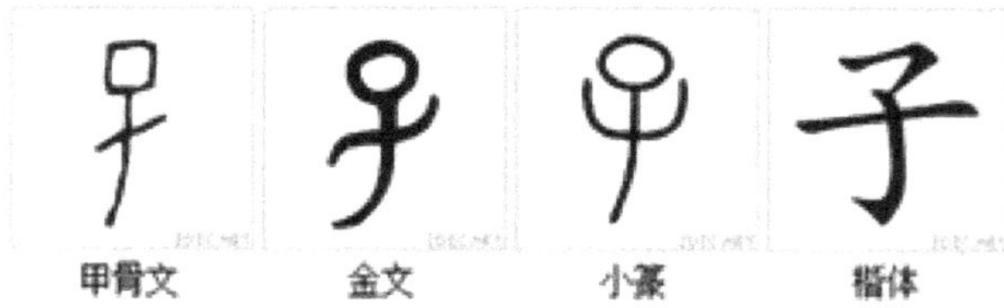

Fig. 19.12. The evolution of 子

Extension of HSEB in Xiaohe to 12 zodiac signs

The male puppet with earrings and red waist rope was extremely prominent in about 300 tombs in Xiaohe Cemetery (Fig. 19.13). This folklore still exists in China

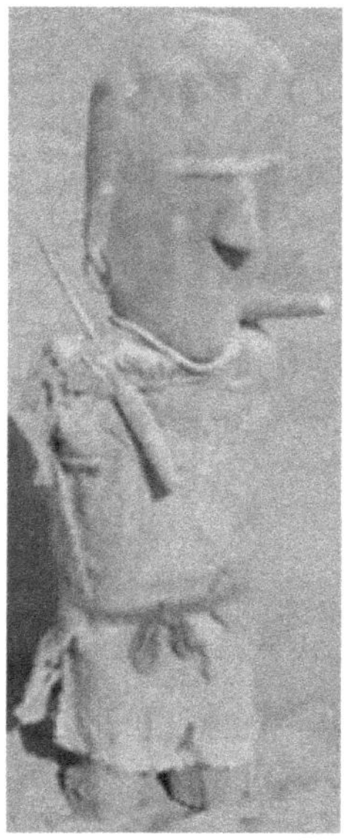

Fig. 19.13. Male puppet

Fig. 19.14. Earthly branches

today, to wrap a red rope on waist or wrist, meaning across the rebirth of the threshold. The ancients named the 12 earthly branches (Fig. 19.14) with the animal names according to shapes of the bright stars in 12 divisions of the sky, called 12 zodiac signs or zodiac animals, the to name the year in turn. The birth year and 12th years after in turn are called threshold years, and a human in the threshold years should wrap some red objects on his body, such as red shorts, red ropes on wrist, ankle or waist, red scarf, and so on.

Xiaohe people called as white Qiang

Xiaohe people belonged to white race as shown in Fig. 19.15 with the Xiaohe Princess who shocked the world.

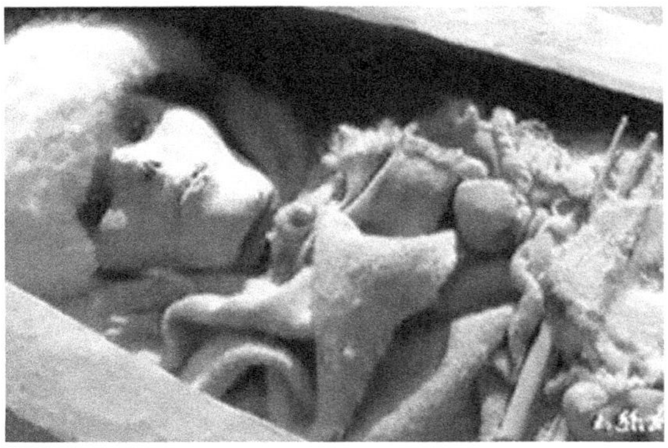

Fig. 19.15. Xiaohe Princess

1) A polished spherical white stone (Fig. 19.16a) was unearthed in a priestess tomb. Qiang worships to the white stones and there are white stones in Qiang's families for worship. This indicates Xiaohe people belonged to White Qiang. Today, every family of Qiang worships white stones on the roofs of their houses, as shown in Fig. 19.16b.

Fig. 19.16. White stones

Ethnicity of the Xiaohe People: Ruo Qiang with Ruomu in HSEB

1. The Xiaohe Cemetery located in the country of Ruo Qiang in Han Dynasty. In Uygur language the meaning of Ruo Qiang is the place to see the spinning wheels, which described the characteristics of this region, i.e., there lived a tribe of Qiang called as Ruo with textile industry. The luxury wool clothing and the sealed coffin leathers in Xiaohe displayed the developed animal husbandry and textile industry. The oracle-bone inscription of Ruo (若),  is a woman kneeling under a tree to tidy wild silk, and its bronze inscription, 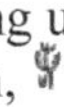, indicated that there was a reeling basin beside her. The oracle-bone inscription of Qiang referred a shepherd. The shape of 若 is like that a priest lifts 3 stems over head representing 3 stars of Big Dipple

2. Ruo Qiang River belongs to Lop Nur water system, originated from Altun Mountains, which showed the trace of migration of Ruo Qiang people from east to west. In ancient documents it recorded that the Ruo Qiang people lived in the northwest of Qinhai and the south of Xinjiang, and some moved to northern Kashmir.

3. Ruo Tree (Ruomu) was the name of Ruo Qiang in HSEB. The Legends of Mountains and Seas recorded that Ruo Qiang people lived in the west and worship to black leaf-shape woods (Fig. 19.17a) and red prismatic-shape (Fig. 17.17b), which is consistent with what were discovered in Xiaohe Cemetery.

Fig. 19.17. Black paddle-shaped and red prismatic

4. The tribe of Ruomu originated at the River Ruo in Sichuan Basin, where was the fief of Chang Yi, the son of Emperor Huang.

5. The leader of Ruomu, An, the grandson of Emperor Huang, was enfeoffed to west in about 2500 BC and established the country of Anxi in about 250 BC, called

Parthia, whose scale replaced the Persian empire. From the enfeoff to Parthia, the offspring of An migrated from Sichuan Bain to northwest in about 2300 years, they might built many different countries and should leave a branch in Tarim Basin.

6. The oracle-bone inscription of An (安) referred a home of woman, and the Xiaohe people belonged exactly to female tribe, because the most tombs buried women, the central tomb buried an old woman and the priestess, and the mud coffin tombs buried women.

The Xiaohe: A blackbird tribe of West Queen

West Queen was a hereditary queen in western China. The Legends of Mountains and Seas recorded that West Queen lived in Jade Mountain (Yushan) with hoopoe and was fed by 3 blackbirds from south.

1. Hoopoe refers to feathers on the out surface of hat. It is a kind of bird with vertical feathers shown in the Fig. 19.18a. Xiaohe people lived by River Peacock (Kongquehe), and peacock was a kind of Hoopoe, therefore, they wore the felt hats with peacock feather (Fig. 18b, c) which was consistent with the tribe of West Queen.

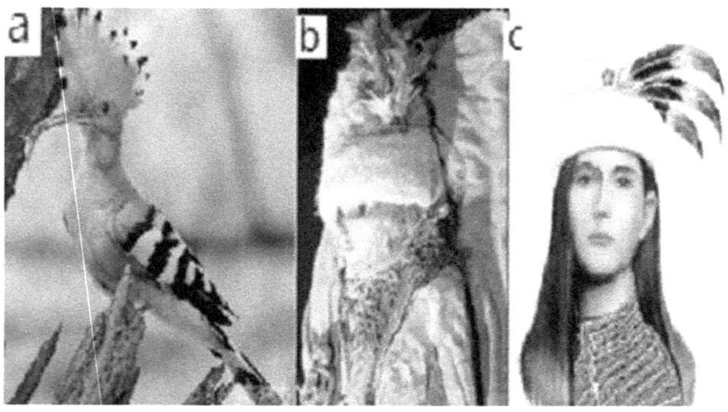

Fig. 19.18. Hoopoe and its symbol in Xiaohe

2. Jade Mountain (Yushan) should be interpreted as place the people mining jade lived, called Yu Tree (Yushu) in HSEB, and it still exists Yushu in Qinghai, meaning the Ruins of King's city in Tibetan language.

3. Yushu referred especially woman tribe because ancients classified male and female as gold and jade respectively, for example, the idiom "golden branch and jade leaf" indicating prince and princess, and "good match of gold and jade" meaning the marriage of a man and a woman.

4. The leaf-shape woods in Xiaohe Cemetery were painted black, and women ornamented with peacock feather, which were consistent with the characteristics of blackbird tribe.

5. The luxury funerals in Xiaohe Cemetery showed their developed agriculture and livestock corresponding to the records of worship to West Queen.

Relatives of the Xiaohe: A Snake tribe

The archeologists found "7 snakes were placed in a man's tomb between his waist and leg. Two wooden rods with carved human faces were at his legs, each of which had seven human faces". The wooden rod was called snake rod by Folke Bergman in his book, Archaeological research in Sinkiang.

1. Placing a rod at a male's leg was called "branch and stick (Zhi Zhang)" in HSEB. The snakes placed on the lower part of a man indicated the totem of human-face and snake-body tribe, and also the offspring of Fuxi and Nuwa, the primary ancestors of the Chinese, whose totem was 2 – snakes combination as a whole with 2 – human faces shown in Fig. 19.19.

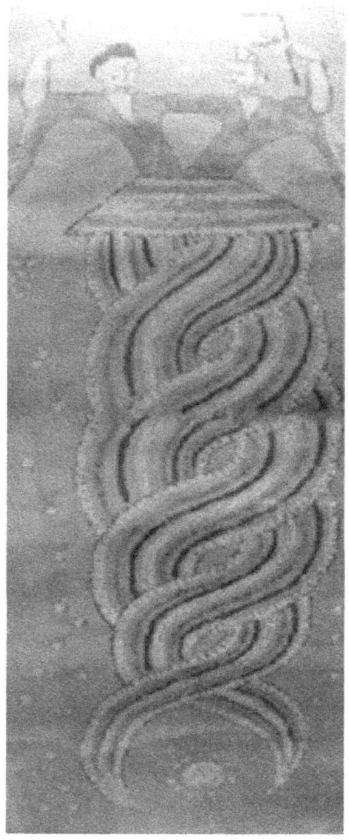

Fig. 19.19. Fuxi and Nuwa

The lizard skulls in the wooden penis (Fig. 19.20) meant that Xiaohe people regarded Fuxi was their ancestor because lizard referred Fuxi due to their similar pronunciations in Chinese. Xiaohe men wore earrings, which was the symbol of snake tribe. In Legends of Mountains and Seas, matching snakes on head meant the snakelike headdress, snakes on ears meant earrings, holding snakes meant wrist bracelets, and stepping on snakes meant ankle bracelets.

Fig. 19.20. Wooden penis

2. 7 – snakes placing on the lower part of the body and the 7 human faces on the wooden rods meant that he was the belonged to the offspring of 7 – snake tribe. Satan recorded in New Testament of Bible first of seven elders in snake tribes, that is, he was the leader of the seven snake tribes. He was defeated in the fight for the place of Polaris. The ancients regarded the Polaris as the center of the heaven; therefore, the fight for Polaris was exactly the fight for the throne of emperor, or the fight to own the world. The ancient Chinese religion was to match the emperors to heaven. Emperor Zhuanxu, the offspring of Emperor Huang, was matched to Polaris. Gonggong, the offspring of Emperor Yan, fought with Zhuanxu for Polaris; therefore, Satan in Bible might be Gonggong.

CLOSE KINS OF THE XIAOHE BASED ON ARCHAEOLOGICAL FINDINGS FROM SICHUAN

The key elements corresponding to the Xiaohe Cemetery were discovered in the archaeological ruins in western China fall into broad categories as follows:

1. Sanxingdui Ruins locate by the south of River Ruo in Chengdu Plain, Sichuan Basin, whose archaeological time was from 2800 BC to 800 BC. They were considered as the center of religious worship as –

a. The statues featured high noses (Fig. 19.21a) and big eyes, in which the high nose pottery man obviously characterized with white race, and most men wearing earrings (Fig. 19.21b), hand and ankle bracelets;

b. A kneeling bronze statue (Fig. 19.21c) wore a headdress of 7 snakes with the face expression of demon. He should be a famous fighter, but his kneeling indicated that he was a defeater of the fight. According to the actual prototypes worshipped in Sanxingdui, this statue should be Satan, or Gonggong, who was the ancestor of the close relatives of Xiaohe people;

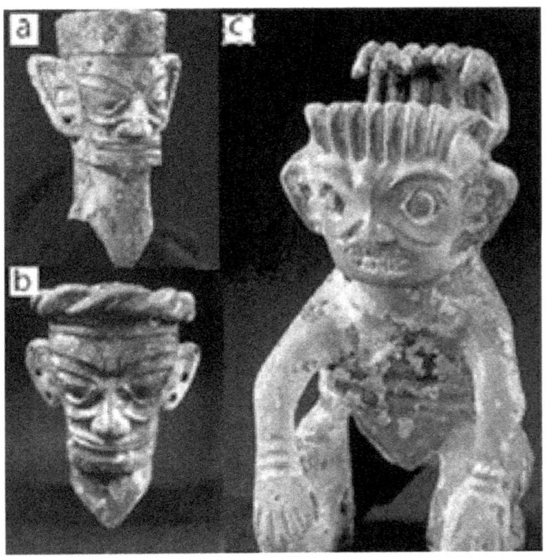

Fig. 19.21. Bronze statues in Sanxingdui

c. Some rituals have 7 – teeth, or 7 – gears, shown in Fig. 19.22a, b respectively;

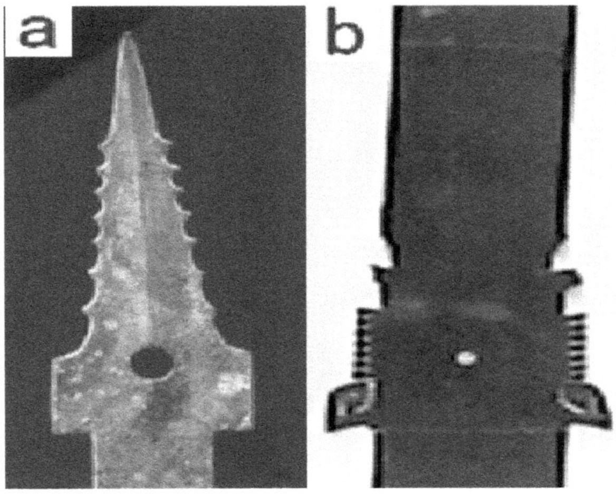

Fig. 19.22. Relics with number 7

d. Hoopoe was shown in the Fig. 19.23;

Fig. 19.23. Hoopoe in Sanxingdui

e. The dresses of Xiaohe people were high pointed felt hats (Fig. 19.24a), skirts (Fig. 19.24b), pointed alice boots (Fig. 19.24c), capes, which were consistent with the woman's dress on a jade from Sanxingdui (Fig. 19.24d). Some pottery and bronze statues in Sanxingdui also showed the priestess, a woman holding the jade, a woman with fists and a high nose woman, all dressed skirts, shown in Fig. 19.25a, b, c, d, respectively.

2. Maiping ruins located at River Ruo (Fig. 19.26c) was speculated as a vast-scale center city by archaeologists and could be traced back to 2500 BC. An aristocratic man wore 7-earrings. The archaeologists discovered that his skull was different from Han ethnic. This area was the place of white people with totem of dog. Some stone coffins were in rectangular shape with cover but no bottom (Fig. 19.26b). The coffins in Xiaohe also had cover but no bottom, in the shape of rectangular in early layer (Fig. 19.26c).

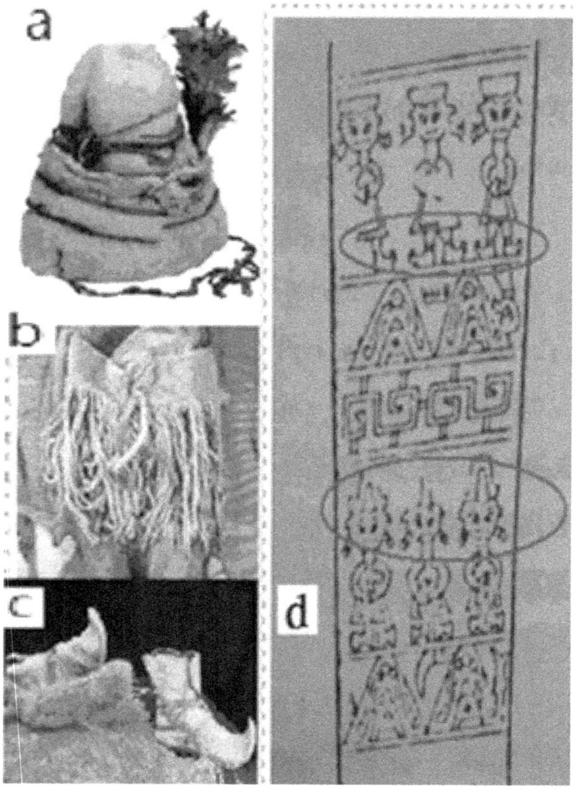

Fig. 19.24. Dresses in Xiaohe & Sanxingdui

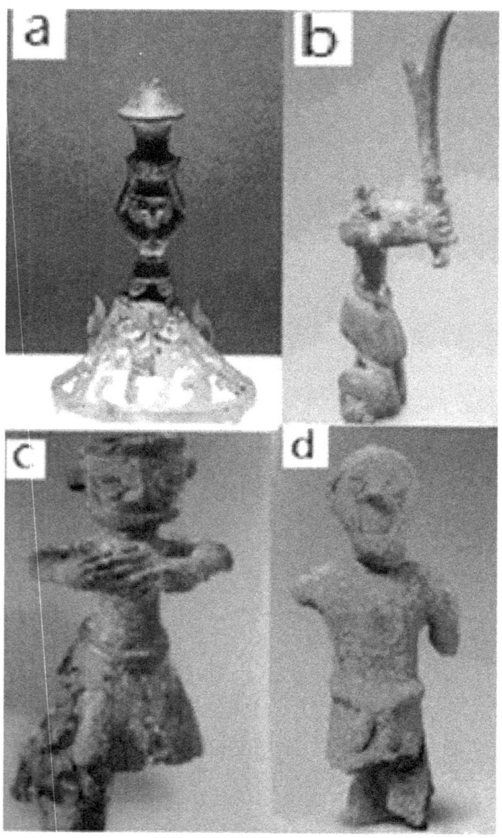

Fig. 19.25. Skirts in Sanxingdui

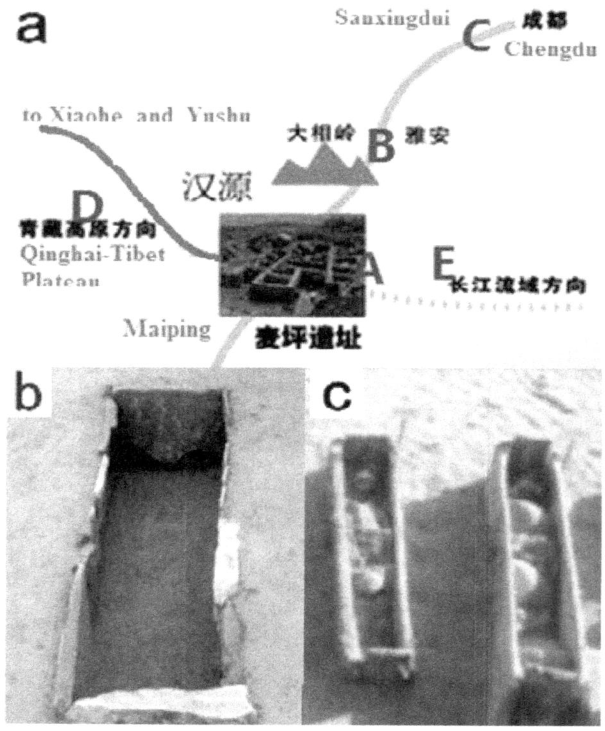

Fig. 19.26. Map and coffins in Maiping and Xiaohe

Remains of the ancient Qiang region

Most ethnics in western China have blood lineage of ancient Qiang, and their folklore and religion was highly consistent with those of Xiaohe people as can be qualified by the following points:

1. Qiang and Xiaohe both people displayed worship of genitals exaggeratedly:

a. Towers were used to indicate penis and pray for newborn boys;

b. The towers were classified male and female, the male were prismatic (Fig. 19.27a), and the female had a slot (Fig. 19.27b) like those in the Xiaoh;

c. The towers were 4 to 20 prismatic constructed by stones (Fig. 19.27c), which were consistent with the Xiaohe Cemetery.

2. Leaves symbolized the female and branches symbolized the male. The male plays pipe, and female plays leaf, therefore their sex could be identified by the sound. The folklore is still kept in ethnics of Qiang, Zang and Yi by River Ruo, including Maiping.

3. Small sticks and branches were inserted in the cliff seams to pray for children as the same reason why Xiaohe people erected the small branch bundles and burr bundles.

4. Qiang people called offspring as seed. The sexual activity of male was called as seeding, bastards as wild seeds, no children as no seeds.

5. Number 7 in folks

a. The coming-of-age ceremony of Qiang was held at the ages of 16 for male and 14 for femal;

b. 7-days memorial ceremony for dead was originated from Qiang in accordance of number 7 in Inner Canons by Emperor Huang, and now become the national ceremony for national martyr. The funerals in Qiang and in Xiaohe were highly consistent with key elements of red and figure 7.

Fig. 19.27. Towers of Qiang

Fig. 19.28. Sticks and branches inserted in cliff seam

6. '7' in ethnic totems and in clothing is seen on:

a. The necklaces of Qiang girls called "seven stars with moon" (Fig. 19.29a), and tippets of Naxi girls was called company of moon and stars (Fig. 19.29b), which consists of a circle or semicircle indicating moon, 7 small circles or semicircles symbolizing 7 stars, and 7 straight bundles meaning light beams.

Fig. 19.29. Girls' dress of Qiang and Naxi

b. Some Zang and Yi girls wear headdress with 7 red coral pendent chains (Fig. 19.30a, b), and smoking pipes of Yi girls have 7 x 2 (14) falling spikes (Fig. 19.30c). Some Zang women wear headdress with 49 red coral rings (Fig. 19.30d).

Fig. 19.30. Ornaments with number 7

7. Totem of Yi has 14 light beams. Totem of Jino (Fig. 19.31) has the inner with the sun and 12 earthly branches, the middle with 14 triangular cups to represent vulva, and the outer with 16 teeth to symbolize the penis.

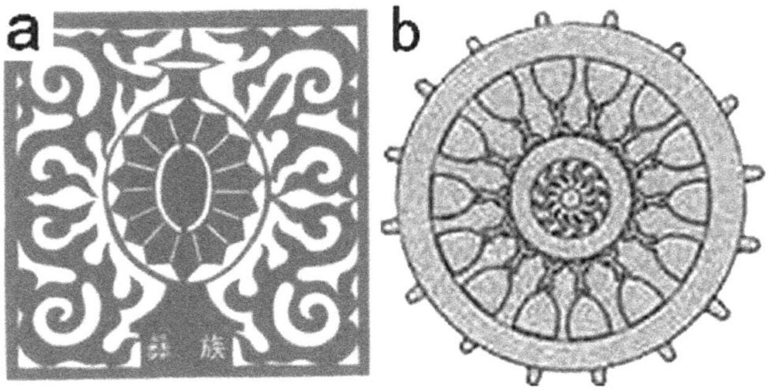

Fig. 19.31. Totem of Yi and Jino

8) High-pointed hat and Phi felt:

a. Yi people's caped (Fig. 19.32-a), snake-shape headdress (Fig. 10.32-b) and earrings are shown in the figure.

Fig. 19.32. High-pointed hat and Phi felt

b. Fig. 19.33 shows that Jino people wear peaked caps.

Fig. 19.33. Peaked caps of Jino

SUMMARY

Naxi women wore high pointed hats until Qing Dynasty. Qiang people wear high pointed hat in a funeral. The Xiaohe cemetery was a place of reproductive worship with HSEB, fabulously displayed the worship for resurrection and the pray for children. The vegetation architecture "sticks and leaves" was used to indicate reproductive process; small twigs, grass thorns and seeds.

To display the reproductive purposes; reproductive elements, red and 7, in Inner Canons by Emperor Huang to show the reproductive ability; numbers 7 and 12+1 to introduce the reproductive idea, to implement the life symbol + and offspring symbol ♀. This reflects a fact that Xiaohe people believed in sun worship and ancestor worship, mastered astronomical knowledge of 12 earthly branches, 12 zodiac signs, and the theory of life sciences of "female blood and male sperm" and "7 of women and 8 of men" in Inner Canons by Emperor Huang.

The study also found that the civilizations of Xiaohe, Maiping and Sanxingdui were highly similar and intersected. At the same time, the folklore in Xiaohe and in Qiang area belonged to same origin. It is suggested that the mankind rebuilt the original religion in the southeast of Qinghai-Tibet Plateau for refuge, used the same astronomical coordinates, 12 earthly branches, to form the highly consistent civilizations. The civilization of Xiaohe and civilization of Chinese originated from same root, and Xiaohe people were the female bird tribe belonged West Queen, called as Ruomu in Xia Dynasty and earlier, as White Qiang in Shang Dynasty, and as Ruo Qiang in the time of the warring states.

References

CUI, HENGSHENG. 2001. *Concise Dictionary of Oracle-bone Inscriptions*. Anhui Education Press.

FENG, XIAOFENG. 2010. Maiping People: the Lead of Cultural Evolution. History. Vol. 11.

FOLKE, BERGMAN. 1997. *Archaeological research in Sinkiang*. Translated by Wang Anhong. Urumqi, Xinjiang People Press.

IDILIS, ABUDURESULA, et al. 2004. The excavations of Xiaohe cemetery. Research of China's Frontier Archaeology Vol. 4.

LUO, GUANGDE. 2010. Hanyuan Archaeology: The Strange Underground Landscape. Ya'an Daily. August 29.

MA, NING. 2006. The Folklore of Red of Qiang under the View of Anthropological Religion. Journal of Xihua University (Philosophy and Social Science). Vol. 6.

NAN, HUAI-CHIN. 2008. *On huangdi neijing and the science of life*. Beijing: Oriental Press.

NAN, XIANGHONG. 2004. Little River Cemetery Reappears. West China. Vol. 6.

WANG, BINGHUA. 2004. Reproductive worship: the core of the spirit culture of ancient mankind. Root Exploration Vol. 4.

20.
IDENTIFICATION OF A BREAST CANCER BRCA1 MUTATION IN WEST BENGAL, INDIA

Abhishikta GHOSH ROY, B.N. SARKAR
Anthropological Survey of India
abhishikta.gr@gmail.com; drbnsarkar@yahoo.com

R. ROY
Cancer Centre Welfare Home and Research Institute

A.R. BANDOPADHYAY
Department of Anthropology, Calcutta University, India
abanthro@caluniv.ac.in

Abstract: *The development of the breast gland takes many years longer than any other organs and breasts more susceptible to cancer, which is one of the leading causes of women's death to the count of 23%, and the BRCA1 mutation carriers are at a lifetime risk on the order of 60-80%. So far, there were a few reports from Asia but none from West Bengal. We noted a significant difference in incidence between urban and rural population of India attributed to socio-cultural and environmental factors. Present study tested 110 patients with 122 controls (no family history) for BRCA1 econ 13 by PCR amplification and then by direct sequencing. We identified a novel (Pro1430Ser) variant among 18 patients and 2 controls. The identification of the novel mutation has implications in developing a database for the population useful for detection and early prognosis of the disease.*

Keywords: *breast cancer, BRCA1, mutation, Bengali, India*

INTRODUCTION

Evolution has been tinkering with the breast all the way down to the cellular level. The survival of a species depends on feeding its young, thereby making the breast critical to the perpetuation of the human race. Making an organ is a very delicate process. The development of the breast gland takes many years longer than any other organs. The special evolutionary features of the breast cells involve estrogen receptors which are places in the cell that receive messages from the estrogen hormone that usually signal the time to start to grow. Therefore, in addition to these biological factors that have made breasts more susceptible to cancer, there are a number of biological reasons for the breast to be more vulnerable to cancer other than any other organ.

Breast cancer has become the most frequent female malignancy with an estimated 422,000 new cases annually being the leading cause of female mortality worldwide (Ferlay et al., 2008). The age adjusted rates vary with different regions and cancer registries with increasing incidences in Indian women with breast cancer (Consolidated Report of Population based Cancer Registries (2001-2004) (Tab. 20.1). Breast cancer being the commonest among women in urban India, is the second most common cancer among Indian women (Samson et al., 2007). BRCA1 mutation carriers have a slightly elevated risk of developing breast cancer. Though it is possible that BRCA1 may interact with estrogen in breast carcinogenesis, but the exact mechanisms are still unclear why BRCA1 mutation carriers develop cancers of these hormonally regulated organs (Lee et al., 2008).

Tab. 20.1. A consolidated report of population based Cancer Registries (2001-2004) on Indian women with breast cancer

City	Age Adjusted Rate
Chennai	26.6
Delhi	29.2
Mumbai	27.5
Bangalore	27.5
Bhopal	22.1
Ahmedabad	7.6
Kolkata	26.8

Approximately 5-10% of breast cancer and 25% of the early onset of breast cancer is characterized by genetic susceptibility resulting from highly penetrant germline inactivation in cancer predisposing genes (Palma et al., 2006). A number of genes have been identified that are associated with breast cancer that has provided means to begin identifying individuals and families with an increased risk. BRCA1 was the most highly anticipated breakthrough in cancer genetics through cloning in the year 1994 (Miki et al., 1994). Mutations in BRCA1 gene results in loss or inactivation of wild-type allele, resulting in expression of a non-functional protein and loss of cell cycle control and DNA repair mechanism in women with tumorigenesis (Welesh and King, 2001).

Estrogen promotes cell proliferation and acts as a genotoxic agent, through its metabolites generating mutagenic DNA damage in breast carcinogenesis (Yager et al., 2006; Cavalieri et al., 2000).

BRCA1 is involved in several cellular functions important in carcinogenesis including DNA damage repair as well as cell cycle checkpoint (Yoshida et al., 2004). In the absence of BRCA1 functioning, the cancer promoting effects of estrogen would be even stronger (Fan et al., 2001). There are likely to be other low risk genes that are distributed more frequently in the population, which interact with environment and lifestyle factors, which results in breast cancer (Samson et al., 2007).

Breast cancer shows uneven geographical distribution in occurrence reflecting the influence of local environmental conditions, lifestyle, hormonal/reproductive pattern and genetic predisposition in the development of the disease (McPherson et al., 2000; Parkin et al., 2001). Breast cancer being second common cancer in the world, it ranks as fifth cause of death, because of its good prognosis, average in developed countries is 73% and 57% in developing countries.

Information on BRCA1 mutation in West Bengal, India is still lacking. Association between breast cancer risk and reproductive and lifestyle variables is still unknown among the Bengali Hindu Breast Cancer patients of West Bengal, India. Therefore, the present study is an attempt to understand the possible risk factors associated with the BRCA1 mutation. This is the first report from West Bengal, thus it will be a hopeful promise in modifying those factors to prevent breast cancer occurrence. Some of the breast cancer risk factors can be modified either through behavioral or environmental changes, whereas genetic and inherited mutations remains unchanged.

MATERIALS AND METHODS

Collection of samples

The subjects included in this study are 108 patients with histopathologically confirmed Breast Carcinoma visiting the Cancer Centre Welfare Home and Research Institute, Kolkata, India, from April, 2010 to December, 2011. Clinical data and detailed family history of each patient were collected with the help of collaborating clinicians after clinical examinations. Ethical approval of the research project using human subjects was obtained from the Institutional Ethical Committee of Anthropological Survey of India, Ministry of Culture, and Government of India. Patient samples were mainly collected from The Cancer Centre Welfare Home and Research Institute, Kolkata and National medical College and Hospital, which are the main referral centers for cases related to Breast Carcinoma. The demographic data (name, age, caste, address, origin, occupation, family history, educational status, etc) and reproductive (age at menarche, regularity of menstrual periods, age at menopause, parity, number of issues) and lifestyle (post menopausal hormone therapy, abortions, use of oral contraceptives, alcohol consumption and smoking practices) data were collected using schedule method. Age at menarche was defined as the chronological age when the women first had her menses; age at menopause was defined as the chronological age when the natural period ceases without any effect of medical or pathological interference. Another 127 controls were collected; those were age sex and ethnic group matched. Controls didn't have any family history of breast cancer.

DNA Isolation: Approximately 5mL peripheral blood samples were collected in BD Vacutainer K2 EDTA (6 mL) with written and informed consent from patients with Breast Carcinoma, and from normal individuals as controls, without any history of breast carcinoma. Genomic DNA was prepared from fresh whole blood by using the conventional phenol-chloroform method (Sambrook and Russel, 2001). Genomic DNA was dissolved in TE (10 mM Tris-HCl and 0.1 mM EDTA, pH 8.0).

PCR Amplification: Polymerase chain reaction (PCR) was carried out to amplify exon 13 and adjacent flanking region in a total volume of 10.0 μL containing 40-100 ng genomic DNA, 0.4 mM of each primer, 0.2 mM of each dNTP, 0.5-1.5 mM of $MgCl_2$ (as appropriate), and 0.2 unit of Taq polymerase (Invitrogen, Carlsbad, CA) in a Thermocycler (GeneAmp-9700; PE Applied Biosystems, Foster City, CA). Annealing temperature is calculated based on Tm of the primer pairs. The exon 13 was amplified using the designed primer:

F: AATGGAAAGCTTCTCAAAGTA;
R: ATGTTGGAGCTAGGTCCTTAC. PCR amplified DNA fragments were analyzed on 2% agarose gel and then visualized by ethidium bromide staining.

Mutation and Polymorphism Detection

The PCR products free of contaminating bands due to non specific amplification were directly sequenced in forward and reverse direction in DNA Analyzer 3730 (Applied Biosystems, USA). Nucleotide changes were detected by comparing sequence obtained in chromatogram with the normal BRCA1 gene sequence using pair-wise BLAST (Tatusova et al., 1999) and *SeqScape* software v2.5.

Appropriate statistical techniques have been applied for analyzing the data using the statistical software SPSS 17.0 version.

Tab. 20.2. Distribution of the SNP

Subjects	CC	CT	Total	P value	OR unadjusted	P value
Patients	93	15	108	0.0053 2 - 7.772	0.2000 (0.0643 to 0.6224)	0.0055
Controls	124	4	128			
	217	19	236			

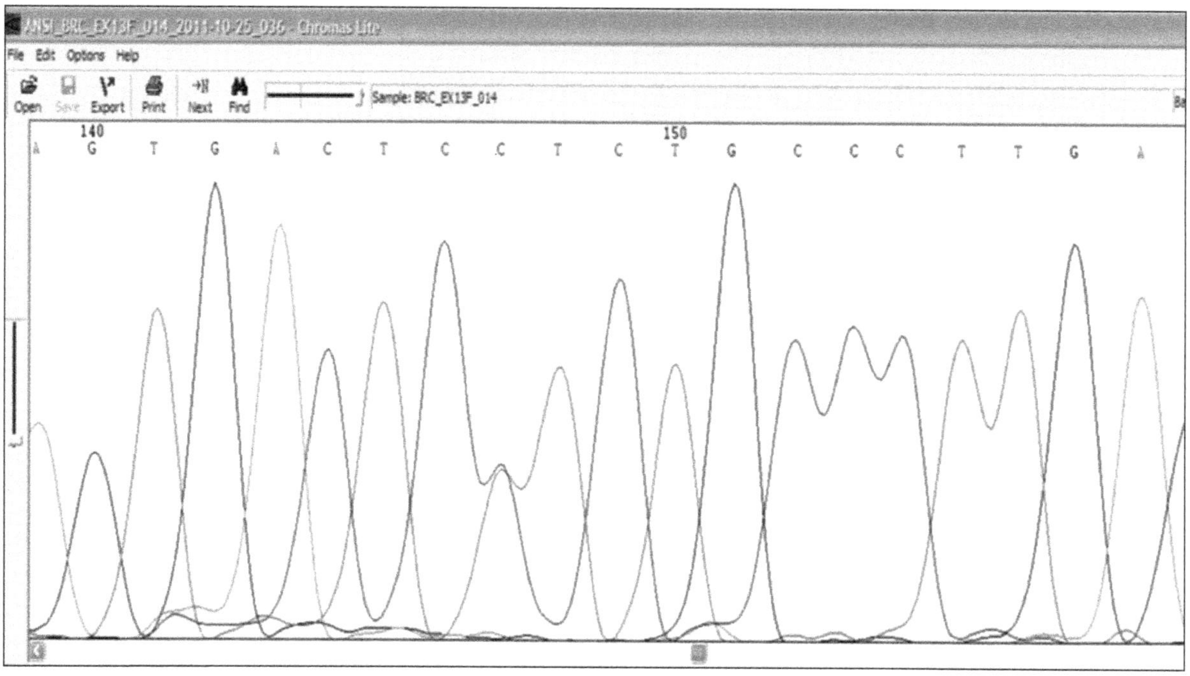

Fig. 20.1. DNA sequencing of BRCA1 gene

RESULTS

A total of 108 breast cancer patients with histopathologically confirmed carcinoma of the breast and 128 controls without any family history of breast cancer were included in the present study. The age of the breast cancer patients ranged from 30 to 72 years with a mean age 54.04 (SD 10.38) years, and mean age of the controls were 54.10 (SD 8.65), ranging between 32 to 72 years. Exon 13 of BRCA1 has been analysed by amplification using designed primer followed by direct sequencing in ABI 3730. DNA sequencing results (Fig. 20.1) confirmed the presence of a variant in axon 13 of BRCA1 relevant number of report exist of BRCA1 germline mutations, with or without a strong family history of the disease. These studies, including the case herein described, underline the importance of mutation screening.

Considering the Indian history and taking into account the multi-ethnic, multi religion status and population structure of this country, the results of this initial study suggest that the spectrum of mutation in exon 13 of BRCA1 gene in this population may differ from what has been observed in other Indian population and elsewhere. Further studies on mutational analysis of the gene will help in identifying the relevant mutation/ polymorphisms and the multiple structure-based alignments of the native structure of these polymorphisms may provide knowledge of the biological effects of the protein corresponding to polymorphisms to help predict the aetiology of breast cancer.

Socio demographic characteristics of the participants have been presented in Tab. 20.3 and the DNA sequencing of BRCA1 gene in Fig. 20.1. A C-T transition was found in codon 1437 which results in the change of amino acid from Proline to Leucine (Fig. 20.1).

Tab. 20.2 demonstrated that 13% of the patients had significant (p=0.0053) change in SNP than the controls (3.125%). The odds of having the SNP for breast cancer patients is 0.2000 times higher (95% CI: 0.0643-0.6224) than those who had no cancer and devoid of family history of breast cancer. This novel variant (ss527767420 has already been submitted to NCBI gene Bank.

The present study revealed no significant differences in age and caste and educational status between cases and controls. But occupation (p=0.0001) and family history (p=0.0001) demonstrated significantly in single women (11.1%) and as well married women (88.9%) among the breast cancer patients in comparison to controls.

Tab. 20.3. Socio-demographic characteristics of the subjects

Socio demographic characteristics	Cases		Controls		P value
	No.	%	No.	%	
Marital Status					
Ever Married	96	88.9	127	100	0.0015
Never Married	12	11.1	0	0	
Occupation					
Home maker	103	95.4	61	47.7	0.0001
service	2	1.9	32	25	
others	3	2.8	35	27.6	
Caste					
Brahmin	23	21.3	30	23.4	0.5058
Kayastha	60	55.6	67	52.3	
Baidya	15	13.9	23	18	
Baishya	8	7.4	4	3.1	
SC	2	1.9	4	3.1	
Educational Level					
Non-literate	3	2.8	4	3.1	
Primary level	11	10.2	9	7.0	0.7902
Secondary	26	24.1	34	26.6	
Hr Secondary	26	24.1	37	28.9	
Diploma/Degree	42	38.9	44	34.4	
Family History					
Yes	72	66.7	0	0	0.0001
No	36	33.3	128	100	

DISCUSSION

We analysed 110 breast cancer patients and 128 controls, we have identified a novel variant (ss527767420) in exon 13 of BRCA1 gene. Many studies have shown that there is a substantial inter-individual variability in cancer risk among BRCA1 mutation carriers. Even among those carrying the same deleterious mutation, there is large variation in age at diagnosis and cancer type. This suggests that there are additional genetic and environmental factors that modify risk; therefore such study attracts medical and genetic attention. So, the present study attempted to understand the possible predictor variables associated with BRCA1 mutation carriers. Inherited mutations in the BRCA1 and BRCA2 tumour suppressor genes are among the strongest genetic risk factors for breast cancer. Therefore, the present study could be taken as one of the early prognosis for the detection of the mutation and modification of the lifestyle factors among the Bengali Hindu females of West Bengal, India.

Acknowledgement

This research work is supported by the grant from Anthropological Survey of India, Ministry of Culture, and Government of India. I am thankful to the patients with Breast Carcinoma and control subjects for voluntarily taking part in this research work and donating their blood samples.

References

ATKINSON, H.G. 2003. Alcohol's "darker side". A drink a day may raise a woman's risk of breast cancer. *Health News* 9(1):4.

COLLABORATIVE GROUP 1996. On Hormonal Factors in Breast Cancer. Breast Cancer and hormonal contraceptives: collaborative reanalysis of individual data on 53,297 women with breast cancer and 100239 women without breast cancer from 54 epidemiological studies. Collaborative Group on Hormonal Factors in Breast Cancer. *Lancet* 347:1713-27.

CONSOLIDATED REPORT 2001-2004. Population Based Cancer Registries National Cancer registry Programme, *Indian Council of Medical Research.*

FAN, S., MA, Y.X., WANG, C. et al. 2001. Role of direct interaction in BRCA1 inhibition of estrogen receptor activity. *Oncogene* 20:77-87.

FERLAY, J., PARKI, D.M., STELIAROVA-FOUCHER, E. 2008. Estimates of the cancer incidence and mortality in Europe in 2008. *Eur. J. Cancer* 46 (4):765-81.

HAILE, R.W., THOMAS, D.C., MCGUIRE, V. et al. 2006. BRCA1 and BRCA2 mutation carriers, oral contraceptive use, and breast cancer before age 50. *Cancer Epidemiol. Biomarkers Prev.* 15(10):1863-70.

LEE, E., MA, H., ROBERTA, M.C. et al. 2008. Effect of reproductive factors and oral contraceptives on Breast Cancer risk in BRCA1/2 Mutation Carriers and Non Carriers: Result from a Population – Based Study. 17(11):3170-78.

LORD, S.J., BERNSTEIN, L., JOHNSON, K.A. et al. 2008. Breast Cancer risk and hormone receptor status in older women by parity, age of first birth, and breastfeeding: a case-control study. *Cancer Epidemiol. Biomarkers Prev.* 17(7): 1723-30.

MARQUIS, S.T., RAJAN, J.V., WYNSHAW-BORIS, A. 1995. The developmental expression of BRCA1 expression implies a role in differentiation of the breast and other tissues. *Nat Genet.* 11:17-26.

MCLAUGHLIN, J.R., RISCH, H.A. et al. 2007. Reproductive risk factors for ovarian cancer in carriers of BRCA1 or BRCA2 mutations: a case-control study. *Lancet Onc.* 8:26-34.

MCPHERSON, K., STEEL, C.M. & DIXON, J.M. 2000. Breast cancer epidemiology, risk factors and genetics. *Br. Med. J.* 321: 624-628.

MIKI, Y., SWENSEN, J., SHATTUCK-EIDENS, D., FUTREAL, P.A., HARSHMAN, K. et al. 1994. A strong candidate for the breast and ovarian cancer susceptibility gene BRCA1. *Science* 266 (5182):66-71.

NAROD, S.A., DUBE, M.P., KLIJN, J. et al. 2002. Oral Contraceptives and the risk of Breast Cancer in BRCA1 and BRCA2 mutation carriers. *Journal of the National Cancer Institute* 94 (23):1173-1179.

NEWCOMB, P.A., STORER, B.E., LONGNECKER, M.P. et al. 1996. Pregnancy termination in relation to risk of breast cancer. *JAMA* 275:321-322.

PALMA, M., RISTORI, E., RICEVUTO, E., GIANNINI, G., GULINO, A. 2006. BRCA1 and BRCA2: the genetic testing and the current management options for mutation carriers. *Crit. Rev. Oncol. Hematol.* 57(1): 1-23.

PARKIN, D.M, BRAY, F., FERLAY, J. & PISANI, P. 2001. Estimating the world cancer burden: GLOBOCAN. *Int. J. Cancer* 94: 153-156.

RAJAN, J.V., MARQUIS, S.T., GARDNER, H.P. & CHODOSH, L.A. 1997. Developmental expression of BRCA2 co localizes with BRCA1 and is associated with proliferation and differentiation in multiple tissues. *Dev Biol.* 184:385-401.

RENADAR, T., PERETZ, T., LIFCHITZ, M. & SHAVIT, L. 2005. BRCA1 and BRCA2 germ-line mutations and oral contraceptives: to use or not to use. *The Breast* 14: 264-268.

SAMBROOK, J. & RUSSEL, D.W. 2001. Molecular cloning: a laboratory manual. In: Irwin N. (ed) Science. Cold Spring Harbor Press, Cold Spring Harbor, NY.

SAMSON, M., SWAMINATHAN, R., RAMA, et al. 2007. Role of GSTM1 (Null/Present), GSTP1 (Ile105Val) and P53 (Arg72Pro) Genetic Polymorphisms and the Risk of Breast Cancer – A Case Control Study from South India. 8:253-257.

TATUSOVA, T. & MADDEN, T. 1999. BLAST 2 sequences a new tool for comparing protein and nucleotide sequences. *FEMS Microbiol. Lett.* 174:247-250.

WELCSH, P. & KING, M.C. 2001. BRCA1 and BRCA2 and the genetics of breast cancer and ovarian cancer. *Hum Mol Genet.* 10: 705-713.

WINER, E.P., MORROW, M., OSBORNE, C.K. & HARRIS, J.R. 2000. Cancer of the breast; Section 2, Malignant tumors of the breast, in: T.V. Devita (Ed.), Cancer*: Principles & practice of* oncology, sixth ed., J.B. Lippcott, Philadelphia, pp. 1651-1659.

YAGER, J.D. & DAVIDSON, N.E. 2006. Estrogen carcinogenesis in breast cancer. *N. Engl J Med.* 354:270-82.

YOSHIDA, K. & MIKI, Y. 2004. Role of BRCA1 and BRCA2 as regulators of DNA repair, transcription, and cell cycle in response to DNA damage. *Cancer Sci.* 95: 866-71.

21.
DEPLETING BIOSPHERE RESERVES: TRADITIONAL AND MODERN CONCERNS IN INDIA

Umesh KUMAR
Anthropological Survey of India, 27, Jawaharlal Nehru Road, Kolkata – 700016, India
umeshkumarpb@gmail.com

Abstract: *Biodiversity is the totality of genes, species, ecosystems and habitats in a region. Environment rich in biodiversity, offers the broadest array of options for sustainable human welfare and for adaptive change. But, increasing human interventions on the ecosystems due to industrialization and urbanization have accelerated biodiversity loss and altered ecosystems. The 'Man and the Biosphere' programme of UNESCO has aimed to facilitate resolution of the increasing conflicts between the people and the Protected Areas. Present study considers that the ideal approach to preserve overall ecosystem health, including species diversity, has to integrate the social fabric, environmental and economic canvas well demonstrated in Indian context where indigenous populations are living within the Biosphere Reserves. This is so as the forest-linked activities of the traditional societies are mediated through a rich traditional ecological knowledge, which has economical, ecological and socio-cultural dimensions. Therefore, societal perceptions of natural resources as a cultural resource – often seen through the conceptual framework of sacred species, sacred groves, sacred water bodies, sacred landscape or even sacred mountain system – need to be integrated in the management strategy. It is suggested that the people' participation through a 'bottom-up approach' involving stakeholders, is very effective in the management of the natural resources and conservation of the ecosystems in Biosphere Reserves.*

Keywords: *biosphere, reserve, India, conservation, conflicts, traditional knowledge, participatory approach*

INTRODUCTION

Biodiversity is the totality of genes, species, ecosystems and habitats in a region. Environment rich in biological diversity, offers broadest array of options for sustainable human welfare and for adaptive change. An environment which is rich in biodiversity constitutes resources upon which humanity depends to fulfil its fundamental needs. Increasing human interventions on the ecosystems have accelerated the process of biodiversity loss which have serious economic and social costs for any country. The experiences of the past few decades have shown that as industrialization and economic development s take place in the society, the patterns of production, consumption and needs change, strain and even alter ecosystems. If one traces the linkages of biodiversity, it is observed that human population and its interventions have been the major factors for enhancing the extinction process. Therefore, the conservation and sustainable use of biodiversity are fundamental to sustainable development. The mapping of mega-diversity zones indicates that most of the mega-biodiversity zones are found in South America, Central Africa and India/South China. An assessment of biological diversity at world level (Mittermeier, 1988; Hannah & Lohse, 1993) reveals that Asia is the most threatened continent.

Approach to understand and preserve biodiversity

The best approach to understand biodiversity is probably to look at it in a hierarchical manner in which biological organisms are organized. The smallest unit starts from the diversity contained in the genetic material of individual organisms and then goes on to encompass the biological communities in which species are organized and on to the ecosystems in which they exist. Biodiversity is recognized at three levels viz., genetic, species and ecosystem (Fig. 21.1); the three components are dynamic in space and time. The genetic composition of the species changes over time in response to the natural as well as to the human-induced selection process. The ecosystems strongly respond to external dynamics and internal stresses.

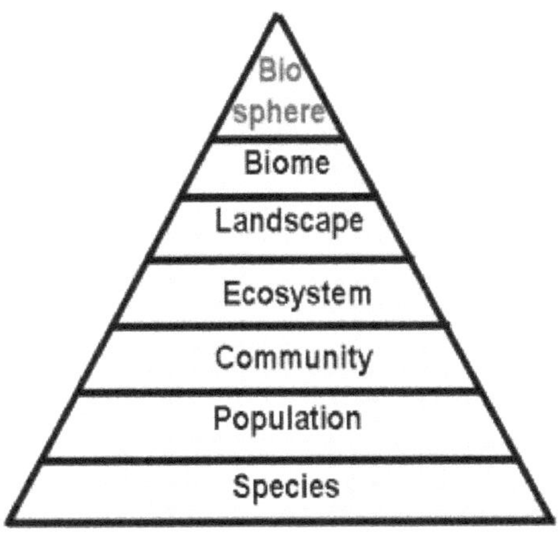

Fig. 21.1. Hierarchy of Biological Organization based on IIRS Report on Biodiversity Characterisation at Landscape Level, 2001

Ecological systems do not exist as discrete units but represent different parts of a natural continuum in the form of landscape. The landscape can also be considered as a higher level of biological complexity, immensely useful for understanding various complex processes. While biodiversity is usually considered at the species level, the maintenance of biodiversity requires management at higher levels of the organization, particularly at the landscape scale, which also includes human beings.

In order to preserve the species diversity very effectively, the management plans must preserve the habitats and landscape structures needed by the target species, rather than preserving the species in isolation from the larger and potentially changing environment. Management practices aimed directly at a particular species run the risk of losing ecosystem functions which might actually be crucial for the target species, but remained unknown when the management plan was created. Furthermore, maximizing benefits for one species may threaten others. Therefore, the idea l approach is to preserve overall ecosystem health, including species diversity, best accomplished by integrating it into the fabric of social, environmental and economic canvas.

So, a process of sustainable development begins when man and environment are integrated together, and leads to a better future for all the living being s. Such considerations gave birth to the concept of Biosphere Reserve in the "Biosphere Conference" organised by UNESCO in 1968. This was the first inter-governmental conference to examine how to reconcile the conservation and use of natural resources, thereby foreshadowing the present-day notion of sustainable development. The conference enabled launching of the *Man and the Biosphere* (MAB) Programme of UNESCO in 1970 which aimed to facilitate resolution of the increasing conflicts between the people and the Protected Areas.

BIOSPHERE RESERVE AS A CONCEPT

Biosphere Reserves are the areas of terrestrial and coastal ecosystems promoting solutions to reconcile the conservation of biodiversity with its sustainable use. They are internationally recognized and nominated by national governments and remain under sovereign jurisdiction of the states where they are located.

The UNESCO's framework stipulates that Biosphere Reserves (BRs) should be representative of a bio-geographic province and should have appropriate zoning system with legally constituted and minimally disturbed core zone. In addition, the site should contain unique and pristine biodiversity, ecosystems and landscape suitable to explore and demonstrate approaches to sustainable development and should be of appropriate size to serve three functions, viz., conservation, development and extension of logistic support. It is aimed to bring at least one representative site in the country within each bio-geographic province under the network of Biosphere Reserves.

BRs are designated to deal with important questions of reconciling the conservation of biodiversity, the quest for economic and social development and maintenance of associated cultural values. Some important aspects of Biosphere Reserves delineate conservation of landscapes, ecosystems, species and genetic variations; economic development, which is culturally, socially and ecologically sustainable and extension of logistic support for research, monitoring, education and exchange of information related to local, national and global issues.

Launched in 1970, *Man and Biosphere* (MAB) programme covers different ecosystem types from mountain to the sea, from rural to urban systems as well as social aspects. MAB has proposed an interdisciplinary research and capacity building aiming to improve the relationship of the people with the environment. In order to achieve the stated objectives it targets study if biodiversity from the ecological, social and economic dimensions. In nutshell, the Biosphere Reserve (BR) is expected to fulfil three basic functions, which are complementary and mutually reinforcing:

1. *Conservation function* – to contribute to the conservation of landscapes, ecosystems, species and genetic variation;
2. *Development function* – to foster economic and human development which is socio-culturally and ecologically sustainable;
3. *Logistic function* – to provide support for research, monitoring, education and information exchange related to local, national and global issues of conservation and development.

BIOSPHERE RESERVES (BR) IN INDIA

The Ministry of Environment and Forest, Government of India has launched the Biosphere Reserve (BR) Programme since 1986 with the specific objectives of this Programme as follows:

1. To conserve the diversity and integrity of plants and animals within the natural eco-system;
2. To safeguard the genetic diversity of species on which their continuing evolution depends;
3. To provide areas for multi-faceted research and monitoring;
4. To provide facilities for research and training, and;
5. To ensure the sustainable use of natural resources through most appropriate technology for improvement of economy and living standard of local people.

Presently there are 18 biosphere reserves in India. These are:

1. NILGIRI BIOSPHERE RESERVE
2. GULF OF MANNAR
3. NANDA DEVI
4. SUNDERBAN
5. SIMILIPAL

6. PACHMARHI
7. NOKREK
8. GREAT NICOBAR
9. MANAS
10. DIBRU-SAIKHOWA
11. DEHANGDIBANG
12. KHANGCHENDZONGA
13. AGASTHYAMALA
14. ACHANAKMAR-AMARKANTAK
15. KACHCHH
16. COLD DESERT
17. SESHACHALAM
18. PANNA

ISSUES LINKED WITH BR DEVELOPMENT

The BRs of India are also the abodes of many traditional societies who have always been an integral part of the ecosystem, functioning there and living close to the nature. Living in the areas rich in natural resource, the traditional societies fulfil many of their livelihood requirements from the rich biodiversity around them. The ecosystems, such as forest not only meets a variety of livelihood needs of the forest dwelling societies by way of timber and non-timber forest products. They also sustain traditional shifting cultivation and agricultural practices contributing to the food security of the people within the BRs.

In the context of India, human populations living within the Biosphere Reserves has always been a matter of conflicting interests and debate. These considerations prompted UNESCO (1996) to move away from the earlier held view of concentric rings of core, buffer and transition zones, to a more flexible way of looking upon the local needs and situations. The flexibility and creativity in the Biosphere Reserve's concept need to be seen from the point of view of both biodiversity as well as the cultural diversity. The benefits of cultural diversity at societal level are comparable to those accruing from the biodiversity. It is important to allude that while the issues of biodiversity have been addressed at length and caught wide attention, the cultural diversity has not been touched upon and left to the point of oblivion.

Building upon the traditional knowledge system

The forest-linked activities of the traditional societies are mediated through a rich traditional ecological knowledge (TEK). TEK has economical, ecological and socio-cultural dimensions. A participatory approach involving stakeholders will prove very effective in the management of natural resources and ecosystem conservation in BR. Such an approach must imply traditional knowledge and traditional conflict prevention strategies in the management of BR resources. For example, in certain parts of North Sikkim, close to the BR, local communities have a system known as zumsha (literally 'gathering of people') *concerned* with the social, developmental and environmental issues of local significance (Kumar, 2013). The local self-governance system of Lachen valley is unique. This system is called the Dzumsa system – a traditional institution entrusted with task of administering and organising activities within a given territory. The Dzumsa general council is *compose*d of heads of the households, headed by a *Pipon*. This system was initially established during the first half of the 19th century to provide structure and cohesion to the societies and activities of the people operative in Lachen and Lachung blocks of Sikkim. In the year 1970s, when Indian Government initiated reorganisation of Sikkim's administration and introduced the Panchyatiraj system, the Dzumsa was officially recognised as system of self-governance in place of Panchyatiraj. This system continues to function even today in the Lachen and Lachung tribes. The zumsha system prescribes and regulates activities such as grazing, tree felling and of medicinal plants and herbs, taking full account of traditional ecological knowledge and imposing penalties for violations. Grazing is permitted only after taking complete care to retain the regeneration capacity of the range land. This includes allotment of a particular piece of land for grazing, determination of the number of the cattle allowed to graze, guidelines about the correct season, the times when cattle be left in/brought out of the forests. The Based on the traditional ecological knowledge, the Pipon recommends the right to fell the trees and collect medicinal plants and herbs considering little danger of over exploitation. Such factors as socioeconomic status of household and systematic allotment of area for collection, considering the capacity for ecological regeneration, are also taken into account in granting the rights of quantity, species, etc. The form of local traditional knowledge can also be integrated with the BR management to make it more efficient. Another such example is restriction on grazing in Nanda Devi BR (the Valley of Flower); it has led to undesirable spread of *Polygonal* sp. Since BRs are home to many traditional societies at various stages of their evolution, building upon TEK is crucial for addressing the issues pertaining to the BR management with a concern for sustainable livelihood needs.

Provisions of Alternative Livelihood

Human population living close to BR is highly dependent on forests in the buffer zones, be it for timber and non-timber forest products. Imposition of Biosphere principles in terms of conservation entails certain restriction on the use of BR resources which translates in to some cost for the people. Studies have shown that people are willing to accept and pay for the sake of conservation if their economic concerns are properly addressed. The studies (Mandal, 2007; Singh & Patil, 2007; Chandra et al., 2013) carried out in various BRs of India reveal that the conflicts crop up only when the pressure of conservation restrictions begins to affect socioeconomic conditions of people living in settlements at the fringes in the absence of viable alternative to their traditional dependence on the BR resources.

An effective conservation policy requires a strategy based on provision of alternative livelihood options and additional resources to defray the pressure and dependence on the BR resources for efficient management and conflict prevention. This can be achieved by capacity building in local communities to enhance the productivity, thereby reducing dependence on biosphere resources that are likely to be restricted.

For the success of BR programmes it is sine-qua-non that local communities must derive direct and indirect benefits. Provision of alternative livelihood options and additional resources to defray the pressure and dependence in the BR resources is also important for the efficient management and conflict prevention. This can be achieved by capacity building in local communities to enhance the productivity, thereby reducing dependence on biosphere resources that are likely to be subject to restrictions. In summary, the most important initiative to be implemented concurrently for effective conflict prevention necessitates the ecosystem protection has to be based on participatory approach, wherein the community be given responsibility for effective monitoring and exercise of traditional rights and forks of resource use.

Sacred Landscape and Conservation

Many traditional forest dwelling societies treat their habitat as sacred. People venerate and propitiate their local and regional deities. The sacredness begins from species level and extends up to the landscape level. It is true in case of the Lepcha and the Bhuti communities of Khangchendzonga and the Shompen of Great Nicobar BRs. It can be explained through an example of Sikkim – known as '*Demazong*', which means a "hidden valley full of treasure, fruits and flowers". It is the land of hydraulic culture where almost every water-body, be it small or big, is treated sacred. In fact, the sacredness begins at species level and reaches up to the landscape level with the Mountain Khangchendzonga acting as the guardian deity (Chandra et al., 2013).

The local deities are related to nature and its components. There are some festivals and rituals to be observed during the year. In the month of August, 'Panglhabsol' festival is celebrated and Buddhist lamas perform rituals and offer prayers to the Mount Khangchendzonga. The monk and devotees go up to Dzongri 'Dhaplha gang' to offer prayers and rituals to Mount Khangchendzonga for good harvest and for the whole region to be free of diseases and natural disaster s. There are altogether 109 existing large and small lakes in Sikkim which are regarded as sacred by the people. Besides the rivers, the springs are also regarded as sacred. People dare not pollute these water bodies. In fact, the people of Sikkim tell the outsiders about the sanctity of these water bodies, and warn them of the nature's wrath if they pollute them. Among these, seven laks, namely, Khecheoparli, Kathok Tso, Bar Cho Marpu, Phu Cho Karpu, Ka Bur La Tso and Dafuk Yum Tso, are considered seven bowls offering to the Mount Khangchendzonga. They are believed to have power of fulfilling the wish, and therefore, called as seven "wishing lakes". Besides the water bodies, there are four holy caves in the four directions of Sikkim, viz, Lhari Nying Phu in the North, Dechen Phu in the west, Khandu Sang Phu in the south and Bas Phu in the east. They are religiously protected by the people who also do not disturb the flora and fauna of the area.

Organic Relation between Man and Nature

Even at meso and micro level, every landscape be it high land or low land, every streams and big trees and caves are believed to be the abode of a guardian deity which (*Yulha, Zibda*) are propitiated with great devotion by the every human habitation and place, which has the footprints of human beings, is decorated with the *wheel of life,* eight lucky signs, prayer flags, chorten, etc. The prayer flags can be seen almost everywhere in Sikkim – on the hills, along the water bodies, beneath a big tree and on the building tops. It seems that sanctity to sacredness flows from the religious rituals. The religion is not an external element to the people but has brought together both man and the nature to interact on day-to-day basis. Consequently, it has led to the emergence and operationalisation of an organic relationship between man and the nature (Fig. 21.2). This relation is world apart from the Newtonian reductionist approach which put emphasis on the importance of parts. On the contrary, it is a holistic relation where man interacts with nature with respect treating it a living entity.

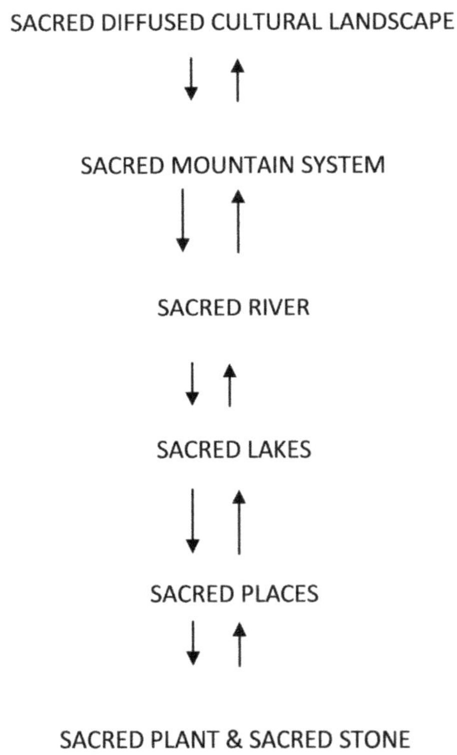

Fig. 21.2. Organic relation between man and nature

CONCLUSIONS

Presently, the management efforts have largely been centred around protection of the natural resources within BRs from the outside impacts. But, at the same time, the use of the resources by local communities residing in the BR, has also been restricted. Yet, the concept of sustainable use of resources has not taken shape due to lack of location specific management understanding. Of late, emphasis is laid on that the eco-development should also combat problems of community participation in forest eco-system management/rehabilitation directly. What more is required for the management of natural resources is the linkage with reinforcement of traditional system of land-based activities based on the traditional value system of local communities.

The BR management, therefore, necessitates not only understanding of ecological issues but also the socio-economic and cultural issues linked with the former. Understanding societal perceptions of natural resources as a cultural resource – often seen through the conceptual framework of sacred species, sacred groves, sacred water bodies, sacred landscape or even sacred mountain system – and integrate it in the management strategy is necessary for the success of the Biosphere Reserve programme. In order to achieve this, conservation-linked development strategy has to be based on a value system that people can understand, appreciate and accept. It, in turn, requires appropriate institutional arrangements for peoples' participation, through a 'bottom-up approach', ensuring that each household takes part in the decision making process at the lowest level in the hierarchy keeping into consideration the gender sensitivities.

Acknowledgments

I am thankful to Dr. A.R. Sankhyan for inviting my contribution and critically going through and improving the manuscript.

References

CHANDRA, R., MANDAL, D.B. & SINGH, A.K. (eds.) 2013. *Man in Biosphere: A Case Study of Khangchendzonga Biosphere Reserve*. Gyan Books Pvt. Ltd., New Delhi.

HANNAH, L. & LOHSE, D. 1993. Annual Report, Conservation Institutions. Washington D.C.

KUMAR, U. 2013. A Study in Ecology and Conservation in Khangchendzona Biosphere Reserve of Sikkim Himalaya. In *Man in Biosphere: A Case Study of Khangchendzonga Biosphere Reserve*. Chandra, R., Mandal, D.B. and Singh, A.K. (eds.). Gyan Books Pvt. Ltd. New Delhi, pp. 31-64.

MANDAL, D.B (ed.). 2007. *Man in Biosphere: A Case Study of Sundarban Biosphere Reserve*. Gyan Books Pvt. Ltd., New Delhi.

MITTERMEIRE, R.A. 1988. Primate diversity and the tropical forest: Case studies from Brazil and Madagascar, and the importance of the mega diversity countries. In *Biodiversity*, Wilson, E.O. (ed.), National Academy Press, Washington, D.C.

SINGH, A.K. & PATIL, S. (eds.). 2007. *Man in Biosphere: A Case Study of Nilgiri Biosphere Reserve*. Gyan Books Pvt. Ltd. New Delhi.

UNESCO. 1996. *The Biosphere Reserve*: The Seville Strategy and Statutory Framework of the World Network. UNESCO, Paris.

22.
ROCK ART IN INDIA: A DATA APPRAISAL

Somnath CHAKRAVERTY
Asiatic Society, Park Street, Kolkata – 700016, India
somraja2008@gmail.com

Abstract: *To delineate the preliterate and prehistoric societies in its totality, the study of rock art is a dependable source to reconstruct history of humankind since the prehistoric era as a continuum from the past to the living reminiscent cultures common among tribal societies living in relative isolation as and when their orthodox cultures maintain life-activities within traditionally set value-system. In the rock art sites of Central India, such as Bhimbetka, Adamgarh as well as in Pachmarhi, the earliest layer of paintings mostly represent larger figures of wild animals either depicted in red ocher, white or grey in colour represent a community which was closely similar to contemporary tribal societies, especially the pastorals and marginal cultivators. In all rock art sites and tribal depictions the figurative and non-figurative motifs are largely symmetrical and confirm to an egalitarian society.*

Keywords: *Rock Art, India, Bhimbetka, Adamgarh, Pachmarhi, Fischer's hypothesis*

INTRODUCTION

Study of rock art is primarily an area for an integrated and multi-disciplinary approach and the focal point is the attempt for retrieval of early rock art unveiling all possible aspects of human activities, ingenuity and gradual modifications developed as an adaptive mechanism. In fact, to delineate the preliterate and prehistoric societies in its totality, the study of rock art is perhaps one of the most dependable sources to decipher an authentic, rational and reconstructed history of humankind existed during the prehistoric era and thereafter. Compared to the direct portray of life and activities that are common in rock art, studies on the other archaeological evidences such as tool typology, analysis of artifacts etc, are indispensable supportive data that are reasonably hypothetic. Study of Rock Art do not necessarily mean to be an exposition only of a prehistoric society and culture but it is rather a cultural trait in continuum from prehistoric past to the living reminiscent cultures. Analogous form of visual imagery is more common among tribal societies living in relative isolation as and when their orthodox cultures maintain life-activities within traditionally set value-system.

Styles and socio-cultural conditions (Fig. 22.1) – both have significant role for application to decipher rock art. It has already been stated that the Rock art research is more successful when its approach is multi-disciplinary. Both in theoretical and technical applications, contribution from several disciplines of earth sciences, physical sciences, life-sciences and social sciences are essential. Objective of such collaborative projects is to develop a holistic view in an integrated form. Anthropological theories and hypothesis evolve from the empirical and regional studies which ultimately project the pan-human principles on varied aspects of society and culture. Art, particularly among the pre-industrial societies is indispensable element from the other aspects of the cultural fabric such as religion, folklore, social organization etc. Therefore, the cultural determinants and other principles may be useful to trace and for logical interpretation or decoding of prehistoric, ancient and other early art.

THE GLOBAL CONTEXT

Rock art research is initiated with successive phases including exploration, recording of the site with documentation of art imagery and other associated archaeological evidences. Then classification of data, analyses and its arrangements gradually reveal its physical realities. Ultimately, a serious researcher with an anthropological background will attempt for a social reconstruction with a note on the palaeoanthropological setting explaining biological and ethnic identities of the concerned groups. In the present scheme, the cultural dimension is only considered with a departure from all biological aspects to further correlate during later phase.

Mention of selected anthropological principles may be relevant in decoding of rock art. Particularly for designing this undertaken study, Robert Redfield's Folk-Urban Cultural continuum model (Redfield, 1956) and J.L. Fischer's hypothesis (Fischer, 1961) relating art Redfield's study helps us to understand the process and the stages of transformation from preliterate folk societies to the urban – based stratified societies. Redfield's bipolar concept of cultural continuum could be applied for analyses of a cultural continuity of some cultural elements of prehistoric art in contemporary performances of ethnic art.

Fischer (1961) had proposed a hypothesis relating art styles and socio-cultural conditions. This proposition had been substantiated by objective statistical tests and incorporating data of Barry's (1957) study in of psychology of child art and Murdock's (1957) "World ethnographic sample".

Fig. 22.1. Petroglyphs of Central India: animals depicted at Pachmarhi (M.P.) and animal foot marks art at Himgir (Orissa)

Tab. 22.1. Re-assessment of Fischer's hypothesis in Indian context (Fisher's Diagram, 1961)

			Rock Art Sites			Tribal Art-Ethnic Group		
	Category	**Identifying Feature**	**Pachmarhi (Madhya Pradesh)**	**Ulap (Odisha)**	**Isco (Jharkhand)**	**Korku (Madhya Pradesh)**	**Juang (Odisha)**	**Soara (Odisha)**
first hypothesis	egalitarian society	visual repeatition of the same element – more conventional and idealized figures	present	present	present	present	present	present
	hierarchical society	varied design – elements						
second hypothesis	egalitarian society	preponderance of maximum empty space	present			present	present	
	hierarchical society	minimum empty space within composition		present	present			present
third hypothesis	egalitarian society	art style more symmetrical	present	present	present	present	present	present
	hierarchical society	art style primarily asymmetrical						
fourth hypothesis	egalitarian society	preponderance of non-enclosed figures	present	maximum			present	
	hierarchical society	enclosed figures		few	present	present		present

A number of hypothetical polar contrasts, available in the art styles of 'different communities under study, have been deduced. It is revealed from the Fischer's hypothesis that certain socio-cultural conditions and functions act as determinants for studying the style of art. The first hypothesis expresses that the art available within an egalitarian tribal society has an association with visual repetition of same element. In other words, in egalitarian societies, figures or motifs become more conventional or idealized, following the artist's intimate experience in group harmony and own involvement in an integrated non-hierarchical society. On the contrary, the varied design elements are represented in the art of hierarchical societies. In the second hypothesis Fischer (1961) had associated the preponderance of maximum empty space with egalitarian societies, whereas the empty space is minimal in the art style present in the hierarchical societies. Postulations have been made in furtherance that the egalitarian societies are more scared about invasion by strangers. They maintain their identity by isolating themselves from other communities and so the inherent idea of isolation is motivated through symbolic creation of enough empty space in the panel as conceived by an artist.

The third hypothesis reveals that the art style of the egalitarian societies are more symmetrical as opposed to the fact that asymmetry is more common to the art of hierarchical societies. The fourth hypothesis points out the preponderance of enclosed figures in the hierarchical societies. Such boundaries are absent in the egalitarian societies. The social stratification or ranking prevails in hierarchical societies influence in the socio-psychological view point of the artist and consciously or unconsciously the idea is reflected in their art.

Besides the four above mentioned major hypothesis, Fisher had also connected the art style to the social organization. Accordingly, the proposition has been made to designate the straight lines, representing the male form, as opposed to curved ones for female. On the contrary, complex, non-repetitive design, representing a hierarchical society, should be associated with societies which strongly favour male solidarity in residence (Fischer, 1961: 84).

Fischer's hypothesis (1961) of course needs further empirical verification. Its proper assessment in practical situation, substantiated with appropriate data, may be verified. It may be an useful model for possible interpretation of rock art. Overall reconstruction of the concerned society may be attempted with the support of art and related archaeological materials of unknown past.

In recent times a significant attempt in decoding the rock art imagery has been attempted by Olsen (1985). In the study on Hovenweep rock art of Rocky mountain area in North America. Olsen had statistically compared the basic stylistic and thematic elements available between the rock art of Hovenweep area near Mesa Varde National Park of Utah state and the traditional art of several numerically dominated American Indian communities of the same region. The study has further traced the genesis, developmental pattern of art elements and chrono-cultural relationship between the art traditions of Hopi, Zuni and Pueblo Indian communities. In Native American reserves or territories in Rocky mountain area, the evidence of art either in the form of material culture or pictographs on rock surface is said to have continuity from prehistoric to modern time. Plenty of relics of art either on pottery or otherwise have been discovered in prehistoric and other archaeological sites. In addition, profuse rock art materials are available from the prehistoric rock shelters and other habitation areas. It is a general trend among the researchers in the field of rock art study in America, to identify and relate rock art with cultural tradition of the present Native American groups. Thus the concept of continuity of art tradition from prehistoric period to present day tribal communities is very much in existence in this part of the continent. Olsen (1985) had analysed the petroglyph and pictographs as a system of visual communications and also observed that initially the rock surfaces had been utilized as a medium to document information about economic ownership, ceremonial ritual, events, status or oral traditions which in due course gradually lost proper significance. *"More information than ownership or identity was included in the prehistoric rock art, such as seasonal information, religious significance or both"* (Olsen, 1985:134). It is also revealed from the same study that a significant number of design-elements of rock art are identical to different clan symbols of particular Native American groups inhabiting in the same area. Qlsen has also discarded the possibility of purposive attempt for sympathetic magic behind the hunting scenes. By making statistical correlations, it has been established that the prehistoric rock art of Anasazi Indian communities have gradually modified and the contemporary Hopi and Zuni art traditions have developed from the Anasazi ancestral tradition. By comparing the prehistoric and living visual communication systems and analysis of the graphic elements with semiotic relationships between parallel cultural traditions, certain similarities have been identified which may ascertain the cultural history of the concerned tribal groups.

ON RECONSTRUCTION

Tool making is the unique capability of man which initially emerged as an adaptive mechanism for success in sustenance. Material culture include all objects including tools and weapons, other artefacts, dress and ornaments, shelter and all household objects, vehicles etc. manufactured artificially with definite shape and purpose for its use. The common arena for research in the field of prehistory, palaeo-anthropology and also in the field of rock art is the reconstruction of prehistoric era. Early human activities and their respective life opportunities during that unrecorded past is viewed in spatio – temporal dimension and it is the central theme for inquiry. In prehistory, generally the indirect evidences such as tools and artefacts, burials etc. are the basic source materials for analyses and analogy. The techno-typological study ultimately converges towards understanding and assessment of their level of skill and probable use and other significances are estimated through logical interpretations and correlations. Such estimations may be 'imposed realities', 'conjectural' and/or may not be bias-free as the observer's view is a predictable entity from a different level of both time and space.

Compared to such indirect and objective archaeological evidences, rock art imagery is relatively more direct, dependable evidence of the past which may reveal the life-opportunities, world view, belief pattern, norms and aspirations of the society. The visual narrative may explain both of overt and covert aspects, natural and supernatural means for their livelihood. Such interpretations revealing their society and culture are relatively uncomplicated, decipherable when the rock art imagery is realistic or naturalistic. Such sites portray arrangement of panels containing visual narratives and episodes containing arrangement and combination of various figures being based on a common theme. But Non-figurative motifs, abstraction of forms, signs and symbols are also coexisting in rock art. Interpretation of such symbolic forms and icon is another stump blockade or challenging hurdle for deciphering of rock art. A general pan-human methodology is to be worked out with a site-wise consideration.

A thorough study on the rock art depictions to identify each of the form of the material culture there, its detail documentation and cataloguing is essential to trace the overall technological level of the society represented in the imagery. The prepared data repository is then compared cross culturally. The comparison with other similar sites as well as contemporary ethnographic data is a meaningful effort to trace the affiliation and identical forms, if any. Apparently and hypothetically, majority of the rock art imagery, particularly in India have close ethnographic parallels and this study being based on visual imagery ultimately reveal its validity.

DEFINING ROCK ART

Rock art may be defined on the basis of its inherent and essential features. Some basic & common features of pan-human rock art are as follows:

1. The representations are carvings, paintings, sculpture, bas relief, paint/s applied on carvings, bruising, and such other visual metaphors and symbolic imagery on any rock surface either in open air boulders, within caves, crevices, tunnels or rock shelters.
2. Technically each identifiable imagery form is composed of minimum possible lines and/strokes, dots, curvatures and this technique attained gradual improvement and became more intricate.
3. Details of the face of the individuals, particularly mentioning eye, nose, lip, and such other detail indicating details for personification could not be traced and/or such features are deliberately avoided by the artists.
4. No sign for preparation of rock canvas as base or space before any art activities.
5. The same rock canvas is utilized in successive generations/periods as evident in superimpositions. But always without erasing the earlier layer.
6. In imagery, different parts of body are mostly proportionate in overall representation.
7. Individual Anthropomorphs and zoomorphs are mostly shown in lateral views.
8. Basic colours and crude pigments are directly used without its further admixtures or combinations by making various shades with the principal colours.
9. Minimum application of colour or pigments particularly in contour lines or as solid block of colour filled in within the marked borders.
10. In later period more elaborate use of colour is evident primarily for ornamentation purpose.
11. In the genesis of rock art simple stick like figures gradually tend to develop figures in detail and descriptive accompaniments with the successive development of more competence, working experiences and skill. Non-figurative signs and symbols gradually become more abstract or simplified in form in course of passage of time.
12. Application of shades to the figures for realistic detail aiming more perfection is rare and mostly such an enhanced technique was adopted much later if at all achieved by that tradition.
13. Initially individual figures thematically mostly unrelated to each other are more common and in later period several figures are thematically and visually articulated to represent a metaphor or episode as a visual narrative.
14. In later development of rock art imagery the metaphors are more common where arrangement of several figures had emerged and a composition seems to be a chronicle primarily functional in an oral tradition as substitutes of scripts. Possibly its purpose was an expression to share or communicate with other viewers of the contemporary period and/or of their future descendants.
15. In initial or formative level rock art display selective few types of repetitive figures or images. The range of variation increases with ongoing maturity of art tradition in subsequent stages.

ROCK ART IMAGERY IN INDIA

Indian rock art scholars like Gordon (1939-40); Wakankar (1976); Pandey (1990, 1993) etc; had suggested chrono-cultural evolution of rock art on the basis of certain morphic criteria such as style, use of colour and superimposition. But the stylistic development can hardly be traced uniformly in actual situation within the entire region. So, for the present purpose priority is given on the quantitative classification of the motifs prevalent in the region. This approach is to trace the thematic priority areas and its range of variables. Depending on the frequency analysis and classification of thematic content, colour-scheme, technique and style of art, the priority area of the artist, and identification of the possible source for thematic inspiration is attempted. Sometimes the repetition of the thematic content reflects the mental obsession of the artist attached to a particular aspect (Chakraverty, 1999). The empirical studies reveal that in preliterate egalitarian societies, artist's perception is generally less individualistic and mostly represents the interpretation of the society (Chakraverty, 2004, 2009; Firth, 1992; Haddon, 1895; Layton, 1991). In the rock art sites of Central India, such as Bhimbetka, Adamgarh as well as in Pachmarhi, the earliest layer of paintings mostly represent larger figures of wild animals either depicted in red ocher, white or grey in colour. In Bhimbetka, within estimated total number of 6214 motifs, a maximum number of 1983 figures or 27.08% of the total number of figures represent different animal species (Mathpal, 1984). Numerically, the human figures rather in association with animal figures, with artifacts or single comprise the majority in the sites. Thus thematically, in central Indian plateau region, the artists were more anthropocentric.

It is also significant that within the total enumerated zoomorphs, a maximum number of 561 figures (33.33%) are of horses. On the contrary, the figures of horse-riders are 510 (89.95%) within the category of anthropomorph in association with zoomorph. No doubt that in the final phase within a particular time frame or for certain generations, the horse and horse-riders had influenced much on mental imagery of the artists. Figures of the archers and armed men are 377 in number, i.e. 54.40% of the anthropomorphic figures are with artefacts. In the battle scenes, individual warriors and other compositions represent group-conflict, violence and inter-ethnic or intra-ethnic struggle. Such idealized figures are thematically predominant in the Central India region.

Arrangement Pattern

The incident and purpose for communication of ideas or transmission of messages through creation of rock art could be ascertained from studying certain norms and practices followed unconsciously during its creation or in its process of manufacture. For such analytical assessment, the selecting criteria for the rock canvas within a site are to be recorded meticulously. In central India, majority of the rock art imagery are located on the naturally available surface of the rock wall in rock shelters, rather than within the inner walls of the caves. The majority of the art imagery, mostly paintings was drawn on open spaces with some rocky covering on the top, as also where maximum daylight is available. In addition to those criterions, majority of paintings are entirely visible from outside entrance part. Whereas, presence of images in inaccessible parts of the natural rock shelters, or in narrow crevices of the rock surfaces are relatively limited in number or totally absent in majority cases. Maximum numbers of paintings are concentrated at a height of eye-levels of an adult visitor remain standing on the floor of any rock shelter. In addition to that selected numbers of images are drawn on the ceiling which could be visible on the top of the head of the visitor only by raising or bending head upwards by a visitor within the painted gallery. In a number of rock shelters, elaborate paintings were drawn on narrow space of the ceilings close to the floors. In such areas the painted panels could only be visible either sitting position or only by laying on floor and keeping the eyes upwards to trace the painted panels above.

Therefore, it may be inferred from the different mode of execution in making art that the images are arranged in open space and visible from outside or from the entrance of the rock shelters reveal that the paintings were prepared with an intention and target so that its anticipated viewers could easily find the painted panels to appreciate or to read or follow its meaningful message-content. Furthermore, it also reveals that the sites containing paintings and engravings within rock shelters and caves has open rocky spaces like boulders and bed rocks in its close proximity. But suck open rocks do not show any sign of engraving. On the contrary, sites where the open rocks contain marks of art activities do not have any natural caves or shelter – like formation in the locality. Such basic prerequisite for art activities identify that mostly such work was created with an intention to preserve. If the motive for preservation had not acted in the artist's mind, then they could have equally explored and executed on adjoining open air boulders and bed rocks as an extension of art activities. But their work was restricted only within a rather protected space under a roof for obvious reason for safeguarding such creations. In such sites where engravings and brushings are only found on open air boulders, essentially the topographic region is without any traces of caves and rock shelters. In such areas, the art-makers had no alternative and better option to preserve their work within limited natural resources.

Study on the angle of visibility may open a new vista for rock art research. The other type of location for visibility of paintings is only when the spectators or audience used to sit or lay down on the rock floor. Such probability is more for viewing when one lay on the floor during leisure period or for sleeping. Therefore the location of the art imagery and its concentration pattern may reveal the possible intention and urge of the artists to communicate with the targeted group of viewers. An elaborate documentation on the distribution pattern of images on the rock surface reveals their scheme, range and limit for space utilization. Such data is helpful to determine the extent of selection and use of rock surface for art activities to a certain height and other limitations for their use. For such deductions, the palaeoclimatic condition of the period is also to be considered. The erosion activity of the rock floor and other possible influencing geological factors are also to be covered for such studies.

In central Indian rock art sites, all images are usually drawn at 10-12 feet height from the floor. Again, up to 4-5 feet from the floor, number of paintings are only few. Either deliberately the painters avoided such spaces in lower level close to the floor understanding greater possibility for early damages from natural factors such as rain, dust and overlapping by clay etc.

From such occurrences it may be observed that the makers of rock art had consciously acted to protect and preserve their works intending uninterrupted clarity for viewing, even in future. During selection of the spot, certain criteria seems to have acted spontaneously and repeatedly, such as – the uneven and broken parts on the surface of the walls were avoided. Such tendency is more common in painted galleries and it indicates that the makers of rock art, had conscious effort to preserve the paintings. Rather, the painters never rejected some spaces which show signs of earlier layer/s of paint, either faded or more or less distinct there.

Therefore, two most common features is marked in rock art, not only in central India but it may be a universal phenomenon. The first of which is – before painting, the natural rock surface was not artificially prepared by application of a base colour or otherwise. The second relatively common phenomenon is – superimposition of paintings. Such incidence is more common in centrally located spaces where natural light is more available and usually attract the viewers' attention more than in remote or dark corners. Such types of analytical approaches need a thorough scrutiny on all rock shelters and caves within a cluster of sites. Study on selection criteria for space on rock has direct relation with seasonal changes of light condition and its direction with varying local climate and other range of factors. Before preparing technical details for developing a suitable methodology for study on selection criteria, a lesson from the ethnological parallels could be more appropriate for understanding its pattern of usage in contemporary but comparable primordial societies (Chakraverty, 1999).

The usual views for rock art representations are profile views for images of both human and animal figures whereas the front or fronto-lateral views are less common. But bodies shown from above or from below are rarely represented. Such conscious or spontaneous attempt by painter to portray a particular figurative view possibly inspired or influenced by the view which is normally observed visually by an individual without any distortion. The profile views require less technical skill to portray a figure than other views. The angle of vision for the best viewing of an image placed on a rock surface, locating the direction of the figure following its base line is essential. Whether the particular placement of the imagery is in hidden crevice or in open and easily visible niche – such occurrences support the deductive logic for understanding the inherent role of rock art. In central Indian sites, paintings are concentrated in such locations where the possibility and chances for its natural protection from rain and wind are more.

In rock art, at least in Central Indian rock shelters, each of the illustrations are either with a single motif or in combinations of several different types connected with a same central theme or episode (Mathpal, 1984; Neumayer, 1983, 1993; Wakankar & Brooks, 1976). But each of the individual clusters of imagery is separated from others mostly by an interval of open space. Often such figures are superimposed when painted in successive period of time occupying the same space in different time period and using different colour, technique, treatment of arrangement etc. But in the entire region, use of distinct borderline to separate each composition is never found. But in mid-eastern region, comprising the states of Jharkhand and Orissa, in certain instances, figurative and non-figurative motifs are bounded by clear-cut boundary lines. Such arrangement of rock art imagery within a discrete frame or outline mark a later qualitative advancement in communicating their visual narratives. Such patterns of enclosed figures illustrate mostly abstract signs and symbols which appeared repeatedly. Similar arrangements of enclosures are common among tribal murals and ritual paintings among Lanjhia Saoras of Orissa, Rathvas and Warlis in Western India (Chakraverty, 2009).

In tribal art, particularly in art practices among thirteen different tribal societies in parts of India, some relevant observations (Chakraverty, 2009) may be mentioned. The analysis is for understanding the selection criteria of spaces for similar art activities in prehistoric art. The tribal wall paintings, more common among Lanjhia Saoras of Orissa (Fig. 22.1), Rathvas (Fig. 22.2) and Warlis (Fig. 22.4) in Western India, paintings are ritualistic drawings depicting the sacred seats of several deities, departed ancestors and their soul, myths of creation, icons and their praises to fulfil their aspirations in mortal life. Such murals, in maximum number of households are painted within the rooms and usually not visible from outside courtyards. Among Saora tribe, such icons are protected in dark areas and covered by hanging seeds, pots containing water, weaponries etc. The Saora icons are secret visual narratives and prayers to their propitiating deities that are strictly maintained only by the family members. All outsiders, including the neighbours are not allowed to locate the exact deity. The idea behind such practices is that – if any evil force from outsiders could identify the deity, then they may divert the deity which may ultimately be harmful for the family. Therefore, for painting of a covert icon for magico-religious purposes, the inner walls of rooms and its protective coverings are essential. On the contrary, among Santals, a numerically dominant tribe in eastern India, wall paintings, mostly decorative motifs are more commonly found on outside walls of their huts signifying their prosperity, wealth and happiness that they prepare or renovate during their annual autumn festival. Traditional Santal society, maintain several customs and taboos related to witchcraft and sorcery. But wall paintings have no immediate connection with such magico-religious practices (Chakraverty, 1987). Therefore, their art is more communicative and open to all section of viewers. In addition to that, Santal art practices sometimes visually narrate their symbols of tribal identity, myths of creation, glorious past and historical episodes of their battle for freedom against forces of British imperial army. It is also relevant to ascertain that in tribal sacred groves in India, terracotta or metal votive figurines are placed as offerings after fulfilment of expectations and hope. The figure of horses and elephants are more common votive figurines present throughout the entire length and breadth of India. Uniqueness of such village sacred grove is that a particular type of figure is repetitively found in cluster within a limited space where the range of variation of form is least. Sometimes, in rock art sites, such overwhelming presence of particular type of figure within a distinct locality or region is not rare.

Fig. 22.2-3. Visual narratives of habitats, social interactions, and battle scenes in Pachmarhi

One such example on contemporary votive figurines and its link with early rock art exists in Kathotia Karad rock shelters (Fig. 22.2) near Bhopal, the capital of the state of Madhya Pradesh in central India. In Kathotia Karad site, more than ten rock shelters have extensive painted galleries. In the same site, a particular rock shelter is considered as the abode of a tribal deity which is believed to cure pains and other various ailments of both hands and legs. The local tribal villagers for such diseases pray in this shrine for mystic cure. On fulfilment of such wishes, they again visit to that shrine and then offer a pair of life-size wooden replica of both hands and legs (Fig. 22.2-3) with other essentials and perform a ritual with the help of tribal shaman or priest. In that particular site dominant numbers of rock art imagery illustrate horse riding warriors and battle scenes. During combats in battles, the possibility for loosing hand and legs or even death is usual. From such trauma of battle as reflected in the rock art, mythical realities had emerged which will act as well as encroaches the mind and belief of indigenous preliterate population. The tribal communities encounter the visual messages from the rock art most frequently and repeatedly during regular collection of forest products which is also a substantial source for their sustenance. In this event, the narrative elements of folklore depicted in rock art motivate and transmitted through generations as living tradition which is not lost in the passage of time and events.

Fig. 22.4. Non-figurative motifs in Pachmarhi (M.P.) and Ghatsila (Jharkhand)

The tribal communities encounter the visual messages from the rock art most frequently and repeatedly during regular collection of forest products which is also a substantial source for their sustenance. In this event, the narrative elements of folklore depicted in rock art motivate and transmitted through generations as living tradition which is not lost in the passage of time and events.

The art of tribal communities and other indigenous populations due to its simplicity in representation provide us certain indications to understand their aspirations and obsession in the mindset of the artist and their community. The tribal artist is conditioned in their social cultural matrix. Thus the work of an individual tribal artist largely represents the idea, belief and sentiment of the group or society (Chakraverty, 2009).

Thematic Range

It is relevant to mention that the reports on empirical studies of contemporary forager and hunter-gatherers in global context reveals that among hunter-gatherers, for organized group hunting, a distinct strategy, experienced leadership and training is essential. In addition to that, the keen observation on behaviour of the targeted animal group, their breeding season and sensitive striking points on body all such information is shared by the matured members of the hunting expedition in their informal group meetings, folklore and related supernatural performances that stimulate the team in preparing prior to their expeditions. In earliest phase of Pleistocene rock art preponderance of zoomorphs is almost a universal phenomenon. Such situation is also evident in central India. In central Indian sites, wild animals are portrayed in visual imagery in isolation. Among tribal hunters, it is customary to mark a particular area with success in availability of suitable hunt. Hunters, identify a particular area with availability of certain wild species. The hunter-foragers store such information in their collective memory through oral tradition. Such indigenous knowledge is shared and inherited by the tribal community for competence in sustenance.

Rock art demonstrate an identical system for maintenance of a non-verbal and visually transmitted arrangement for communication within a generation and even beyond that time limit. Rock art acted as a unique indigenous knowledge resource for better sustenance in the forest habitat even valid to their successive generations. Hunting-gathering activity does not necessarily confirm the presence of Palaeolithic or Mesolithic period. Foraging had continued as an alternative source for livelihood, particularly during crisis. Such activity is still valid among marginally located tribal communities. Hunting is also a popular sports and annual ritual for initiation of youths. Such practices are common among several tribal communities in India.

Chrono-culturally, the next stage demonstrates wild animals in association with anthropomorphs, primarily as hunter, either individual or organized in groups. Such activities give some clue for successful mode of technique adopted by the group. Visual narratives depict urge for sharing of experiences and anecdote for each expedition. To early preliterate societies, rock art was a system of virtual data-repository created and developed by successive generations for the benefit of themselves as well as to their successors. It may be recapitulated that in central region four successive and partly overlapping cultural phases exit in rock art imagery. The stages are:

1. Earliest level of figurines representing hunting-gathering, foraging groups with large figures of wildlife.

2. Intermediate phase indicating marginal stage of food collection, group-hunting by improved technological means and beginning of pastoral economy.

3. Full-fledged pastoral society with magical belief and aspiration for breeding of domesticates. This phase illustrate a wide range of group activities and aspects on life opportunities.

4. Presence of extensive battle scenes and use of metal weapons by two rival groups confronting each other in varied contexts.

In rock art of central India, a distinct but overlapped phase is marked by the presence of pastorals in rock art (Fig. 22.4). With display of essential activities for animal domestication, providing suitable security to domesticates against threats indicate the major issues for concern for their community. As it is revealed from the images, the threat from wild carnivores was in initial period. The pastoralists had to remain prepared to combat attacks from other rival groups, particularly of plunderers who raided on their herd. Such periodic insecurity and conflict, both seems to be the prime motivating factors for creation of rock art. In rock art of early historic period, descriptive accounts of various battle scenes, mostly feud (Fig. 22.2-3) had influenced the artists. Rather, such theme or view had turned to an obsession affecting the mental imagery of its creators (Chakraverty, 2003). Sometimes, intimate visual records in association of more realistic contextual assemblages convincingly support the view that possibly the painters had witnessed such conflict and genocide by themselves directly. No distortion of tool and weapons are observed in art compositions illustrating such battle-scenes. Use of wheel and chariots are illustrated (Fig. 22.3) signifying organized effort by urban influences.

The ultimate or final stage of rock art display well articulated scripts like 'Sankh-lipi' or conch shell variety and early 'Brahmi' scripts sometimes found superimposed over layers of earlier figurative paintings. But in Indian context no evidence is reported for evolving a proto-script from its ancestral rock art imagery. Battle-scenes are rarely present in the rock art of northern region, eastern region or in the south. From a recently discovered site in Nilgiri hills, Tamil Nadu of South India, portrayal of warriors and similar battle scenes are common. The maximum presence of a particular cultural element within a specific eco-cultural area suggests that the context, significance and functional use of the rock art vary from one region to the other. It follows the anthropological concept and principles of cultural-area. Either the artists had direct historical experiences about the battle or they indirectly learned about the episode through folklore and other modes of oral tradition. The figures of horse or horse-riders are prevalent in tribal and folk art of central and western parts of India. Various votive figurines representing horse or horse-riders are integral part of rituals, belief and practices among the tribal societies like Gond, Muria, Bison-horn Maria, Korku and Baigas living in the almost similar eco-cultural areas within the state. Among those tribal societies, horse is considered as the most suitable seat for an honourable person and so it is the symbol of might, authority, power and prestige.

The pre-urban economy as represented in the rock art of the region has two overlapping stages. The greater number of depictions is of forager and pastorals. The facets of life-opportunities among the food-gatherers were elaborately represented whereas; the glimpses on the warriors do not contain any further details. There are two possibilities, either the warriors were alien, invaders, mythical characters of the semi-nomadic group. Due to historical factor the marginal ethnic communities had joined in a battle as hired force. But nowhere in the rock art within the entire region one can find the technologically superior type of weapons, particularly use of any fire arms or presence of warriors in association to any household scenes. But in several cases, the depiction of umbrellas is associated to dance-scenes, rituals (Fig. 15) and it sometimes cover the head of the horse-riding warriors. The economy of the society was possibly in transition. Besides food gathering, they reared selected types of domesticates such as cow, water-buffalo, goat or sheep etc. In majority of the rock art sites, no direct pictorial evidence for settled agriculture or use of plough exists. Fowling, collection of honey and seasonal fishing were their other sources for subsistence as evident in the rock paintings. In Chambal river valley and also in selected few sites of Betwa source region, the representation of bullock cart and use of animal power for ploughing is manifested in rock art.

Individual and group-hunting both were practiced simultaneously. Snares and nets were frequently used for hunting. In Bhimbetka and its adjoining sites of Betwa source region several painted motifs represent use of harpoon heads most possibly equipped with multi-barbed microlithic blades. Digging sticks are also depicted in the same layer enclosed by hunting scenes.

Presence of animal foot marks and figure of large mammals were attempted for planned group hunting. A clear division of labour is represented where the womenfolk were principally engaged in household activities. Responsibilities allocated to womenfolk included-grinding of fruit and cereals, cooking, nourishment of children, keeping of household utensils and other personal belonging etc. Figurines are rarely represented as carrying pots or with heavy head load of goods. Women are often shown as dressed in a short skirt but with projected breasts with a view to identify them discretely. Breast-feeding scenes are also frequent.

In hair style, both male and female figurines are more or less identical. Men had long hair and they arranged their hair either in a bun or a knot of hair at a nape of one's neck. Mostly all warriors had similar hair style.

The rock art is a unique source for studying the material culture of the society in its eco-cultural milieu. The bow and arrows were indispensable for every household. The baskets were of different types. Clay pots, club, four kinds of axes, including hunting axes and water containers. Rock shelters were their usual dwellings. Sometimes they also constructed lean-to type semi-permanent huts. Variety of musical instruments including horns, pipes, different types of drums and other percussion instruments were popular to them. Horns and pipes are frequently associated to battle scenes.

The horse and elephant-riding figures point out political hierarchy existed in the society. But above all, hunter-gatherers and animal-rearing groups were possibly more egalitarian in type in their socio-political structure. Their group-solidarity is evident in the religious practices of the community. Scenes represent group dances during rituals, magic and other supernatural performances. It is interesting to note that their dance-style was more or less similar to the Gond-speaking tribal groups inhabiting the region. In Mahadeo area of Pachmarhi, descriptive account of certain rituals in rock paintings has identical phases similar to the performances of local tribal communities. Their annual autumn festival is related to the belief in fertility cult rendering more benefit from successful breeding and trouble-free maintenance of domesticates. During such festivities and ritual hand impressions in red ocher and other colours is essentially imprinted on the body of cattle and other domesticates.

Similar belief in fertility cult is not rare in the rock art, where x-ray paintings of cubs are delineated within the womb of quadrupeds. In rock art, palm prints, thumb impressions, hand stencils and finger markings are plenty. Such symbolic representations are also common to the local tribal households that are believed to protect themselves from evil eye, sorcery, witchcraft and similar other malevolent supernatural forces.

The other features present in the rock art point out the presence of animistic type of religion. Head hunting was practiced in the society. Head-hunting scenes are depicted in the Rajat Prapat rock shelters of Pachmarhi (Neumayer 1985). The personal adornment and hair style of victims as well as the assailant are similar. Possibly it portrays in-group feud or rivalry. Representation of two vultures on the tree depicts their presence as birds of prey who would exploit the carcass. They were waiting for the carcass as source of food within the battlefield. The composition signifies the artist's idea and perception to communicate the horrified eventualities of a battle.

Sometimes, the attitude of the artist towards a particular motif is reflected in their nature of implementation. Sometimes, in the compositions of the rock painting, particularly the dimensions of certain motifs are consciously exaggerated. The disproportionate motif, mostly larger than its other associated parts point out the purposeful attempt for attracting its viewers. In India, the early artist did not paint all available rock shelters within a locale. They had intentionally selected very few but suitable rock shelters which were more accessible for communication and display to attract its possible viewers. The visual language of the early society had gradually transformed into pictorial narratives for self identification. In Jharkhand state the rock art sites are related to a local legend of 'Kohbar' or marriage booth of a tribal king and the newlywed couple spent the night of their marriage in the rock shelter already decorated with paintings. Some of the sites, located near the Kiul River and in the bordering areas between Bihar and Jharkhand are also known to the local villagers as – *Kohbar* and local rivulet is identified as – Rani Gadar or after the unknown queen. The figurative motifs, often stylized horse riders, single or in groups, two persons carrying another within a palanquin or marriage-litter, symbolic designs and other unidentified motifs are common. In Jharnawa Rani Gadar site two persons armed with sword-like object hanging from their belts are shown encircled by two or three different rows of rounded objects, possibly be a row of stone boulders and the scene most possibly represent the visual account of a megalithic burial commemorating two warriors. Similar composition of painted motifs representing burial practice has recently been reported from Chik Rampur, Karnataka (Sundara, 2006). In the same site, in a different panel painting signify a schematic map-like drawing showing households of a village-like settlement with its distinct boundary. There several horse riders and other villagers in defence against other horse riding invaders are engaged in protecting their territory. The village territory with arrangement pattern of conical thatched huts, cultivated land and the well-planned entrance of the village almost resemble existing village settlements in that locality. In the entire area a significant proportion of non-figurative sign and symbols coexist with figurative rock art imagery. The motifs are mainly monochrome and represented in red ochre.

Rock Art & Tribal Art: a Comparative approach

In rock art of Pachmarhi, the majority of the figurative representations are solid-bodied (79.67%), stick figure or line drawing (14.49%) and outline or contour (3.79%). The figurative motifs are mostly naturalistic (90.63%) with the exception of few (0.67%) figures of zoomorph and anthropomorph which have been identified as more stylized. Within naturalistic figures an anthropomorph is rarely shown with traces of projected nose, distinct eyes and in general all human and animal figures are devoid of such recognizable facial details. On the contrary, several human figures, mostly depicted within battle-scenes are provided with distinct dress, other forms of personal adornment including ornaments and hairstyle. In both sectors it is observed that the initial drawings of anthropomorph and zoomorphic figures constitute arrangements of triangles to form the required figure.

In Korku society, horse-riding figures signify possession of wealth, power, prestige, higher status and overall dominance of the man in the society. Such figures as well as armed man in battle-scenes are frequent in Central Indian rock art sites. Such representations are quite common in Bhimbetka and other neighbouring rock art sites of Betwa source region and in Adamgarh hillock located in the southern bank of Narmada River.

In both sectors of art, technique of figural representations are mostly standardized and idealized, and repetitive forms are frequent. In both categories of art, the artist's creative impulses have seldom been expressed. The activity and experience of the ethnic group have largely been represented rather than the expression of individual perception.

A total number of 31 common elements (Chakraverty, 1999) of morphic types have been identified in both rock art and tribal art of Pachmarhi. It comprises 4 types of zoomorphs including depictions of unidentified quadruped mammals, deer, horse and cow. Within anthropomorphs, two morphic types are respectively group dance and man guiding horse-rider. In the category of anthropomorph with zoomorph and material trait, a total number of three types of pictorial elements are common which are – horse rider holding artefact, hunting with bow and arrow and scene of animal husbandry. In the category of anthropomorph with material trait, a maximum number of 11 motif-types in both sectors are identical. Such type of figures include man holding knife and dagger, holding umbrella, drummer, holding stick or club, musicians playing flute or horn, holding bow and arrow, holding only arrow, holding only bow, and figurative drawing of man and woman shown under a shelter. In material trait, identical items are respectively; – umbrella, sword, knife or dagger, pot or basket, rock shelter or trait for habitation, drum, bow and arrow, club or stick and flute or horn. One generalized figure of a plant in the category of phytomorph and one solitary motif of floral-like design under the category of non-figurative motif is identical in nature.

As the comparative study reveals, in both sectors, the pictorial glimpses on the pattern of life, activities and material culture are mostly indistinguishable. 31.96% of the total motif-elements identified in the rock art are thematically identical to those of tribal art. Therefore it is not unlikely that the Korku art tradition had been influenced by the rock art persisting in the same eco-culture region.

In the present issue there are other observations pertaining to hypothesis, especially of Fischer (1961). It is based on the data of world ethnographic samples. Fisher's first hypothesis entails that in the art of egalitarian tribal societies visual repetitions predominate. In hierarchical societies, varied design elements are more conspicuous. The data on rock art localities and tribal societies (Chakraverty, 2009) under this study have been examined (Fig. 22.1) and found that in all rock art sites as well as those of tribal societies repetition of art form is a common feature.

Fischer's second hypothesis relates egalitarian societies with presence of maximum empty space. On the contrary in hierarchical societies minimum empty space is left in art compositions. Considering the rock art localities in Pachmarhi maximum empty space exist within the painted area of the rock canvas. Such feature is also common in the art of the three tribal societies under study. Exception is met with in the rock art of Ulap, in Qrissa and Isco in Jharkhand. Therein the art motifs are more congested and inter-spaces are minimum pointing to a relatively advanced form of society with hierarchal phenomenon (Fig. 22.1).

Testing the third hypothesis, it has been found that in all rock art sites and tribal depictions the figurative and non-figurative motifs are largely symmetrical which confirms the egalitarian society as their origin. Fischer (1961) had also related enclosed figures with the egalitarian societies, and in opposition, enclosed figures are predominant within hierarchical societies. In rock art, of Pachmarhi, in Juang society and partly in Ulap, art motifs are non-enclosed. On the other hand, Isco rock art and in art forms of the Korku and Saora tribal society enclosed figures is preponderant. Considering the above noted hypotheses, it may be inferred that rock art of Pachmarhi is a product of egalitarian society, whereas in Ulap and Isco sites, some elements of developed form are found to exist. Among the Korku and Lanjia Saora societies their art illustrates some features which are placed in transition between egalitarian and full-fledged hierarchical societies. These inferences are to be checked with the utilization of other forms of data for the purpose of confirmation (Fig. 22.1) (Chakraverty, 2009). In central India, the early rock art represent a community which was closely similar to contemporary tribal societies, especially the pastorals and marginal cultivators. Both similarities and differences are important for reconstruction of the early society and understanding their art as an integral cultural element. The basic similarity between the arts of two different chronological periods may be explained in terms of common economic pursuits, social organization and material culture. The cultural continuity of certain elements is prevalent in the cultural-area under consideration.

On the basis of superimposition and techno-stylistic consideration, it is confirmed that the rock art of the region had developed through successive chrono-cultural periods. On the contrary, in the same region, the tribal art tradition is significantly present (Elwin, 1955). The analogy between the two sets points either to a relationship or to a connection between the two art forms (Chakraverty, 1999). It may be mentioned that the similarities are often observed between the rock art and tribal art with regard to theme, forms, styles and motifs. In addition to the same, there is a significant overlapping on the distribution pattern between the two sets. Based on the above argument, it may be hypothesized that the tribal art is a continuity of the total tradition of rock art.

It is indeed essential to make attempt for establishing the continuity between the two sets on the basis of various forms of data information with the methodological standardization and application of rigorous analysis.

Among preliterate tribal Communities, the oral tradition including myths and legends is the major source for reconstruction of ethno-history. In order to understand the culture continuum, the comparison between rock art and ethnographic account of relevant tribal societies from the same region may be attempted for a more rational approach in deciphering rock art. In addition to the study of tribal art, different aspects of material culture, social organization and the cultural expressions as reflected in rock art could be a useful source to trace the ethnic milieu in rock art.

Acknowledgment

I am very grateful to Dr. A.R. Sankhayan for providing me opportunity of including my contribution.

References

ANDERSON, R.I. 1979. *Art in Primitive Societies*. Englewood Cliffs: Prentice-Hall.

ARCHER, W.G. 1947. *The Vertical Man: a Study in Primitive Indian Sculpture*. London: George Allen and Unwin Ltd.

BARRY, H. 1957. Relationships between child training and the pictorial arts. *J. Abnormal and Social Psychology*. 54: 380-383.

BOAS, F. 1955. *Primitive Art*. New York: Dover Publication.

CHAKRAVERTY, S. 1987. Tribal murals of Chhotanagpur. *The India Magazine* 6 (2): 20-27.

CHAKRAVERTY, S. 1995. Isco: Rock Art Site in Hazaribag. In *Recent Perspectives on Prehistoric Art in India and Allied Subjects,* R.K. Sharma & K.K. Tripathy (Eds.), pp. 74-99. New Delhi: Aryan Books International.

CHAKRAVERTY, S. 1999. Primeval Rock Art and Ethnographic Analogy in India. *Bollettino del Centro Camuno di Studi Preistorici* (31-32): 213-222.

CHAKRAVERTY, S. 2003. *Rock Art Studies in India: a Historical Perspective*. The Asiatic Society. Kolkata.

CHAKRAVERTY, S. 2004. Mystical Icons of Lanjhia Saoras: a Primordial Tribal Art Tradition of India. In *New Discoveries, New Interpretations, New Research Methods,* pp. 155-163. Brescia: Centro Camuno di Studi Preistorici.

CHAKRAVERTY, S. 2009. Indigenous Tribal Art: Strategy for Conservation and Promotion. *Journal of Social Anthropology* 6 (1-2): 87-102.

CHAKRAVERTY, S. 1995. Rock Art and Ethnographic Data Retrieval in India. In *Rock Art vis-à-vis Ethnography*. Rabindra Nath Agrawal (ed.). Nagpur, Maharastra: Gupta Publishers.

DALMIA, Y. 1988. *The Painted World of the Warlis: Art and Rituals of Warli Tribes of Maharastra*. New Delhi: lalit Kala Akademi.

DAS, A.K. 1979. *Tribal Art and Craft*. New Delhi. Agam Kala Prakashan.

DAS, A.K. 1992. North East India, in *Tribal India: Ancestors, Gods and Spirits*. Pp. 3-22. Bombay. Marg Publications.

ELWIN, V. 1948. Notes on the Juang. *Man in India*. 28 (1-2): 1-182.

ELWIN, V. 1951. *The Tribal Art of Middle India: a Personal Record*. Bombay; Oxford University Press.

ELWIN, V. 1955. *Religion of an Indian Tribe*. Bombay: Oxford University Press.

ELWIN, V. 1959. *The Art of North-East Frontier of India*. Shillong: North-East Frontier Agency.

FAWCETT, F. 1901. Notes on Rock Carving in the Edakal Cave, Wynaad. *The Indian Antiquary* 30: 409-421.

FIRTH, R. 1951. The Social Framework of Primitive Art. In *Elements of Social Organization*. Edited by R. Firth., pp. 155-182. London: Watts.

FIRTH, R. 1992. Art and Anthropology, in *Anthropology, Art and Aesthetics*. Edited by Coote, J. & A. Shelton, pp. 15-39. Oxford: Clarendon Press.

FISCHER, J.L. 1961. Art Styles as Cultural Cognitive Maps. *American Anthropologist* 63: 79-93.

GHOSH, A.K. & CHAKRAVERTY, S. 1988-89. Genesis and changes of Tribal Art -examples from West Bengal. *Indian Museum Bulletin* 23 & 24: 136-141.

GHOSH, A.K. & CHAKRAVERTY, S. 1995. Rock arts of eastern India and their bearing with local Tribal arts: a cross-sectional study on continuity in India at the dawn of history. *In India at the Dawn of History* Asok Datta, Asok K. Ghosh and C. Margabandhu (Eds.). Delhi: Agam Kala Prakashan.

GORDON, D.H. 1939. Indian Rock Paintings-II; the Rock Paintings of Kabra Pahar, Raigarh State. *Science and Culture* 5(5): 269-275.

GORDON, D.H. 1940. Animals and demons in Indian cave art. *Science and Culture* 5(11): 662-667.

GORDON. D.H. 1940a. Warfare in Indian cave art. *Science and Culture* 5 (10): 662-667.

GORDON. D.H. 1958. *Prehistoric Background of Indian Culture* Bombay: N.M. Tripathi Publication.

HADDON, A.C. 1895. *Evolution in Art*. London. Walter Scott Ltd.

HASELBERGER, H. 1961. Methods in studying ethnological art. *Current Anthropology* 2: 341-384.

JAIN, J. 1984. *Painted Myths of Creation: Art and Ritual of an Indian Tribe*. New Delhi: Lalit Kala Akademi.

LAYTON, R. 1991. *The Anthropology of Art*. Cambridge: Cambridge University Press.

MATHPAL, Y. 1984. *Prehistoric Rock Paintings of Bhimbetka, Central India*. New Delhi: Abhinav Publications.

MATHPAL, Y. 1995. *Rock Art in Kumaon Himalaya*. New Delhi: Indira Gandhi National Centre for the Arts.

MILLS, J.P. & HUTTON, J.H. 1929. Ancient Monoliths of North Cachar. *Journal of the Asiatic Society of Bengal* (N.S.) 25: 285-300.

MURDOCK, G.P. 1957. World Ethnographic Sample. *American Anthropologist* 59: 664-687.

NEUMAYER, E. 1983. *Prehistoric Indian Rock Paintings*. New Delhi: Oxford University Press.

NEUMAYER, E. 1993. *Lines on Stone – the Prehistoric Rock Art of India.* New Delhi: Manohar.

PANDE, B.M. 1972. 'A Neolithic Tectiform from Burzahom, District: Srinagar'. *Journal of Indian Anthropological Society* 7: 175-177.

PANDE, B.M. 1971. Neolithic Hunting Scene on a Stone Slab from Burzahom, Kashmir. *Asian Perspective* 14: 194-198.

PANDEY, S.K. 1993. *Indian Rock Art.* New Delhi: Aryan Books International.

REDFIELD, R. 1956. *Peasant Society and Culture.* Chicago: University of Chicago Press.

SUNDARA, A. 1968. Rock Shelter, Hire Benakal Hills, Raichur. *Indian Archaeology 1967-68 – A Review*, p. 67.

SUNDARA, A. 2006. Karnataka Rock Art: Some Distinctive Aspects. *Purakala* 16: 35-49.

WAKANKAR, V.S. & BROOKS, R.R.R. 1976. *Stone Age Paintings in India.* Bombay: D.P. Taraporevala Sons & Co.

23.
ASTRONOMICAL ORIENTATION OF THE TREPANNED NEOLITHIC WOMAN OF BURZAHOM, KASHMIR

Iharka SZÜCS-CSILLIK
Romanian Academy, Astronomical Institute, Astronomical Observatory Cluj-Napoca,
Cireşilor 19, 400487 Cluj-Napoca, Romania
iharka@gmail.com

Alexandra COMŞA
Romanian Academy, Archaeological Institute, "Vasile Pârvan" – Center of Thracology,
Calea 13 Septembrie 13, 050711, Bucharest, Romania
alexcomsa63@yahoo.com

Anek R. SANKHYAN
Palaeo Research Society, Ghumarwin, H.P.-174021, India
arsankhyan@gmail.com

Abstract: *We present a very interesting skeleton No. 7 of a 25-30 years female from the Neolithic site of Burzahom in Kashmir Valley, India, dated to about 4000 BP which exhibits multiple trepanations on the skull. Archaeological and anthropological investigations had revealed that the woman was suffering from a brain anomaly and head injury and was insane or epileptic and looked 'different'. Among the ten excavated skeletons at Burzahom, the present study observed that the skeleton No. 6 was astronomically 'different' from the rest of all. Our results corroborated with those from a very recent scrutiny of craniometric data (by ARS) had shown that the skeleton 6 was of a 'very robust' man who belonged to a round-head, broad-faced, broad-nose and medium tall phenotypically very distinct population compared to the long-head, narrow face, long nose and tall Burzahom population This enabled us to conclude that the 'robust man' was a shaman-cum-surgeon from a distant place who treated the woman spiritually for long and later performed trepanations. He was accepted by the Burzahom people with whom he had lived for good, died and buried there within their habitat but with a different orientation to mark a difference.*

Keywords: *Burzahom, Neolithic, India, multiple trepanations, astronomical orientation, the shaman-cum-surgeon*

INTRODUCTION

Burzahom as a prehistoric site has been much different from other Neolithic sites of India. During early Neolithic times the people living in this area had adopted a strange way of living to avoid the cold weather of Kashmir. They used to dig up pits inside the ground which served as homes for these nomadic people, called as 'pit-dwellers'. The area is full of deep and wider pits with narrow opening at the top. The deep pits included a few steps as staircases at the entrance, but the smaller pits had no such steps and the people could easily jump in. These pits were plastered from inside with mud and were either oval or round shaped structures, barring a few rectangular or square shaped.

In due course of time during the later phase of Neolithic, the underground pits gave way to mud-brick dwellings built on the top of the pits or nearby. The people also started the use of simple earthen pots of different shapes, sizes and colours generally grey, red or brown. There is also evidence of polished stone and bone tools of animal bones, horns and antlers, besides harpoons, arrowheads, spear heads and daggers used for hunting. Some bone needles discovered there indicate knowledge of sewing and making of clothes from animal hides. The excavated burnt potteries and pieces of wood reveal that they fired the pottery and cooked food. The human burials were found in the habitation site, under their dwelling units. A few burials had animals too, like dog, wolf and ibex.

Besides Burzahom, several Neolithic sites were discovered in Kashmir valley, viz., Begagund, Gofkral, Hariparigom, Olchibag, Pampur, Panzgom, Sombur Waztal, and Brah, but skeletal evidence is yet to be found from them.

BURZAHOM CEMETERY IN TIME & SPACE

Burzahom Archaeological site is located about 10 km northeast of Srinagar in the Kashmir Valley. The Burzahom burial site (22°57'N: 79°31'E) is on a slightly raised terrace formed of the dried-up lakes, known as Karewa in Kashmiri. The entire Kashmir Valley is a cup-shaped flat surrounded by the Pir Panjal mountain range with tall birch trees widely spread in the valley, which give the name Burzahom, meaning the "place of birch". For its scenic natural beauty enhanced by forests and lakes, the Kashmir valley is a world famous tourist resort and regarded as paradise on earth for its cold climate in

Tab. 23.1. Graves's data at Burzahom Archaeological Site

SKL No.	Orientation	Azimuth	Position	Depth	Sex	Grave goods
SKL 1	WE	270	Crouched	5'-7'11"		Earthen goblets
SKL 2	WE	270	Foetal	–	Child	–
SKL 3	NW-SE	315	Crouched	5'10"		Skull of a dog
SKL 4	EW	90		5'6"		Red ochre
SKL 5	SE-NW	135		6'		Red ochre, earthen pot, small barrel shaped paste bead
SKL 6	NS	0		3'9"	Male	Animal bones
SKL 7	NE-SW	45	Crouched	7'4"	Female	Red ochre, animal bones, antler, horn pieces, soap stone circular disc
SKL 8	SE-NW	135		10'4"		5 carnelian barrel beads
SKL 9	–	–		–		–
SKL 10	NW-SE	315		10'7"		Circular stone bowl

the summer months. This is why the early Neolithic settlers had selected the place for their living. The forests and mountains were a continuous source of food and water.

The Burzahom site was excavated between 1961 and 1968, and yielded 10 human skeletons of different cultural stages Neolithic (Period 1), Neolithic-Megalithic (Period 2), and early historic cultures. The Burzahom Neolithic period is divided into two phases: Phase I (lower) and Phase II (upper). The Phase I belongs to the early pit-dwellers, which has not yielded any human skeleton, and yield the earliest C^{14} date of 2375±120 BC (>4375BP). It is the Phase II which has yielded seven skeletons, No. 4-10 (Tab. 23.1). It includes the trepanned skull (skeleton 7) referred here as the *Burzahom skull*, and dates around 2000 BC (=ca. 4000 BP). The remaining three skeletal burials (1-3) are of the Neolithic-Megalithic cultural complex (Basu & Pal, 1980; Sankhyan & Weber, 2001).

The Position of the Burials

Among the ten Burzahom skeletons, five are of adult males, three of adult females, one of a juvenile and one of a child. All the interments had their orientation close to the NW-SE directions, except the skeleton 6, oriented to N-S direction. Some of them had grave goods, and some others had not (Tab. 23.1). The phase II Neolithic burials had been discovered in the habitation area below the dwelling's floor, at a considerable depth, which varied from 4 feet to 9 feet and 6 inches. When considering those seven interments of the Neolithic phase II, we should bear in mind that 5 of them have been found in a "primary position", which means that the corpse was buried in a complete form, without missing parts. On the other hand, there are also two "secondary burials" of totally or partially disarticulated skeletons likely resulted from the excarnation of the individuals. In this case, as a consequence of the mentioned method of disposing of the dead, the decomposition of the body has taken place. Some smaller parts of the skeleton had been carried away by birds or animals and the others remained in place, but disturbed. Both, the extended and crouched positions of the skeletons were observed. A special characteristic of the Neolithic phase II was the use of the ochre, spread both on the human and animal bones (Basu & Pal, 1980: Table 1). One of the burials contained an earthen pot (SKL 5) and a barrel-shaped paste bead, another had barrel-shaped carnelian beads (SKL 8) and another one had a circular stone bowl (SKL 10). One of the burials (SKL 3), which was assigned to the Neolithic-Megalithic Phase, also contained a dog skull buried together with the human individual, considered to be an Early Chinese custom.

After a detailed anthropological study it resulted that the respective burials were part of a long-headed and tall homogeneous population, being more related to the mature Harappan in the Cemetery R37 than to other contemporary Neolithic populations elsewhere in India. Also, given the presence of the 'Charles facet' and of the 'squatting facet' on the distal femora and tibia of some individuals they are possibly closer to the today's Punjabi people in Northwest India.

The Skeleton of the Trepanned Skull

The skeleton 7 bearing the trepanned skull was primary in articulated position found in a crouched position, with a Northeast-Southwest orientation. The individual was reportedly accompanied by animal bones, antler and horn pieces including a stone soap-disk. The burial was not disturbed and was placed inside the pit with the right side of the skull lying downward and the trepanned surface upwards. It is interesting to note that the human bones were reportedly covered with red ochre.

Anthropological data on Trepanned Skull

The anthropological study of the skeleton by Basu & Pal (1980) had revealed that it belonged to a woman, aged at 26-30 years, with an ovoid skull, having marked dolichocrany and high calvaria. Its cranial capacity was of 1353 cc. The skull is elongated, with cryptozygy. It has some missing parts, like: the basal part of the cranium, part of the occiput, the greater wing of the sphenoid, the nasal bones, as well as part of the right maxilla. It could be observed that a hemihypertrophy and asymmetry existed on the left side of the skull, with the left mastoid,

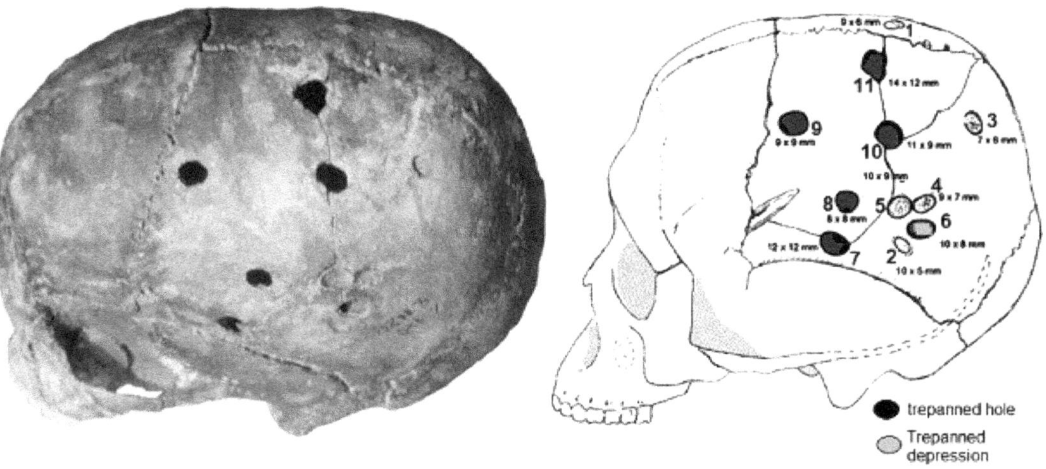

Fig. 23.1. Trepanned Burzahom skull and the re-sequencing of the trepanation process (after A.R. Sankhyan, 2015)

orbit and parietal oversized, associated with a deformity of the prelambdoid region. The forehead is low, slightly receding and with a distinct post-glabellar depression. The nasal aperture is pyriform. Occipital squama is protruding and the occiput narrows and converges downward with lambdoid ossicles present at the left lambdoidal suture. The cranium is hyperdolichocranic with L/B index 68.7 and hypsicranic with length/auricular height index 64.3. The face is very narrow, with medium height, hyperlepten, with superior facial index 57.1. The skull has a gracile look.

The Speciality of Burzahom Trepanation

For a detailed general account on the practice of trepanation worldwide the reference may be made to the article No. 18 in this volume by Comşa & Sankhyan. We may not repeat on the practice of trepanation in general, but in consideration of similar cases elsewhere we will discuss the significance of the rare multiple trepanations done on Burzahom skull. When we see multiple trepanations on human brain case we feel scared and amazed to know that such a case could ever exist, especially if the individual had lived for some time with so many surgeries on the skull! The skull under reference presents six small but completed holes and 5 tiny shallow depressions as well. Allchin and Allchin (1969) were the first to notice the presence of trephination, yet the first study of the interventions was done by Roy Chowdhury (1973) – a medical expert who considered that trepanations were done upon a living person for a medical purpose. After seven years, the trepanation was reviewed by two physical anthropologists, Basu & Pal (1980) who contested this conclusion and considered that it was a posthumous intervention for the sole purpose of taking out bone roundels to be worn as amulets. About twenty years later, the skull was thoroughly examined/re-analysed by Sankhyan & Weber (2001) and Sankhyan (2008) taking in consideration both earlier studies. They assumed it a surgical intervention and conceptualized a sequence of trepanation events classifying the 11 attempts made in three major stages at different times.

Sankhyan (2015) has recently reviewed the earlier sequence of the attempted trepanations in the *Encyclopaedia of Surgery* (in Press) from where his edited version and the image of trepanned skull and the revised sketch of the perceived trepanation process is reproduced here (Fig. 23.1) along with his revised observations. He re-classified the 11 perforations into: depressions (1-5) and trepanned holes (6-11) assuming that in the first sitting the surgeon started his treatment of the patient with a gentle stroke given on the top of the head with a view to see if the patient is relieved of the pain and chronic ailment and turns normal. This stroke caused a small depression (1) on the top of the cranium. The left hypertrophy of the cranium is indicative of some brain anomaly/tumour which might have turned the patient insane or epileptic. With the same intentions the surgeon attempted the second stroke but this time down below on the left parietal bone (2). The third stroke was tried on the posterior extreme (3). All the three strokes were probably given in a single sitting after brief intervals, which resulted in three impressions or shallow depressions (2-3) on the cranial vault. In the second sitting, when the patient could not get any relief from earlier treatment, the surgeon gave two greater stokes on another site which caused deeper depressions (4 & 5), followed by another and bigger one that removed the outer table of the bone into the diploe space by perforating the skull slightly but leaving the inner table intact (6). Again, when the patient showed no sign of relief, the surgeon thought of real deeper interventions in the third sitting, which perforated the cranium and formed two complete holes (7 & 8) made close by possibly for bloodletting and taking out the degenerated blood tissue. They are almost of the same size and outline and were made very neatly and carefully by the same instrument. Finding no relief to the patient, the surgeon made another trepanation in the fourth sitting, which was probably his last effort on the living patient when a single neat rounded hole (9) was made. However, the patient perhaps succumbed to death in this attempt, so that the remaining two holes (10 & 11), which are bigger in size and uneven in shape, were most likely a post-mortem study trial to ascertain the cause of the pathology, and for a simple demonstration/training to the future surgeon. There are fracture lines emanating from these big perforations indicating that the patient had suffered from a trauma.

Sankhyan (2015) further clarified that no instruments that could have been used in trepanation were found in the trepanated burial site except some animal bones, antler, horn pieces, and a soap stone circular disc. It is therefore possible that the antler piece or animal bones could have been used as drills of various diameters for trepanation. Erdal & Erdal (2010) observed that only the drilling technique with flint or bone drills was practiced during Neolithic times, whereas the cutting and scraping techniques were developed later with bronze instruments during the Bronze Age, followed by iron instruments during the megalithic period. In this background, I am of the opinion that we may rule out the possibility of inviting an Indus surgeon speculated by George Weber earlier (Sankhyan & Weber (2001)).

Sankhyan (2015) also concurred with Basu & Pal (1980) about the extraction of the roundels from the Burzahom skull except in the post-mortem trepanation. We further share his argument that trepanation at Burzahom skull was NOT done for extracting roundels which were possible through a cutting technique, whereas Burzahom trepanation indicates drilling rather than cutting. The drill would rather damage and push the bone piece inside the skull thereby making it difficult to retrieve until the skull is hollow and devoid of brain. In this context Sankhyan and Weber (2001) had also previously considered that the trepanned woman was a solitary case in the necropolis of Burzahom and that there was no direct evidence of wearing human bone roundels as amulets in North Indian tribes.

Who performed the Burzahom operation?

It is difficult to believe that a small group of the Burzahom nomads had the skills of performing such a surgical operation. It seems that the Neolithic populations, at least those European ones, from Moldova, a region of today Romania, took over the trepanation technique from nomadic populations, the so-called bearers of the Pit Grave Culture. Perhaps, these tribes, given their mobility, had two main reasons to improve their ways of treating diseases: they came in contact with many other communities, so that they also had contacts with many pathogens and also had developed immunity. They could refine their technique and instruments through inter-cultural exchanges; they learned and were receptive to everything about medicine. Moreover, the health condition of the people was very important, as the ill ones had to be taken care of and even transported from one place to another, facts that would have increased the efforts of the community. Many hunter-gatherers in India are snake charmers, traditional medicine men and shamans who render good services to the agricultural communities; some of the members of the latter have also learnt such skills from the nomads. But, it is interesting to observe that the hunter-gatherer Massai of Africa still do trepanations. The rare and interesting way of living in underground pit dwellings is also a testimony to the innovative capability of the Burzahom people. But, the cases of multiple trepanations are rare while the single or double trephinations in vivo are in thousands in various periods of the world history (Comşa, 2008). As in the Romanian Neolithic there are no multiple trephinations, we could mention a skull from the Early Bronze Age site of Cetăţenii din Deal (Argeş County) (Russu & Bologa, 1961), as well as another one from Zimnicea-Cemetery I (Teleorman County), both with traces of four unfinished trephinations (Necrasov, 1977). On the American continent, a much later skull from Cuzco (Peru), dated ca. 1000 AD, had seven holes on its calotte (Greenblatt, 1997:31). There is a mention of cases of multiple trepanations in the Inca (Verano 2003). Two skulls have four big trepanned holes each; in one skull these were done by the boring and cutting technique and in the second by the circular grooving technique (Erdal & Erdal, 2010). The well-healed holes are a striking evidence of four surgical operations carried out in vivo in Peru.

ASTRONOMICAL ORIENTATION

It seems that the Old World was one of the most important "centres" of the trephination practice, one of them being situated in the region of today's France. The largest number of trepanned skulls in that country had been found in collective burials and were dated to Late Neolithic-Chalcolithic, being concentrated in two principal material cultures (the Treilles group on the Grande Causse and the Seine-Oise-Marne Culture from the Paris basin (Beyneix 2013: 54). The present case from Burzahom is very interesting, given the several trephinations found, either completed or not upon a single subject.

Astronomically, a population which practices agriculture observes the sky, constellations in different manner than the hunter-gatherer and fisher populations, as they have different purposes. Their observations about the sky are reflected in burial orientations. In the present case the orientation is just guidance, and reflected no solar cult; it showed that the Burzahom population was a homogeneous nomadic community. The epileptic woman was buried in the same manner used for the other members of the little group.

In Neolithic time the sky was enthralling, the night sky was dark, with a lot of bright stars. At the foot of the Himalaya Mountain it must have been a very clear sky, perfect for daily observations. This Neolithic sky must have been so fascinating that the ancient man had a lot of cults conceived by observing the sky. Moreover, those people had introduced the sky rituals in their life (in every day concerns, in burials, in constructions, etc.). The people have used the astronomical phenomena in order to orientate in time and space, for calculating the ideal weather corresponding to their agricultural occupation, for hunting success, etc. Many Neolithic civilizations existed on the Indian Territory.

Our astronomical investigation refers to the Burzahom Neolithic site (latitude: 22°57'N) with 10 skeletons from two phases, dated around 2000 BC (4000 BP). For orientation in space, the ancient man from Burzahom used the Sun's apparent motion on the sky. He observed that the Sun rises and sets, and he must have also observed that the sunrise and sunset locations on the horizon varied in one year. He must have observed that the day and night are not of the same lengths. At equinoxes, in spring and autumn, there is a period when the length of the day is equal with the length of the night. At solstice, in summer, there is a period when the day is very long, and in winter, there are few days when the night is very long. These oscillations must have been known by a daily ancient observer. In a year, the sunrises and the sunsets describe an arc at the horizon, called *solar arc* with the endpoints at the summer and the winter solstices. It is very interesting that the Neolithic man preserved his daily concerns, beliefs, conceptions, religion in the burial methodology, such as grave goods, position of the skeleton, burial placement in the cemetery or near the dwelling, etc. Astronomically, he kept the orientation of the skeleton.

Studying a lot of Neolithic cemeteries worldwide, one can realize that the skeletons are astronomically orientated. Historians say that the Neolithic man believed in afterlife which is why he put the grave goods near the deceased person, who could use them in next birth. People made the graves towards the sunrise, sunset direction or towards the cardinal points as per their spiritual conceptions. From archaeo-astronomical point of view, if a necropolis had the orientation between the solstice points for most of its skeletons, i.e. on the "solar arcs", then it results that the burials were arranged towards Sun direction, and that that Neolithic community practiced the solar cult (Comşa & Szücs-Csillik, 2013).

THE SOLAR ARC AT BURZAHOM

The orientation of skeletons (skull to pelvis) or the axis of the graves was derived from the published records. The azimuthal distribution of orientation of the 10 skeletons from Burzahom Site is given in Tab. 23.1 and Fig. 23.2. What we can see that the main orientation was NW-SE. In fact, large sites preserved better traces of astronomical orientation during Neolithic time or other periods. We calculated the solar arc at Burzahom Neolithic Site by using the fact that the local apparent positions of the Sun, which depends on local geographical coordinates, are depicted by horizontal coordinates (h – altitude, A – azimuth). The transformation relations between equatorial and horizontal coordinates are given by (Smart, 1986):

$$\sin(\delta) = \sin(\varphi)\cdot\sin(h) + \cos(\varphi)\cdot\cos(h)\cdot\cos(A),$$

where φ = geographic latitude, δ = the declination of the Sun (Wittmann, 1979). If we consider h = 0 (zero horizon) we obtain the theoretical azimuths:

$$A = \arccos\left(\frac{\sin(\delta)}{\cos(\Phi)}\right)$$

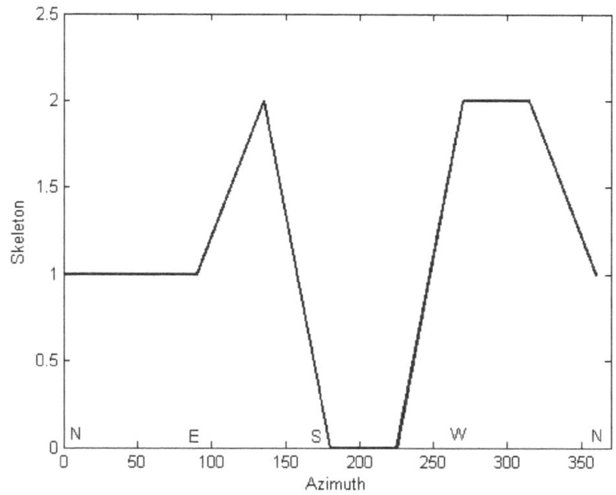

Fig. 23.2. Azimuthal distribution at Burzahom Neolithic Site

The real solar arc at Burzahom Archaeological site (2000 BC) is not at zero horizons, because the place situated at an altitude of 1800 m above the sea level is surrounded by high mountains, which shifted the solar arc towards the South direction (see Lăzărescu & Turcu, 2013). The main orientation of the Burzahom Archaeological Site was NW-SE, according to the published records (Fig. 23.2). Inside the solar arc at Burzahom (2000 BC) there are 3 skeletons (1 at East and 2 at West). Based on the small sample for burial orientation, it remain inconclusive whether the solar cult was practiced by the Neolithic people of Burzahom.

The Burzahom skull (SKL7) goes fairly well in morphometric with three other skeletons, except the male Skeleton 6 which is much robust and different from all the rest. The Burzahom population was physically more homogeneous and anthropologically the SKL7 is not different from the others.

DISCUSSION

In Neolithic people at Burzahom lived in small groups. Each group had individual characteristics, habits, beliefs, which were specific to the group. If someone was accepted from outside the group, then it was either fully assimilated within the group, or kept with at least part of its customs and beliefs. In the European necropolis of Basatanya from Hungary we have found skeletons differentiated by astronomical orientation. Most of the skeletons were aligned to the apparent motion of the Sun (Solar Cult) but some of them (anthropologically different) were orientated to the North (Csillik et al., 2004).

Elsewhere, in some archaeological sites there are known cases where the deceased was buried with the face downward. In doing so its astronomical orientation fell out of the solar arc. Such persons were punished for something and marginalized even after death too (Szücs-Csillik et al., 2010). In the case of the SKL 7 of the

Tab. 23.2. Theoretical and estimated solar arc at Burzahom (4000 BC)

Solar arc	A_{rise}		A_{set}	
	Summer solstice	***Winter solstice***	***Summer solstice***	***Winter solstice***
Theoretical	63°51'	116°08'	243°51'	296°08'
Estimated	74°40'	127°52'	232°40'	285°52'

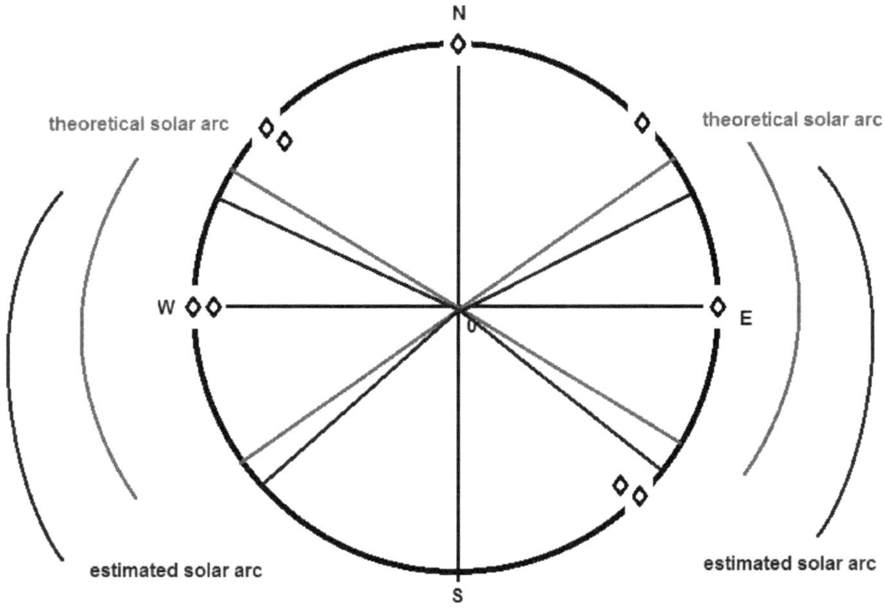

Fig. 23.3. Theoretical and estimated solar arc with skeleton azimuths

woman of Burzahom with multiple trepanations we are observing a different special case.

The Burzahom woman with, at least, slightly abnormal skull, might have been insane, epileptic, or otherwise "different", and such people were regarded with awe and fear in most early societies (Sankhyan & Weber, 2001). The word "different" applied to the woman of the trepanned skull is due to expected different behaviour and not as a distinction due to morphology, despite her left side hypertrophy.

Studying the astronomical orientation of the Skeleton 7 with the trepanned skull from Burzahom, we realized that it has an orientation that falls out from the main orientation of Burzahom Archaeological Site, namely NW-SE orientation, and moreover, falls outside of the given solar arc. Four skeletons are orientated towards NW-SE or SE-NW direction, three on the EW axis, one with no data, and two are orientated N (skeleton 6) and NE-SW (skeleton 7). The skeleton 6 belongs to a robust man, and skeleton 7 to a gracile sick woman lying on her right side with the trepanned side of the skull placed upwards. Because skeleton 7 is lying crouched on the right side, the view of the deceased person is to the North. It is important to notice the position and the orientation of the skeleton, the grave goods position and orientation and the trepanned skull position and orientation.

The Neolithic man had the desire to preserve his belief by performing the burial ceremony. Every position of the skeleton reflects the qualities or attributes of the dead person (Szücs-Csillik et al., 2010). We, as three different specialists, working together as a team upon the Burzahom trepanned skull, made a pioneer work with the present study. Our astronomical investigation shows that in Neolithic a small group of ten semi-nomadic people lived separately from the other local inhabitants. Astronomically, we may infer that somebody (likely the skeleton 6) well mastered in medicine, treated the sick woman (skeleton 7) got fully involved in the process and was eventually embraced by the group.

Astronomically, the North orientation of the skeleton 6 is an important direction for travellers, the nomadic people. In Neolithic the North direction could have been determined in two specific ways. During the day it could be followed the movement of the Sun and its shadow (at the equinoxes). During the night it could be observed the spot around which the starry sky used to spin (the so-called world pillar). Many Neolithic communities who relied upon the concept of world pillar as a major belief were shamanistic ones. In this case, the North point represented the spot that assured the communication with the after world, a task that could have been accomplished just by the shaman (Comşa & Szücs-Csillik, 2013). In recent times, the North point is in the direction of the

Polaris star, of the Ursa Minor (Smaller Bear) constellation. In the past, for instance, in the year of 4800 BC the polar star was Edasish (iota Draconis), while in the year of 3000 BC the polar star was Thuban, from the Dragon constellation.

Who trepanned the woman?

Our assertions are documented by a recent scrutiny of the anthropological data on skeleton 6 and other individuals of the Burzahom series by Sankhyan (2015), which concluded that the Trepanner was indeed the man of skeleton 6. Sankhyan (op.cit.) argued that anthropometrically the man belonged to a distinct population unlike the rest of the Burzahom skeletal population. The skull of skeleton 6 is of a very 'robust' man who died in the age of 51-55 years. The man had a rounded head as is apparent from the mesocranic skull with LBI=74.9. His face was very broad with large bizygomatic breadth (150 mm) and so was his very broad nose, hyper chamaerrhine with nasal Index 68.0. His forehead (frontal) was much receding and low. All other skeletons of Burzahom collection belong to a homogeneous long-headed (dolichocranic) to very long-headed (hyper dolichocranic) population having long to longer noses, and narrow vertical faces; they were also taller than the 'robust' man. The Burzahom skeletal series finds its nearest phenotypic parallel with the Harappan R37 series.

Basu & Pal (1980) had attributed the cranial differences of the skeleton 6 simply to individual variation than to the migration. However, the recent scrutiny by Sankhyan (op.cit.) revealed that the skull 6 resembled the NGK 7 skull from Nagarjunakonda (Andhra Pradesh) of the contemporary Neolithic series of Southern India. Hence, both astronomically and morphologically our conclusion is that the 'robust man' was a new entrant from a distinct outside population and we identify him as the Trepanner of the Burzahom woman. Although for a Neolithic man from Southern India, Kashmir is too distant to visit, but Kashmir must have been an attractive heavenly place in the Himalayas ever since the first man discovered that. This is why a lot of people came and settled there. Those days medical surgeons were presumably very uncommon, and the nomadic people were at a greater advantage to have more information about such specialists than the settled people. So, the Burzahom Trepanner might have been brought there as a medical specialist.

But, again we may argue why he was buried there with others? Probably, the treatment of the woman continued for long which necessitated his longer stay as the young beautiful lady was suffering from some mysterious ailment/epilepsy/insanity, presumably due to a brain tumour apparent from the left-sided cranial hypertrophy. In those days such diseases were primarily attributed to infliction of some evil spirits and there was little difference between the spiritual specialists or shamans or the medicine men. Trepanations were done as treatment for removing the evil spirits; there is even a horrifying recent instance of self-trepanation done for liberating the spirit and enhancing self-enlightenment by Joey Mellen and Amanda Feilding mentioned in *Eccentric Lives & Peculiar Notions* by John Michell [Elliott's Home Page].

But, at Burzahom the sick woman needed prolonged life-saving spiritual and ultimately surgical intervention and was therefore in the hands of this shaman-cum medical man. She had to bear the torture of so many strokes on her vault for trepanations, obviously for treatment and not for bone roundels as argued by Basu & Pal (1980). But, every likely she could not survive the 9th operation; the last two interventions could have been as post-mortem research experiment. Nevertheless, the surgeon was indispensible and might have prolonged the life of the sick woman, and therefore accepted by the community. He lived his life there and after his death his body was buried within the habitat of the group, but oriented differently as a distinction from the rest of the burials.

References

ALICHIN, B., ALICHIN, R. 1968. *The Birth of Indian Civilization.* Penguin Books, Bombay.

BASU, A., PAL, A. 1980. *Human Remains from Burzahom*, Memoir, No. 56, Anthropological Survey of India, Calcutta.

BEYNEIX, A. 2013. Néolithique à l'aube de la médecine occidentale, Archéologia 12 (506): 53-60.

COMŞA, A. 2008. *Aspects of Archaic Medicine: Human and Animal Trephination on the Territory of Romania.* Cetatea de Scaun, Târgovişte. Geschichte der Medizin und der Naturwissenschaften, B. 45, H. 1, Wiesbaden: 45-47.

COMŞA, A., SANKHYAN, A.R. 2014. Amazing skills: practice of trepanation around the world, In A.R. Sankhyan (ed) *Recent discoveries and perspectives in human evolution*, p. 171-180. BAR: BAR Publishing, England (This volume).

CSILLIK, I., MAXIM, Z., BARLAI, K. 2004. *The archaeoastronomical work on the database of the Basatanya burial site, Hungary.* Annals of the Tiberiu Popoviciu Seminar of Functional Equations, Approximation and Convexity, 2:157-170.

ERDAL, Y.S., ERDAL, O.D. 2010. A Review of Trepanations in Anatolia with New Cases. *Int. J. Osteoarchaeol.* Published online in Wiley InterScience (www.interscience.wiley.com) DOI: 10.1002/oa.1154: pp. 1-30.

GREENBLATT, S.H., FORCHT DAGI, T., EPSTEIN, M.H. (ed.) 1997. *History of Neurosurgery in its Scientific and Professional Contexts.* The American Association of Neurological Surgeons, Park Ridge.

LĂZĂRESCU, V.A., TURCU, V. 2013. *Seasons of Death. Towards an Archaeological-Astronomical Method for Studying Ancient Burial Rites.* BAR International Series, edited by Vasile Cotiugă and Ştefan Caliniuc. BAR Publishing, Oxford.

NECRASOV, O. 1977. *Nouvelles Données sur la Pratique de la Trépanation chez les Populations Préhistoriques et Protohistoriques de Roumanie*, Mitteilungen der

Anthropologischen Gesellschaft in Wien 108:130-137.

ROY CHOWDHURY, A.K. 1973. Trepanation in Ancient India. *Journal of the Asiatic Society of Bengal* 15: 203-204.

RUSSU, I.G., BOLOGA, V. 1961. Trepanationen im Gebiet des heutigen Rumänien, Südhoffs Archiv für Geschichte der Medizin und der Naturwissenschaften, B. 45, H. 1, Wiesbaden:45-47.

SANKHYAN, A.R. 2008. Surgery in Ancient India. In *Encyclopaedia of the History of Science, Technology, and Medicine in Non-Western Cultures* (ed. Helaine Selin), Springer Link, part 19, pp. 2060-2063.

SANKHYAN, A.R. 2015. Surgery in ancient India. In *Encyclopaedia of the History of Science, Technology, and Medicine in Non-Western Cultures* (ed. Helaine Selin). DOI:10.1007/978-94-007-3934-5_9727-2. Springer Science + Business Media Dordrecht 2015, pp. 1-9.

SANKHYAN, A.R., WEBER, G.H.J. 2001. Evidence of Surgery in Ancient India: Trepanation at Burzahom (Kashmir) over 4000 years ago. *International Journal of Osteoarchaeology* 11: 375-380.

SANKHYAN, A.R., ROBBINS SCHUG, G. 2011. First evidence of brain surgery in Bronze Age Harappa. *Current Science* 100 (11): 1621-22.

SMART, W.M. 1986. *Textbook on Spherical Astronomy*, Sixth edition revised by R.M. Green 1977, first edition 1931, reprinted 1986. Cambridge, Cambridge University Press.

SZÜCS-CSILLIK, I., COMŞA, A., MAXIM, Z., SZÜCS, I. 2010. Case studies of Archaeoastronomy in Romania, Archaeologia e Calcolatori, 21: 325-337.

VERANO, J. 2003. Trepanation in Prehistoric South America: Geographic and Temporal Trends over 2.000 Years. In R. Arnott, S. Finger, and C.U.M., Smith (eds), Trepanation*: history, discovery, theory.* Swets & Zeitlinger B.V. Kisse, Netherlands, pp. 223-236.

WEBER, J., WAHL, J. 2006. Neurosurgical aspects of trepanations from Neolithic times. International Journal of Osteoarchaeology16: 536-5545. DOI: 10.1002/oa.844.

WITTMANN, A. 1979. The Obliquity of the Ecliptic, Astronomy & Astrophysics, 73, p. 129-131.

www.ingramcontent.com/pod-product-compliance
Ingram Content Group UK Ltd.
Pitfield, Milton Keynes, MK11 3LW, UK
UKHW061847041026
15591UKWH00015B/956

9 781407 313726